1.
B
a

2

CONTROL PROCESSES IN
MULTICELLULAR ORGANISMS

CONTROL PROCESSES IN MULTICELLULAR ORGANISMS

A Ciba Foundation Symposium

Edited by
G. E. W. WOLSTENHOLME
and
JULIE KNIGHT

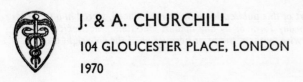

J. & A. CHURCHILL
104 GLOUCESTER PLACE, LONDON
1970

First published 1970

Containing 107 illustrations

Standard Book Number 7000.1431.4

Contents

Membership

Symposium on Control Processes in Multicellular Organisms, held by the Ciba Foundation at the India International Centre, New Delhi, 17th–21st March 1969

F. G. Young *Chairman*	Department of Biochemistry, University of Cambridge
B. K. Anand	Department of Physiology, All-India Institute of Medical Sciences, New Delhi
P. M. Bhargava	Regional Research Laboratory, Hyderabad
Suzanne Bourgeois	The Salk Institute for Biological Studies, San Diego, California
D. P. Burma	Department of Biochemistry and Biophysics, College of Medical Sciences, Banaras Hindu University, Varanasi
R. W. Butcher	Department of Physiology, University of Massachusetts School of Medicine, Worcester, Massachusetts
M. Cohn	The Salk Institute for Biological Studies, San Diego, California
G. M. Edelman	The Rockefeller University, New York
C. Fortier	Département de Physiologie, Faculté de Médecine, Université Laval, Québec
C. Gopalan	Nutrition Research Laboratories, Indian Council of Medical Research, Tarnaka, Hyderabad
H. Harris	Sir William Dunn School of Pathology, University of Oxford
V. Jagannathan	Biochemistry Division, National Chemical Laboratory, Poona
A. Korner	Department of Biochemistry, University of Sussex, Brighton
R. B. Livingston	Department of Neurosciences, School of Medicine, University of California at San Diego, La Jolla, California
F. Lynen	Max-Planck Institut für Zellchemie, Munich
J. F. A. P. Miller	Experimental Pathology Unit, The Walter and Eliza Hall Institute of Medical Research, Melbourne
J. Monod	Institut Pasteur, Paris
A. S. Paintal	Vallabhbhai Patel Chest Institute, University of Delhi
M. R. N. Prasad	Department of Zoology, University of Delhi

V. Ramalingaswami	Department of Pathology, All-India Institute of Medical Sciences, New Delhi
R. I. Salganik	Institute of Cytology and Genetics, Siberian Department of the U.S.S.R. Academy of Sciences, Novosibirsk
P. S. Sarma	Department of Biochemistry, Indian Institute of Science, Bangalore
O. Siddiqi	Tata Institute of Fundamental Research, Bombay
G. P. Talwar	Department of Biochemistry, All-India Institute of Medical Sciences, New Delhi
J. R. Tata	National Institute for Medical Research, London
A. White	Department of Biochemistry, Albert Einstein College of Medicine of Yeshiva University, New York
V. P. Whittaker	Department of Biochemistry, University of Cambridge and Department of Neurochemistry, Institute for Basic Research in Mental Retardation, Staten Island, New York

The Ciba Foundation

The Ciba Foundation was opened in 1949 to promote international cooperation in medical and chemical research. It owes its existence to the generosity of CIBA Ltd, Basle, who, recognizing the obstacles to scientific communication created by war, man's natural secretiveness, disciplinary divisions, academic prejudices, distance, and differences of language, decided to set up a philanthropic institution whose aim would be to overcome such barriers. London was chosen as its site for reasons dictated by the special advantages of English charitable trust law (ensuring the independence of its actions), as well as those of language and geography.

The Foundation's house at 41 Portland Place, London, has become well known to workers in many fields of science. Every year the Foundation organizes six to ten three-day symposia and three or four shorter study groups, all of which are published in book form. Many other scientific meetings are held, organized either by the Foundation or by other groups in need of a meeting place. Accommodation is also provided for scientists visiting London, whether or not they are attending a meeting in the house.

The Foundation's many activities are controlled by a small group of distinguished trustees. Within the general framework of biological science, interpreted in its broadest sense, these activities are well summed up by the motto of the Ciba Foundation: *Consocient Gentes*—let the peoples come together.

Preface

ELEVEN out of twelve of the Ciba Foundation's conferences are held in the Foundation's house in London. Approximately once in two years a symposium or study group is arranged in some part of the world where we have not been before, with four particular purposes in mind: to pay respect to medical, biological and chemical scientists in the host country, to bring them into informal contact with some of the leading relevant research workers from other parts of the world, to provide them and their colleagues in the governmental administration of research with the encouragement of an international occasion, and together to try to forge bonds of personal friendship and professional cooperation which hopefully may endure through the stresses and strains of this troubled world.

India's response to our proposal to hold a symposium in New Delhi was outstandingly positive. Reference is made in the following pages by our chairman, Professor F. G. Young, to the great helpfulness and generosity of many of our hosts, which was supremely personified in the gracious, scholarly and moving welcome given by the then President of India, Dr Zakir Husain, when he came to open the conference. Professor Jacques Monod, in a brilliant introductory public lecture (not published here), prefaced his speech with the comment that more than most scientists he had taken part in conferences all over the world but he had yet to meet the President of the United States or even the President of France; and he went on to say, amid the approval of the densely crowded audience, that he found it highly significant that the President of one of the largest countries in the world should see fit, and find time, to inaugurate a small, highly technical, scientific symposium of this international character. It was with all the deeper regret, therefore, that we learnt of the sudden death of Dr Zakir Husain some six weeks after our meeting.

The Ciba Foundation was even more than usually indebted to Professor Young for his part in the initiation, formulation and direction of this symposium. He will undoubtedly share our hope that this book will be regarded by the biochemists, endocrinologists, immunologists, neurobiologists and others who find it of value, as an appropriate scientific tribute to our Indian hosts and their colleagues.

THE EDITORS

CHAIRMAN'S OPENING REMARKS

Professor F. G. Young

THE idea of holding this conference took shape in conversations between Professor Pran Talwar and myself during the last days of 1965. At that time I was visiting India at the invitation of the Society of Biological Chemists of India to take part in their annual scientific meeting, which was being held in New Delhi in that year. Professor Talwar must take much responsibility for the initiation of this idea, which was effectively transmitted to Dr Gordon Wolstenholme early in 1966. After the necessary preliminary negotiations had taken place, the title of this symposium took its place among the conferences to be held under the auspices of the Ciba Foundation in 1968. But the problems of crossing international frontiers in the arrangements for a conference of this sort are not inconsiderable and in 1967 the decision had to be taken to postpone the conference from March 1968 to March 1969. In the meantime the indefatigably peripatetic Dr Wolstenholme was able to come to Delhi to discuss the arrangements on the spot. Here I would like to express my gratitude, on behalf of many of us, for the labours of Professor Talwar and his colleagues for all they have done to ensure that the arrangements for this symposium should proceed smoothly. We are very grateful indeed to them for all their hard work in this connexion over a very long period.

The title of this symposium is a broad one, and to some this symposium may seem to cover ground which is very similar to that of other meetings held during the past few years. But I myself believe that by discussing Control Mechanisms in Multicellular Organisms in a different environment, and among an international group which has never before met as a group, we may sow seeds that will bear fruit in the scientific years to come. Moreover, this symposium is of the special sort that the Ciba Foundation has made its own: a group of 25 to 30 active members who can all see each other during the discussions and get to know each other during the three and a half days of the debate. They will get to know each other in a way that would certainly not be possible if the membership were larger or the time of the conference shorter.

This symposium includes members from eight different countries which are, in alphabetical order, Australia, Canada, France, Germany, Great Britain, India, the Soviet Union and the United States of America. Four of the five great continents are represented among our members and the members come from many different scientific disciplines, though all have an interest in common in medical science.

The exchange of ideas and of factual information, face to face, by scientific colleagues who cross international boundaries in order to meet together inside and outside a conference room, plays an important part in advancing knowledge in these modern times, when the aeroplane has facilitated quick travel to such a remarkable degree. Many of us will be making visits to medical and scientific institutions in other parts of India after this conference in Delhi is over. I know that we shall all benefit in this way from the strengthening of old contacts and the making of new ones.

In this symposium there is a departure from the normal pattern in that scientific observers are present who have the opportunity of hearing papers and discussions, but not of participating in the proceedings. This procedure has been followed in a small number of Ciba Foundation conferences held outside Great Britain, and some years ago, at the conclusion of such a conference of which I had been Chairman, I was able to thank all the members of it for the vigorous part they had played in the discussion and at the same time to thank the observers for their silence. I am sure that I shall have the opportunity to do likewise at the end of this symposium.

Finally, may I beg nobody to think that any question is too simple to ask on an occasion such as this. We are from very different medical and scientific backgrounds and often a simple question directed to an expert can prove to be a most penetrating and fruitful one.

IN VITRO STUDIES OF THE LAC OPERON REGULATORY SYSTEM

SUZANNE BOURGEOIS AND JACQUES MONOD

*The Salk Institute for Biological Studies, San Diego, California, and
Département de Biologie Moléculaire, Institut Pasteur, Paris*

IN the present paper we wish to summarize and discuss some recent *in vitro* studies of the interaction between operator (O), repressor (R) and inducer (I) in the *lac* system of *Escherichia coli*. We feel that such a discussion may not be altogether irrelevant to the subject of this symposium, devoted to control systems of higher organisms. For several years now the question has been posed whether or to what extent the conclusions and concepts derived from genetic-biochemical studies of regulation in bacteria might apply to the fundamental problem of the control of gene expression in higher organisms.

The scheme of genetic control based on the analysis of the *lac* system of *E. coli* has been widely used both in theorizing on the subject of differentiation, and as a working hypothesis in experimental studies of, for example, the mode of action of certain hormones. No doubt these questions will be discussed again during this symposium. In the present paper, however, we do not intend to theorize or speculate on these wide problems, but rather focus attention on the molecular properties of the *lac* repressor, as they begin to be known from studies of the purified protein, and on the kinetics of its interactions with the operator and inducer. Owing in particular to the development of some very sensitive methods, these interactions can now be studied in detail, allowing one to define in precise, quantitative terms such problems as the specific "recognition" of a DNA segment by a protein, and the nature of the transition undergone by the repressor-DNA complex in the presence of inducer.

While it is reasonable enough to believe that the very simple logic of the *lac* system need not apply, as such, to the control of gene expression in higher organisms, the nature of the elementary molecular interactions involved at each step in a logically complex system may well turn out to be

3

fundamentally the same as in a simple one. At this very basic level, therefore, generalizations may be justified and useful. The methodology of
these *in vitro* studies may also, in part, be applicable to the analysis of control
systems of higher organisms. We have, for this reason, elaborated somewhat the section on methods.

EXPERIMENTAL DATA

I *The membrane filter assay for repressor–inducer (RI) and repressor–operator
(RO) complexes*

In vitro work on the *lac* operon regulation was initiated by Gilbert and
Müller-Hill (1966) who developed an equilibrium dialysis assay for the
lac repressor (R) based on the binding of a labelled inducer, isopropyl-
thiogalactoside (IPTG). Using this test, they could achieve a partial
purification and identification of R. They also demonstrated (Gilbert and
Müller-Hill, 1967) specific binding of R to DNA carrying a *lac* operator
region by showing that repressor labelled *in vivo* migrated with λφ80
dlac DNA in a glycerol gradient, a technique previously worked out by
Ptashne (1967) for the λ phage repressor.

A very convenient and sensitive technique has been developed to assay
for either RI or RO complexes, based on the observation that R would
stick to membrane filters and retain the bound ligand, ^{14}C-labelled
IPTG in the first case, or ^{32}P-labelled λφ80 *dlac* DNA in the latter (Riggs
and Bourgeois, 1968; Riggs *et al.*, 1968). The RI or RO complexes trapped
on the filter are stable enough for a constant high percentage of them to be
retained even after some washings which are necessary to eliminate the
background of free radioactive ligand.

It is worth pointing out the possible general usefulness of the membrane
filter technique as we understand it today for measuring binding of ligands
to a great variety of proteins. Membrane filters were first used to retain
nucleic acid–protein complexes like ribosomes (Nirenberg and Leder,
1964) or RNA polymerase–DNA complex in conditions in which the
RNA polymerase by itself was not retained (Jones and Berg, 1966). Kuno
and Kihara (1967) have since shown that it is a general property of proteins
to stick to membrane filters. Retention to filters is affected by ionic strength,
and magnesium is sometimes required. Membrane filters have been used
to study the binding of t-RNA to amino acid synthetase (Yarus and Berg,
1967) and recently the binding of C-carbohydrate from *Pneumococcus*
(Cohn, Notani and Rice, 1969) and of DNA (Schubert, Roman and Cohn,
1969) to antibodies. Conditions have also been established for adsorption

on filters of L-lactic dehydrogenase and polynucleotide phosphorylase (Thang, Graffe and Grunberg-Manago, 1968) and it has been shown that the first of these two enzymes is inactive when bound to filters, whereas the latter is still active. Some alteration of protein upon sticking to filters may stabilize the bound ligand to washing and therefore need not be a drawback.

The sensitivity of the assay will be limited by the specific activity of the radioactive ligand used and by its capacity to be readily washable from the filter when in free form. When using native DNA as a ligand, it is most useful to work in the presence of 5 per cent dimethyl sulphoxide (Legault-Démare et al., 1967) to reduce the background of [^{32}P]DNA. As it is applied here to the *lac* repressor, the membrane filter technique can assay

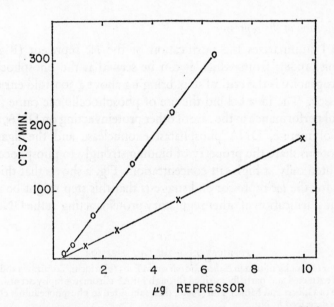

FIG. I. Antiserum assay (-×———×-). Fifty μl of antiserum was added to 0·96 ml of TMS buffer (0·2 M-KCl, 0·01 M-magnesium acetate, 10⁻⁴ M-EDTA, 0·001 M-mercaptoethanol, 0·01 M-Tris, pH 7·4) containing 10⁻⁷ M-[¹⁴C]IPTG and the indicated amount of repressor. The mixture was incubated overnight at 5°C and filtered through a 13-mm HA Millipore filter. Then, without washing, the filter was dried and counted for [¹⁴C]IPTG. Background was 7 counts/min and has been subtracted.

Millipore filter assay (-O———O-). One ml of TMS buffer containing 10⁻⁶ M-[¹⁴C]IPTG and the indicated amount of repressor was filtered under vacuum through a 13-mm HA Millipore filter. The filter was washed 3 times with 2 drops of TMS buffer, dried, and counted for [¹⁴C]IPTG. A background of 7 counts/min has been subtracted. For both of these assays, the amount of repressor was calculated from the equilibrium dialysis values for IPTG-binding activity.

as little as 10^{-16} moles of RO complex. Its extreme sensitivity and rapidity have allowed accurate kinetic study of the RO interaction (Riggs and Bourgeois, 1969a, b).

II *Binding of inducer to repressor*

As shown in Fig. 1 the amount of [^{14}C]IPTG bound to the filter increases linearly with repressor concentration in the range from 0·5 μg to at least 6·5 μg. A limiting factor in the use of this technique to assay for repressor in crude extracts is that the filter quickly saturates with contaminating proteins. The labelled RI complex can be specifically precipitated from crude extracts by a rabbit antiserum obtained by injecting purified repressor. The linearity of the antiserum assay is also illustrated in Fig. 1.

III *Purification of the repressor*

Table I summarizes the purification of the *lac* repressor (Riggs and Bourgeois, 1968), from which it can be seen that the phosphocellulose chromatography is the critical step, bringing about a 300-fold enrichment in repressor. The idea behind the use of phosphocellulose came from its successful performance in the case of other proteins acting on DNA, namely DNA polymerase, DNA phosphatase-exonuclease and the ligase. All these proteins share the property of binding strongly to phosphocellulose, being eluted only at high salt concentration. Fig. 2 shows that this is also the case for the *lac* repressor and suggests that this step might be a useful one in the purification of other regulatory proteins acting at the DNA level.

TABLE I

PURIFICATION OF THE *LAC* REPRESSOR

Repressor from 1 kg of frozen *Escherichia coli* strain E203 (Horiuchi, Tomizawa and Novick, 1962) was extracted and partially purified through DEAE chromatography according to the procedure of Gilbert and Müller-Hill (1966), then submitted to phosphocellulose chromatography, as described in Fig. 2.

Step	Total protein in repressor fraction (mg)	Specific IPTG-binding activity*
Crude extract	70000	0·1
23–35% (NH₄)₂SO₄ cut	7000	1·1
DEAE chromatography	3200	2·6
PO₄-cellulose chromatography	4·5	800·0
G-200 gel filtration	0·9	1800·0

*Percentage excess [^{14}C]IPTG binding per mg protein per ml.

One can see that the IPTG-binding activity elutes off the phosphocellulose column as two separate fractions, A and B. Although these are

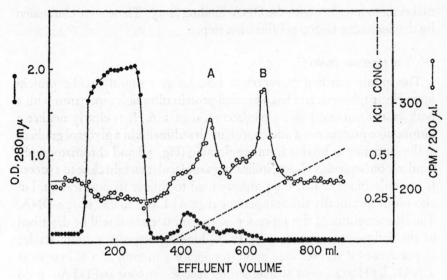

FIG. 2. After DEAE chromatography 180 ml of partially purified repressor fraction were dialysed against LLK buffer (0·01 M-mercaptoethanol, 0·01 M-K$_2$HPO$_4$, pH 7·0), and were applied to a 2·5 × 60 cm column of phosphocellulose (Bio-rad, 0·95 mequiv/g, lot No. B2231), washed and equilibrated with LLK buffer. The column was eluted with a linear 0 to 1 M-KCl gradient in LLK buffer. IPTG-binding activity was determined by a very simple, convenient technique. The column was equilibrated with 10^{-7} M-[^{14}C]IPTG, and the sample and eluting buffer also contained the same concentration of [^{14}C]IPTG. Samples of each fraction were counted directly for ^{14}C. Fractions containing binding activity as measured by this technique were pooled and always checked by equilibrium dialysis. (-●——●-), O.D.$_{280}$; (-○——○-), IPTG-binding activity; (---), KCl concentration.

present in about equal amounts in the experiment shown in Fig. 2, fraction A is most often present in much smaller amount than B and sometimes it is absent.

When the membrane filter technique is used to analyse IPTG-binding activity along a glycerol gradient, fraction B sediments at 7s while fraction A sediments around 3 to 3·5s. Moreover, the IPTG-binding activity in fraction A is precipitated by an antiserum directed against B. These observations suggested that the 3·5s protein represented a subunit of the 7s repressor. This interpretation was later confirmed when these IPTG-binding subunits were obtained from a purified 7s repressor (see Section IV).

Although essentially pure, the 7s repressor preparations often contain minor contaminating proteins that could not be eliminated by phosphocellulose chromatography and exhibit some affinity for DNA which makes them interfere with the DNA-binding assay. These were eliminated by the Sephadex G-200 gel filtration step.

IV *The repressor protein*

The highly purified 7s repressor behaves as a slightly acidic protein upon electrophoresis and has a normal protein ultraviolet spectrum with a peak at 277 nm (mμ) and a 280/260 ratio of 1·6. It is clearly neither a histone-like protein nor a nucleoprotein. It sediments in a glycerol gradient at the same rate as bovine immunoglobulin (Fig. 3a) and chromatographs similarly on Sephadex G-200, indicating a molecular weight close to 150000. It not only binds IPTG, the property used to follow its purification, but also binds specifically the *lac* operator region of a λφ80 *dlac* phage DNA. The characteristics of this repressor–operator interaction will be described in the following sections. The IPTG-binding activity is quite stable; 50 per cent of it was retained upon storage for 5 months at 0°C in 0·01 M Tris-HCl, pH 7·4—0·01 M-magnesium acetate—0·0001 M-EDTA— 0·06 M-mercaptoethanol. By contrast, the DNA-binding activity was entirely

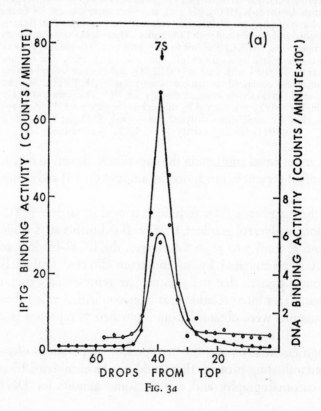

FIG. 3a

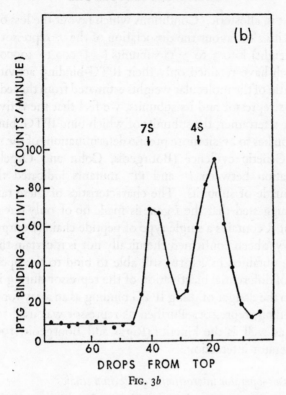

FIG. 3. Repressor was layered on top of 5–30 per cent glycerol gradients in 0·01 M-Tris-HCl, pH 7·4—0·01 M-magnesium acetate—0·0001 M-EDTA—0·0001 M-dithiothreitol and centrifuged for 9 hours at 250000 × g. Fractions were assayed for IPTG-and for DNA-binding activities by the membrane filter technique described in the legends of Figs. 1 and 4, respectively. The arrows indicate the position of 7s γ-globulin and 4s ovalbumin used as markers.

(a) Fresh active repressor preparation. ●——●, IPTG-binding activity. ○——○, DNA-binding activity.

(b) Repressor stored as described in the text. ●——●, IPTG-binding activity. No DNA-binding activity could be detected.

lost after storage in these conditions and the IPTG-binding activity now sedimented in a glycerol gradient as two peaks, the major one around 3·5s (corresponding to a protein of molecular weight about 35000 to 40000) and the second one at 7s (Fig. 3b). No DNA-binding activity could be detected in either peak.

The differential decay of IPTG and operator-binding activities indicates that they correspond to two distinct, presumably non-overlapping, binding sites, suggesting that the interaction between these sites must be indirect—that is, allosteric. Conditions which favour the loss of operator-binding activity also favour the dissociation of the 7s repressor ($\sim$150000 molecular weight) into 3 to 3·5s subunits ($\sim$35000 to 40000 molecular weight) which have retained only their IPTG-binding activity. On the basis of the ratio of the molecular weights estimated from the sedimentation data of the 7s repressor and its subunits, we feel that the active repressor is likely to be a tetramer, the subunits of which bind IPTG, but obviously this conclusion has to await more precise determinations of the size of these molecules. Genetic evidence (Bourgeois, Cohn and Orgel, 1965) of complementation between i^+ and $i^s i^-$ mutants indicated that the *lac* repressor is made of subunits. The characteristics of these rare complementations suggested that the *i* gene is made up of only one cistron and therefore that R contains a single type of peptide chain (Bourgeois, 1966). This has not yet been confirmed chemically nor is it certain that subunits could not be obtained in a form still able to bind to the operator. The observation of differential inactivation of the repressor during its purification points to the danger of using IPTG binding as an assay for active, that is, DNA-binding, repressor. Purified 7s repressor was used to study the equilibrium as well as the kinetics (Riggs and Bourgeois, 1969b) of the repressor–operator interaction.

V   *The repressor–operator interaction: equilibrium studies*

The basic experiment showing specificity of binding of the *lac* repressor to the *lac* operator region of the DNA is illustrated in Fig. 4.

When increasing amounts of purified 7s *lac* repressor are added to a constant amount of ^{32}P-labelled λφ80 *dlac* DNA, increasing amounts of [^{32}P]DNA are retained on the filter until a saturation level is reached. By contrast, *lac* repressor does not cause retention of DNA without the *lac* region (wild type λφ80). Moreover, a higher concentration of repressor is needed to achieve retention of DNA from a phage carrying an operator constitutive mutation (λφ80 *dlac o^c*). The *o^c* mutant used here has a base level of β-galactosidase equal to 5 per cent of the fully induced level and, as expected, its operator region has retained affinity for the repressor, except that it is weaker than for wild-type operator.

What is the affinity of the *lac* repressor for the wild-type operator? The first estimate (Riggs and Bourgeois, 1968) has turned out to be incorrect not only because the greater lability of the DNA binding compared to that of the IPTG-binding activity had not yet been recognized, but also because the measurements were carried out at concentrations of R and O too high to get an accurate equilibrium constant. Of the two reactants, operator

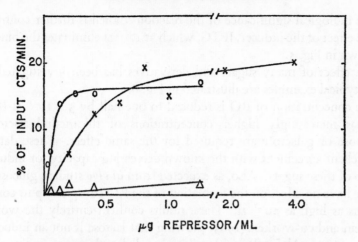

FIG. 4. ^{32}P-labelled DNA was isolated from a heat-inducible *lac* trans-ducing phage λφ80 *dlac* (λ$_{857}$ h80 *dlac*, originating from E. Signer) as described elsewhere (Riggs *et al.*, 1968). Five μl of a 10 μg/ml solution of ^{32}P-labelled DNA (in 0·015 M-NaCl—0·0015 M-sodium citrate, pH 7·0) were added to 0·095 ml of TKID buffer (0·01 M-magnesium acetate—0·01 M-KCl—0·0001 M-EDTA—0·006 M-mercaptoethanol—5 per cent dimethyl sulphoxide—0·01 M-Tris-HCl, pH 7·4) containing purified repressor at 25°C. The mixture was filtered through a Schleicher & Schuell 13 mm B$_6$ membrane filter, washed with TKID buffer, dried and counted. O——O, o^+ DNA. ×——×, o^c DNA. △——△, wild-type φ80 DNA.

and repressor, the operator concentration can easily be calculated from the specific activity of the DNA preparation used and by knowing from the genetics that each 30 million molecular weight phage DNA carries one operator region. The active repressor can then be titrated against a known operator concentration when working under conditions where the binding is essentially stoicheiometric—that is, at concentrations 100 times above the dissociation constant. After such titration of the active repressor, a binding curve in which repressor is varied against a known low concentration of DNA can be used to determine the equilibrium constant. A series of control experiments has also been done to ensure that the presence of the membrane filter does not perturb the equilibrium of the reaction. All these experiments (Riggs and Bourgeois, 1969*b*) are consistent with the stoicheiometry of the repressor–operator interaction written as follows:

$$R + O \rightleftharpoons RO$$

Under our experimental conditions one molecule of 7s repressor is bound per operator region with a dissociation constant of ~10^{-13} M. The binding becomes less tight with increasing salt concentrations and is very little affected, if at all, by temperature between 0°C and 35°C.

The biological significance of the reaction studied is further confirmed by the effect of the inducer, IPTG, which at 10^{-3} M eliminates the binding, as shown in Fig. 5.

The effect of many sugars and derivatives has been investigated and some typical examples are illustrated in Fig. 6.

The concentration of RO is reduced to one half by 5×10^{-6} M-IPTG, whereas increasingly higher concentrations of thiomethylgalactoside, melibiose or galactose are required for the same effect. These data are in excellent agreement with the known decreasing capacity for induction *in vivo* of these sugars. Also, as expected from *in vivo* studies, glucose and lactose have no effect on the formation of the RO complex up to concentrations as high as 10^{-1} M. These results confirm entirely the work of Burstein and co-workers (1965) showing that lactose is not an inducer of the *lac* operon. Altogether, the striking parallelism between *in vitro* results and the *in vivo* characteristics of induction gives confidence in the meaning of the system observed.

An important bonus of the membrane filter technique of assay of RO complex is the fact that it can be scaled up to purify operator regions (Riggs and Bourgeois, 1969a) and provides an assay for the *lac* operator by competition between unlabelled and radioactively labelled O or R. This competition assay has been used to follow the purification of the operator as well as to show that the repressor does not bind detectably to denatured λφ80 *dlac* DNA (Riggs and Bourgeois, 1969b).

Finally, the rapidity of the filtration technique has allowed kinetic studies (Riggs and Bourgeois, 1969a) which give insight into the mechanism of inducer action.

VI *Binding of repressor to operator: kinetic studies*

Equilibrium studies suggest that the repressor-operator interaction can be expressed by the simple equation:

$$R + O \underset{k_b}{\overset{k_f}{\rightleftharpoons}} RO$$

This interaction has been further characterized by measuring the rate constants for the forward and back reactions, k_f and k_b. Formation of RO complex is very fast but its rate can be measured quite accurately at concentrations low enough for the attainment of equilibrium to take several minutes while stopping the reaction by filtration takes about 30 seconds. Even better accuracy is achieved by stopping the reaction by adding, before filtering, a hundred-fold excess of unlabelled λφ80 *dlac* DNA. Any RO

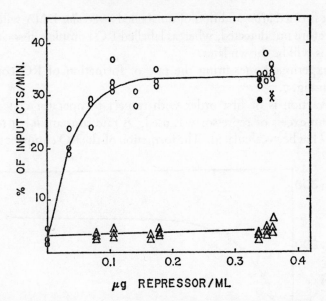

FIG. 5. Experiment performed as described in Fig. 4. ○——○, no IPTG added. △——△, 10^{-3} M-IPTG present in the reaction solution.

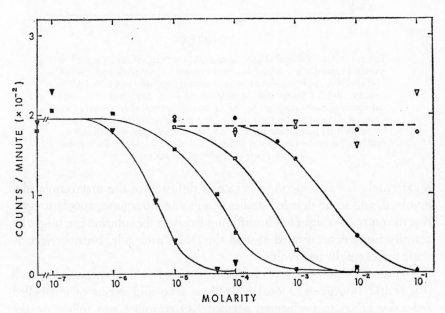

FIG. 6. Experiment performed as described in Fig. 4. ▼, IPTG added to the reaction mixture; ▣, thiomethylgalactoside added; □, melibiose; ●, galactose; ○, lactose; ▽, glucose.

complex formed after addition of unlabelled λφ80 *dlac* DNA will be cold and therefore not detected, whereas labelled RO complex dissociates very slowly, as will be shown later.

The experiment measuring the rate of formation of RO complex is shown in Fig. 7.

The reaction is of first order with respect to operator concentration because an excess of repressor was used. A rate constant $k_f = 4$ to 8×10^9 $M^{-1} sec^{-1}$ has been calculated. The formation of the RO complex therefore

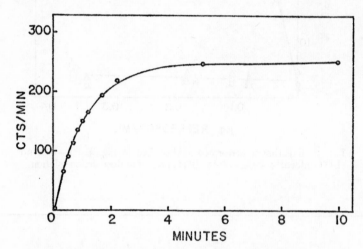

FIG. 7. $10^{-12}M$-^{32}P-labelled λφ80 *dlac* DNA (0·033µg/ml) and $3 \times 10^{-12}M$ purified repressor are mixed in a final volume of 3·1 ml of buffer (0·01 M-Tris-HCl, pH 7·4—0·01 M-magnesium acetate—0·01 M-KCl—0·0001 M-EDTA—0·0001 M-dithiothreitol—5 per cent dimethyl sulphoxide) containing 50 µg ml of bovine serum albumin at 25°C. One ml aliquots of this mixture are filtered through 25 mm Schleicher & Schuell B_6 membrane filters which are washed once with the above buffer without bovine serum albumin, dried and counted. Each point is the average of triplicate filtrations.

is extremely fast considering the rate of diffusion of the macromolecules involved, and some detailed studies (Riggs and Bourgeois, 1969*b*) suggest that the repressor might not be diffusing freely in the solution but might be actually more concentrated around the DNA molecule, because electro-static forces might be involved.

The rate of dissociation of the RO complex has been measured as well (Riggs and Bourgeois, 1969*a*) by adding a 20-fold excess of unlabelled λφ80 *dlac* DNA to preformed labelled RO complex and following the disappearance of labelled RO with time. In those conditions, the un-labelled DNA will compete with the labelled one and the rate-limiting reaction is the dissociation of labelled RO. As a control, addition of excess cold λφ80 DNA which does not carry a *lac* operator should not promote the dissociation of labelled RO complex. The result of such an experiment is shown in Fig. 8.

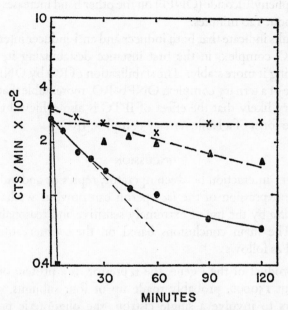

FIG. 8. Purified repressor is added to achieve half saturation of $1 \cdot 7 \times 10^{-11}$M-^{32}P-labelled λφ80 *dlac* DNA ($0 \cdot 5$ µg ml) in buffer ($0 \cdot 01$ M-Tris-HCl, pH $7 \cdot 4$—$0 \cdot 01$ M-magnesium acetate—$0 \cdot 01$ M-KCl—$0 \cdot 0001$ M-EDTA—$0 \cdot 006$ M-mercaptoethanol—5 per cent dimethyl sulphoxide) containing 50 µg/ml of bovine serum albumin at $25°$C. At time zero 10 µg/ml of unlabelled DNA is added and $0 \cdot 1$ ml aliquots are filtered in triplicate at the times indicated through 13 mm Schleicher & Schuell B_6 membrane filters which are washed once with the above buffer without bovine serum albumin, dried and counted. Each point is the average of three filtrations. ×——×, unlabelled wild-type λφ80 DNA. ●——●, unlabelled λφ80 *dlac* DNA. ▲——▲, unlabelled λφ80 *dlac* DNA + 10^{-3}M-ONPF. ■——■, unlabelled λφ80 *dlac* DNA + 10^{-3}M-IPTG.

From the initial rate of decay of RO, the half-life of the complex is of the order of 25 minutes; that is, a rate constant for the back reaction $k_b = 4$ to 6×10^{-4} sec^{-1} at $25°$C.
The values obtained for the rate constants agree within experimental error with the previously mentioned dissociation constant:

$$K_{diss} = \frac{k_b}{k} = \frac{4 \text{ to } 6 \times 10^{-4} \text{ sec}^{-1}}{4 \text{ to } 8 \times 10^9 \text{ M}^{-1} \text{ sec}^{-1}} = \sim 10^{-13} \text{ M}$$

VII *Inducer and anti-inducer interaction with RO complex*

The experiment described in Fig. 8 shows that the half-life of the repressor–operator complex, which is 25 minutes in the absence of inducer, is less than 1 minute in the presence of 3×10^{-5} M-IPTG. The anti-inducer ortho-nitrophenylfucoside (ONPF) on the other hand increases the half-life of RO to about 120 minutes.

These results indicate that both inducer and anti-inducer interact directly with the RO complex, in the first instance destabilizing it, and in the second making it more stable. The stabilization of RO by ONPF indicates the existence of a ternary complex, ONPF–RO, more stable than RO. This makes it very likely that the effect of IPTG is also achieved through the intermediate form of a labile ternary complex, IRO.

DISCUSSION

The ternary interaction between operator, repressor and inducer which controls the expression of the *lac* operon can now, as we have seen, be studied *in vitro* by the use of extremely sensitive and reasonably accurate methods. The main conclusions based on these observations may be summarized as follows:

(*a*) The product of the *i* gene, the repressor, is a protein of molecular weight about 150000, probably made up of four subunits. Since the *i* gene appears to involve a single cistron, the oligomeric protein must contain several (probably four) identical polypeptide chains.

(*b*) The protein attaches to the operator segment of competent native DNA with an extremely high affinity constant.

(*c*) In the presence of inducer, the operator–repressor complex dissociates very rapidly.

These observations may be considered to prove directly the validity of the conclusions drawn some years ago (Jacob and Monod, 1961; Willson *et al.*, 1964) from genetic-biochemical evidence and from the *in vivo* study of induction kinetics. Let us consider one by one the different steps involved in this mechanism, as they are now clearly defined by the *in vitro* studies.

The high stability and extreme specificity of the RO interaction poses the problem of the nature of the configurational features of the DNA segment which are recognized by the protein. Because of the apparent (to us) monotony of the Watson and Crick structure, the possibility has been suggested by Gierer (1966) that the operator might involve self-complementary strands and be able, therefore, to assume a tertiary structure comparable to that of t-RNA. If so, the protein might recognize this structure rather than actually a sequence of base-pairs in the conventional structure. Moreover, since such a tertiary structure should be less stable than the Watson-Crick helix, the latter would prevail in the absence of repressor (that is, in the presence of inducer), thereby allowing transcription to proceed.

This ingenious scheme is not favoured by the *in vitro* results. If the association involved, or had to be preceded by, a transconformation of the fairly rigid Watson-Crick structure, one would hardly expect this reaction to proceed at such a high rate. Moreover, as we have seen, the stability of the complex appears to be largely independent of temperature, indicating that the ΔH term in the equilibrium is negligible, while a transconformation such as suggested by Gierer should involve a significant enthalpy contribution.

It seems more likely, therefore, that the structure recognized is, and remains, in the Watson-Crick configuration. If so, the repressor must presumably recognize the sequence itself.

Assuming the interaction of each base-pair with the repressor to contribute an average of $0 \cdot 5$ to 1 kilocalorie to the interaction (a reasonable figure, for an elementary non-covalent interaction) a sequence of 10 to 20 pairs could be considered sufficient to account both for the specificity and for the stability of the association, while the deletion of one or two base-pairs would decrease the association constant by a factor of 10 or 20, as observed with the o^c operators (presumed, on strong evidence, to result in general from small deletions).

The geometry of the RO complex presents a curious and interesting problem, because of the tetrameric structure of R. If the tetramer is a regular (symmetrical) one, as is the case, in general, of oligomeric enzymes (Schachman, 1963; Reithel, 1963), then it bears four identical recognition sites, of which it could not, for obvious geometrical reasons, use more than one or two at a time in its association with the operator. One is led to speculate on the possibility that the protein undergoes a major reorganization of its quaternary structure when it binds to the operator, and also that the latter may entail a repetitive sequence of base-pairs.

Several observations reported in the experimental section suggest strongly that the operator–inducer interaction, mediated by R, is indirect, therefore presumably due to an allosteric transition of the protein. First, as we have seen, there is evidence that the inducer analogue and antagonist ONPF (Müller-Hill and Rickenberg, 1965) associates with (and stabilizes) the RO complex, most probably by binding at the inducer site, which must therefore be considered as free from direct, steric, hindrance. Hence the competition between O and I for R appears not to be a direct one.

Moreover the protein may lose all its affinity for O, while retaining it towards I. Such differential loss of affinity towards different ligands has often been observed with allosteric enzymes, giving rise to the phenomenon of "desensitization", the interpretation of which, in fact, was largely responsible for the development of the theory of allosteric interactions (Monod, Changeux and Jacob, 1963; Monod, Wyman and Changeux, 1965). Desensitization of an allosteric protein is very often, if not as a rule, found to be correlated with an alteration of the stability of its quaternary structure. This is also the case of the *lac* repressor, as we have seen.

The kinetics of the interaction, however, raise a number of interesting and difficult problems. As we pointed out, a comparison of the *in vivo* inducing activity of different galactosides with their *in vitro* capacity to dissociate the RO complex leads to entirely parallel results. However, *in vivo*, the relation between the concentration of inducer and the differential rate of enzyme synthesis is strongly sigmoidal (Herzenberg, 1959), suggesting (although evidently not proving) cooperative interactions of *several* inducer molecules for binding to the repressor. No evidence of such cooperation has been observed, so far, *in vitro*. This is disturbing (even apart from comparison with the kinetics *in vivo*) because of the probable tetrameric structure of the repressor.

Direct measurements of the forward and backward rates of the dissociation reaction:

$$RO \rightleftharpoons R + O$$

in the presence and absence of inducer reveal a most interesting fact. As we have seen, while in the absence of inducer the association is very fast, the dissociation is extremely slow, and the measured rates actually account for the high equilibrium constant of 10^{13} M^{-1}.

In the presence of inducer, the equilibrium is displaced in the direction of dissociation, due mainly, it would seem, to a considerable *acceleration* of the rate of dissociation. To account for this fact, it seems inevitable to assume that the inducer not only stabilizes differentially the two states, but actually allows the reaction to follow a pathway different from the one it takes in the absence of inducer. To be more precise: the inducer must be considered to decrease the activation energy of the transition. In other words, the inducer appears to act, in part, as a *catalyst* in the transition of R from the bound to the free state. There is no parallel to this observation in the case of the (so far very few) allosteric enzymes whose transitions have been studied using fast kinetic methods. But the possibility that specific ligands might play such a role in other physiologically significant control systems obviously is of very great interest.

Taking this remark as a lead, we may perhaps at this point indulge in some speculation. It seems likely that certain control systems in higher organisms will turn out to involve, as the *lac* system does, "pure" regulatory proteins, devoid, by themselves, of any activity and serving exclusively as mediators of chemical signals. It is also a reasonable guess that in many instances the logic of the system will be similar to that of the *lac* system, namely: the activity of a macromolecular (or multi-macromolecular) constituent will be under the control of a micromolecular compound (metabolite or hormone) acting through the intermediacy of a specialized protein, able to recognize both the controlling ligand and the active macromolecular constituent. Because the molar concentration of the latter component, as well as that of the control protein, will in general be quite low, the efficiency of such a system will require that the complex formed between the two be of high stability. In the *lac* system we find a K_A of 10^{13} M^{-1} for the RO complex, amounting to about 10 Kcal of stabilizing energy, resulting in the fact that the spontaneous dissociation of the complex is quite slow (as shown by its measured half-life of about 30 minutes) whereas induction *in vivo* takes place within seconds. Non-covalent complexes between macromolecular components may easily attain much higher stability. A degree of stabilization amounting to 25 Kcal need not involve more than 15 to 30 "elementary" interactions (between individual residues) and yet corresponds to a K_A of the order of 10^{18} M^{-1} for the complex, whose half-life would be measured in terms not of seconds or minutes but of centuries. In order usefully to control the dissociation of such a complex, a micromolecular ligand would have to accelerate the rate of dissociation by decreasing the activation energy of the transition; that is, it should act as a catalyst.

We shall not go so far as to propose, on this basis, a purely imaginary system of control involving a macromolecular component to which (say) two states would be thermodynamically available, each preferentially stabilized by alternative compounds, but separated by a high-energy barrier. Transition from one to the other state would then virtually not occur except in the presence of a ligand able specifically to *catalyse* the transition (by binding to the activated intermediate state). Yet it may be useful to remember that any phenomenon of differentiation must represent, fundamentally, the *kinetic* stabilization (following a transitory period of "induction") of thermodynamically unstable (metastable) states. Our speculative scheme of control is intended only to illustrate the fact that "differentiation" at the level of single macromolecules or macromolecular complexes is not *a priori* to be excluded as a possibly very important mechanism of long-term control.

SUMMARY

Some recent *in vitro* studies of the interaction between operator (O), repressor (R) and inducer (I) in the *lac* system of *E. coli* are summarized and discussed.

The *lac* repressor is a protein of molecular weight about 150000, probably made up of four subunits but very likely containing a single type of peptide chain. It has distinct operator and inducer binding sites.

It binds the operator specifically with an extremely high affinity constant $K_{diss} = \sim 10^{-13}$M. This constant agrees with the measured rate constants which show a very fast association ($k_f = 4$ to 8×10^9 M^{-1} sec^{-1}) and slow dissociation ($k_b = 4$ to 6×10^{-4} sec^{-1}) of the RO complex. In the presence of inducer the RO complex dissociates very rapidly while anti-inducer stabilizes the complex.

The nature of the molecular interactions involved in this simple system might fundamentally be the same as in the control of gene expression in higher organisms and the methodology may also be of use in the analysis of more complex control systems.

Acknowledgements

Work at the Salk Institute was supported by National Science Foundation Grant No. GB 5302 and National Institutes of Health Training Grant No. 1 TO1 CA05213-01, both to Dr Melvin Cohn.

Research in the Department of Molecular Biology at the Pasteur Institute has been aided by grants from the U.S. National Institutes of Health, the Centre National de la Recherche Scientifique, the Délégation Générale à la Recherche Scientifique et Technique, and the Commissariat à l'Energie Atomique.

REFERENCES

BOURGEOIS, S. (1966). *Sur la nature du répresseur de l'opéron Lactose d'*E. coli. Thèse de doctorat de l'Université de Paris.

BOURGEOIS, S., COHN, M., and ORGEL, L. E. (1965). *J. molec. Biol.*, **14**, 300–302.

BURSTEIN, C., COHN, M., KEPES, A., and MONOD, J. (1965). *Biochim. biophys. Acta*, **95**, 634–639.

COHN, M., NOTANI, G., and RICE, A. S. (1969). *Immunochemistry*, **6**, 111–123.

GIERER, A. (1966). *Nature, Lond.*, **212**, 1480–1481.

GILBERT, W., and MÜLLER-HILL, B. (1966). *Proc. natn. Acad. Sci. U.S.A.*, **56**, 1891–1898.

GILBERT, W., and MÜLLER-HILL, B. (1967). *Proc. natn. Acad. Sci. U.S.A.*, **58**, 2415–2421.

HERZENBERG, L. A. (1959). *Biochim. biophys. Acta*, **31**, 525–538.

HORIUCHI, T., TOMIZAWA, J., and NOVICK, A. (1962). *Biochim. biophys. Acta*, **55**, 152–163.
JACOB, F., and MONOD, J. (1961). *J. molec. Biol.*, **3**, 318–356.
JONES, O. W., and BERG, P. (1966). *J. molec. Biol.*, **22**, 199–209.
KUNO, H., and KIHARA, H. K. (1967). *Nature, Lond.*, **215**, 974–975.
LEGAULT-DÉMARE, J., DESSEAUX, B., HEYMAN, T., SEROR, S., and RESS, G. P. (1967). *Biochem. biophys. Res. Commun.*, **28**, 550–557.
MONOD, J., CHANGEUX, J. P., and JACOB, F. (1963). *J. molec. Biol.*, **6**, 306–329.
MONOD, J., WYMAN, J., and CHANGEUX, J. P. (1965). *J. molec. Biol.*, **12**, 88–118.
MÜLLER-HILL, B., and RICKENBERG, H. V. (1965). *Biochem. biophys. Res. Commun.*, **18**, 751–756.
NIRENBERG, M., and LEDER, P. (1964). *Science*, **145**, 1399–1407.
PTASHNE, M. (1967). *Nature, Lond.*, **214**, 232–234.
REITHEL, F. J. (1963). *Adv. Protein Chem.*, **18**, 123–226.
RIGGS, A. D., and BOURGEOIS, S. (1968). *J. molec. Biol.*, **34**, 361–364.
RIGGS, A. D., and BOURGEOIS, S. (1969a). *Biophys. J.*, **9**, A84.
RIGGS, A. D., and BOURGEOIS, S. (1969b). In preparation.
RIGGS, A. D., BOURGEOIS, S., NEWBY, R. F., and COHN, M. (1968). *J. molec. Biol.*, **38**, 365–368.
SCHACHMAN, H. (1963). *Cold Spring Harb. Symp. quant. Biol.*, **28**, 409–430.
SCHUBERT, D., ROMAN, A., and COHN, M. (1969). In preparation.
THANG, M. N., GRAFFE, M., and GRUNBERG-MANAGO, M. (1968). *Biochem. biophys. Res. Commun.*, **31**, 1–8.
WILLSON, C., PERRIN, D., COHN, M., JACOB, F., and MONOD, J. (1964). *J. molec. Biol.*, **8**, 582–592.
YARUS, M., and BERG, P. (1967). *J. molec. Biol.*, **28**, 479–490.

DISCUSSION

Harris: I have three sources of uneasiness about your scheme, Professor Monod. The first derives from the re-mapping of the *lac* locus. Until very recently we thought that the order was IOPZYA; that is, that the promoter region was distal to the operator in the direction of transcription. We now believe that the promoter region is proximal to the operator, and that therefore the operator region must be transcribed under normal conditions, unless there is a curious hop, skip and jump motion of the polymerase. This implies that the templates from this region contain the operator region. If this is so, there are two alternatives. Either these templates are translated, in which case the operator region forms part of some structural gene for some protein; or, if they are not translated, the template contains an untranslated operator region. That second possibility seems to me to put the whole thing back into the melting pot. If all templates for this region contain an untranslated operator region (or homologue of it), what is this region doing on the template except acting as a recognition site? How do you know that in the intact cell the "repressor" proteins which you assay *in vitro* against DNA would not react with the operator region on RNA?

Monod: You raise the point that if the operator is transcribed and a copy is present on the messenger, why isn't it recognized by repressor? I think there is a simple answer to this; the repressor does not bind to single-stranded DNA; and so it would have no reason to bind to the transcribed RNA piece.

Harris: I cannot accept this explanation that the repressor would not bind to RNA because RNA is single-stranded. The *in vitro* binding experiments are done with DNA which, when denatured, is a tangled mess; but you can be sure that the templates for protein synthesis will not be a tangled mess in the intact cell.

My second point is that I do not understand why the binding of a protein to the beginning of the *lac* region should block subsequent transcription elsewhere in that region. In the case of the tryptophan region where, again, there is a "promoter" site proximal to an "operator" site, Imamoto and Ito (1968) have shown that transcription can begin at internal sites in the "operon". Why should the *lac* region be different from the tryptophan region?

Monod: Transcription can start anywhere in the tryptophan region only in certain genetic conditions; it is extremely doubtful whether it does so normally, *in vivo*. If transcription could start anywhere in any operon all sorts of stupid pieces of protein would be made, which I do not think exist.

Harris: But you do get "stupid" pieces of RNA made by the tryptophan region. That is just what is found.

Monod: You may have stupid pieces of RNA in the deleted mutants.

Harris: My third point is a technical matter. Nobody has given any detailed information about the extent to which the DNA preparations used for the "repressor" assays are contaminated with protein. All these DNA preparations are made by very simple phenolic extractions which I would expect to give a 1–2 per cent contamination with protein. I wonder to what extent these extraordinarily high affinities, which vary between different laboratories, might not be due to contaminant proteins.

Monod: We would say that there may be 1 per cent protein but that is irrelevant. Why should the presence of some protein, presumably an irrelevant impurity, account for the association constant?

Harris: The question is whether it is an irrelevant impurity or whether at least some of the radioactive "repressor" molecules which you are assaying might not be binding to a protein which contaminates the DNA preparations.

Monod: If a specific protein were present we would have found mutants of that protein and they should modify the regulation. We have looked hard for genes non-allelic to the I gene or the O gene which would modify the regulation, and have never found any among thousands of mutants.

Bourgeois: All attempts to find suppressors of operator constitutive mutants have failed (Bourgeois, 1966; Bourgeois, Cohn and Orgel, 1965). This is evidence that the operator gene is not translated. Furthermore, when you denature DNA the binding to the repressor is lost; when you renature it you recover the activity. This is in essence a purification step to get rid of the postulated cryptic protein.

Monod: I think that the evidence against any specific protein having a role is extremely strong.

Bourgeois: The crude extract contains many proteins which bind to DNA and thereby stick it to the filter. However, this binding does not require the presence of the operator region. As purification proceeds the non-specific binding (neither IPTG-sensitive nor operator-dependent) is lost. No evidence has been obtained that any protein other than repressor is necessary for the repressor–operator interaction.

Talwar: May I ask about the possible mechanisms by which the repressor protein recognizes the DNA locus? If the binding is stereo-specific, is there reason to believe that there would be so many specific configurations in the DNA molecule where such proteins can fit in?

Monod: This is still a matter of speculation, but two or perhaps three possibilities have been suggested. The first is that what is recognized is a given DNA sequence of bases in the intact Watson–Crick double helix type of structure. Secondly, as Gierer has suggested, the structure of the operator region could be such that it can exist either as a Watson–Crick double helix, or, because the sequence of bases is such that each of the two DNA fibres is self-complementary, in the form of a loop. If Gierer's idea is right we might postulate that what is being recognized is not a sequence but a secondary structure in the DNA. There is no evidence yet on which to choose between these possibilities; however the transition from the Watson–Crick helix state, which should be the most stable state, to the loop structure would require a high activation energy, which seems to be incompatible with the extremely high association constants. I therefore disfavour Gierer's scheme unless one modifies it to say that the stable structure is always there and what is recognized is some secondary structure, but there is no transition between the two.

Bhargava: Gierer's model would allow for the initial recognition by the repressor of a specific sequence in the Watson–Crick helix state. The binding could then lead to the loop structure (which cannot be transcribed), so that the transition would be obligatorily dependent on the presence of an active repressor molecule.

In your experiments, Dr Bourgeois, what was the minimum size of the DNA molecules which could be bound to the inducer? And if you broke the DNA pieces down further, did you get progressively reduced binding or was the binding abruptly lost at some stage?

Bourgeois: In our binding experiments we use intact phage DNA of molecular weight 30 million. We have also shown binding to 1 million molecular weight pieces of that DNA, obtained by sonication. We did break the DNA down further using spleen DNase and followed the killing curve of operator (assayed by its binding activity) compared with the reduction in molecular weight of the DNA (determined by sedimentation equilibrium). From such a preliminary experiment the operator appears to be very small indeed, of the order of 20 base pairs, the only assumption being that the DNase cleaves DNA at random.

Monod: There are other difficulties; according to these results of Dr Bourgeois and also some unpublished work of W. Gilbert, the repressor is a tetrameric molecule and so presumably has four recognition sites. But if it is a globular protein with the usual structure for protein tetramers it cannot use its four sites at the same time, which seems absurd, especially given the very high dissociation constant. This is a nice problem of geometry.

Lynen: I was struck by the fact that you can filter the repressor on a Millipore membrane and can wash the filter several times; what was the technique used? In that respect, what is the dissociation constant of the complex between inducer and repressor?

Monod: The equilibrium constant for the repressor–inducer complex is 5×10^{-6}, which is the right order of magnitude; it fits with the findings *in vivo*.

Bourgeois: For both the repressor–inducer complex and the repressor–operator complex the mixture is filtered slowly and the filter always washed with the same small volume of buffer. A detailed description of the technique as well as control experiments and a discussion of its validity for measuring the dissociation constant in the case of the repressor–operator complex will be published soon.

Lynen: Would you not expect that if you washed extensively enough you would lose all the inducer?

Bourgeois: If one goes on washing one does slowly lose counts, which is why we wash only twice, usually, and with standard volumes, and we get reproducible results.

Monod: I think the results are incomprehensible unless one assumes that there is a denaturation of the protein on the filter itself, so that it melts, as it were, around the ligand. If the dissociation constant of the complex on the membrane remained the same as the dissociation constant in the solution you couldn't use the technique.

Talwar: What is the effect of ageing on the repressor?

Monod: According to Dr Bourgeois' results, on ageing the repressor protein dissociates into four subunits and this is accompanied by loss of affinity for the operator site. There are good reasons, mostly genetic, for there being only one type of subunit, although one might have expected two kinds, one with the operator site and the other with the inducer site. But there are no complementing mutants and therefore there is presumably a single type of polypeptide chain. This is not an isolated observation. Many allosteric enzymes lose their recognition for one of the receptors but not the other even though their subunits consist of a single polypeptide chain. It is one more case of what we call "desensitization", which was one of the initial reasons for assuming that the two sites are distinct and don't even overlap.

Harris: May I make a more general point? Professor Monod has suggested that we should attempt to make a bridge between bacteria and eukaryotic cells in discussing the regulation of concatenated sequences of enzymes. The best study of such a sequence in any eukaryotic cell, as far as I am aware, is the nitrate reductase system in *Aspergillus nidulans* (Pateman and Cove, 1967). Here, a sequence of enzymes is involved in the reduction of nitrate. The enzymes are regulated coordinately, and there is both "negative" and "positive" control. But each of these genes in *Aspergillus* maps at a completely different locus. They are not clustered together at all. But the enzymes are regulated coordinately nonetheless. This suggests that the clustering of the genes at the *lac* locus in *E. coli* is not required in order to achieve coordinate regulation of the related group of enzymes. It seems more likely that this clustering is required by the sexual mechanisms of *E. coli*. In *E. coli*, where the sexual mechanism is the transfer of the chromosome in an ordered sequence, it would presumably be a selective advantage to keep groups of genes involved in a single metabolic pathway together; because, if the chromosome breaks in the middle of such a group, a lethal situation could be produced. In higher organisms, which do not have this kind of sexual mechanism, there would be no selective advantage in keeping such genes together.

The "operon", the cluster, may really have little to do with coordinate regulation; it may simply be a device for safe sex.

Monod: I think I agree with you. In bacteria also there are systems where the coordination of regulation is not as strict and stoicheiometric as for an operon. The best example is the arginine system of *E. coli.* It is a very important question whether there are operons of more than a single gene in higher organisms. There are a number of examples where this is probable. But the genetics is so poor that it is difficult to decide. But I would agree that the presence of many operons with many genes, up to eight or nine in bacteria, may have something to do with the sexual process.

Siddiqi: I would like to raise here the possibility of controlling the formation of active enzyme in the absence of protein synthesis. Such an activation might involve a change in the covalent structure of the enzyme. We have been looking at the genetic regulation of aryl sulphatase in *Aspergillus nidulans* (Siddiqi, Apte and Pitale, 1966). This enzyme is repressed by inorganic sulphate or by amino acids that contain sulphur. Mutations in one of several different genes lead to constitutive enzyme synthesis.

An interesting fact has emerged. If cycloheximide is added some time after derepressing the enzyme, enzyme formation continues undiminished for several hours. The incorporation of labelled amino acids into protein, on the other hand, ceases immediately upon the addition of cycloheximide. Thus during a period when there is no detectable incorporation of amino acids into protein, enzyme activity increases fifty-fold. If cycloheximide is added soon after derepression, only a limited amount of enzyme is formed; that is to say, there is an early cycloheximide-sensitive step in the formation of active enzyme and a subsequent cycloheximide-insensitive step. This may be a trivial variation of the basic Jacob–Monod system. On the other hand an additional control at the cycloheximide-insensitive step may offer some advantages.

Monod: This type of phenomenon might be analogous to the activation of pancreatic enzyme, for example; or it may be a more sophisticated kind of system similar to what has been found by Stadtman and his co-workers with glutamine synthetase, where the enzyme is allosteric to a vast number of ligands, and it has two "states", one with a covalent AMP group and one without. Here again allosteric systems seem to be involved because whether the modifying ligand is covalent or non-covalent may depend on the physiological advantages. We have also the old example of phosphorylase *b* which can be activated either by phosphorylation with a specific enzyme, phosphorylase kinase, or by adding AMP, and the evidence from studies by H. Buc is that kinetically and in many other respects phosphorylase *b* activated by AMP has exactly the properties of phosphorylase *a*. These different types of regulation need not be mutually exclusive, therefore.

REFERENCES

BOURGEOIS, S. (1966). Thesis, Université de Paris.
BOURGEOIS, S., COHN, M., and ORGEL, L. (1965). *J. molec. Biol.*, **14**, 300–305.
IMAMOTO, F., and ITO, J. (1968). *Nature, Lond.*, **220**, 27.
PATEMAN, J. A., and COVE, D. J. (1967). *Nature, Lond.*, **215**, 1234.
SIDDIQI, O., APTE, B. N., and PITALE, M. P. (1966). *Cold Spring Harb. Symp. quant. Biol.*, **31**, 381–382.

MODULATION OF ENZYME ACTIVITIES BY METABOLITES

F. LYNEN

Max-Planck Institut für Zellchemie, Munich

IN his book *Wasserstoffübertragende Fermente* published in 1948 Otto Warburg, one of the great pioneers of modern enzymology and biochemistry, wrote: "In einem noch so komplizierten Gemisch von Substanzen und prosthetischen Gruppen wird jede Fermentreaktion so ablaufen, als ob alle in ihr nicht teilnehmenden Stoffe nicht vorhanden wären. So kann im Leben keine physiologische Fermentreaktion die andern stören."* Only about 20 years have passed since then, but our views about enzyme reactions occurring in a complex mixture of substrates and coenzymes, as present in living systems, have completely changed. We are now fully aware of the fact that many metabolites, not directly involved in a chemical reaction catalysed by a specific enzyme, can influence that enzyme's activity to a considerable extent. Modulation of enzyme behaviour by variation in the concentration of specific metabolites has been observed in very many cases and the important role of these effects in biological regulation is now generally accepted.

The first observations of such metabolite effects go back more than 20 years. As an example, already in the forties the two Coris (Cori and Cori, 1945) had found that muscle phosphorylase *b* requires the addition of AMP for full activity. But the general importance of metabolite modulation in metabolic control processes was first recognized in connexion with studies on biosynthetic pathways in microorganisms. It was observed that mutant strains of microorganisms blocked late in a biosynthetic sequence tend to accumulate an intermediate prior to the blocked step, but in many cases this accumulation is abolished if the normal endproduct of the sequence is supplied at a level allowing rapid growth

* "No matter how complicated a mixture of substances or prosthetic groups [coenzymes], every enzyme reaction will take place as if all the other substances, not necessary for it, were not present. In life therefore no physiological enzyme reaction can disturb another."

28

(cf. Umbarger, 1961). This fundamental observation was further supported by tracer experiments which demonstrated that the incorporation of endogenous carbon into an amino acid was specifically depressed when the amino acid was available in the medium (Roberts *et al.*, 1955). In order to explain such effects Umbarger (1956) and also Yates and Pardee (1956) developed the concept of feedback control mechanisms: the inhibition of enzymes early in a biosynthetic pathway by the end-products of this pathway. Finally, when it was found that feedback-controlled enzymes such as aspartate transcarbamylase (Gerhard and Pardee, 1961) or threonine dehydrase (Changeux, 1961) lost sensitivity to their inhibitors under various conditions while retaining catalytic activity, it became apparent that such enzymes might possess separate but interacting sites for substrates and inhibitors. Emphasizing the fact that the inhibitor need not be a steric analogue of the substrate, Monod and Jacob (1961) introduced the term "allosteric inhibition" to describe such interactions. In the following years this new term was mainly accepted.

In order to explain "allosteric inhibition" Gerhard and Pardee (1962) suggested in 1962 that binding of inhibitor leads to a conformational change in the enzyme, with a concomitant alteration of affinity for substrate at the catalytic site. The ground for this suggestion, which has served as the point of departure for later discussions of regulatory enzyme kinetics, was well prepared by Koshland's "induced-fit" theory of enzyme action. In his theory, first published in 1958, Koshland (1958) had already postulated that small molecules could induce conformational changes on binding to enzymes.

The new element introduced with Koshland's "induced-fit theory" as compared to Emil Fischer's "template" or "lock and key" hypothesis (Fischer, 1898) of enzyme action may be seen in the fact that in the new theory the "flexibility" of proteins was fully recognized. If one looks back, it seems very astonishing that enzymologists at that time had generally accepted the hypothesis that substrates are flexible and are fixed in a specific conformation when bound to enzymes. But the enzymologists were not at all aware of the fact that enzyme proteins, when binding the substrates, ultimately may also be fixed in a specific conformation. At that time enzyme proteins were generally considered to possess a rather rigid structure. From the recent X-ray analysis of a number of enzyme proteins (cf. Perutz, 1969) compared with the corresponding enzyme-substrate or enzyme–inhibitor complexes it is now well established that the final conformational state of a protein is the sum of the complementary interactions between ligand and protein (Koshland, 1968).

2*

As a consequence of conformational changes of enzyme proteins, activation or inhibition of enzyme activity may occur. For the molecules interacting at the "allosteric sites" of enzymes, and thereby inducing these effects, the terms negative and positive "effectors", "modifiers" or "modulators" were introduced (Atkinson, 1966). It was also found that any such modulation of enzyme activity may result from effects on the two parameters of enzyme kinetics: either on the Michaelis constant (K_M) or on maximal velocity (V_{max}) (Monod, Wyman and Changeux, 1965).

CONTROL OF FATTY ACID SYNTHESIS

I would like to restrict the further discussion to results which were obtained in studies on fatty acid synthesis and its regulation, where modulation of enzyme activities by metabolites was found to be important. Fatty acid synthesis from acetyl CoA is achieved in two steps. In the first step acetyl CoA is combined with carbon dioxide to form malonyl-CoA (equation 1). This reaction, which is linked to the cleavage of ATP into ADP and orthophosphate (P_i), is catalysed by acetyl CoA carboxylase. In the second step fatty acids are formed from malonyl CoA. This synthesis, catalysed by fatty acid synthetase, requires NADPH as a reducing agent and small amounts of acetyl CoA, which serves as "primer" of the process (equation 2).

(1) Acetyl $CoA + CO_2 + ATP \rightleftharpoons$ malonyl $CoA + ADP + P_i$

(2) Acetyl $CoA + 7$ malonyl $CoA + 14$ NADPH $+ 14$ H$^+ \rightarrow$
    palmitate $+ 8$ CoA $+ 7$ $CO_2 + 14$ NADP$^+ + 6$ H_2O

Equation 2 holds for the process catalysed by the fatty acid synthetase from animal tissues (Hsu, Wasson and Porter, 1965). The analogous enzyme isolated from yeast cells catalyses the formation of fatty acyl CoA derivatives like palmitoyl and stearoyl CoA (Lynen, Hopper-Kessel and Eggerer, 1964).

With respect to the control of fatty acid synthesis in whole animals and in tissue slices, it has been known for many years that dietary and hormonal conditions may exert pronounced effects (Fritz, 1961; Masoro, 1962). The main observations are summarized in Table I. Starvation, fat feeding or diabetes, induced by the injection of alloxan, decrease fatty acid synthesis greatly. In the latter case it is restored on giving insulin to the alloxan diabetic animals. Refeeding of starved animals restores fatty acid synthesis, if the diet is not high in fat. A diet high in carbohydrate and low in fat leads to the highest rates.

TABLE I

RATE OF FATTY ACID SYNTHESIS UNDER VARIOUS CONDITIONS

Metabolic or hormonal state	Rate of fatty acid synthesis
Starvation	decreased
Diet high in fat	decreased
Diet high in carbohydrate	increased
Diabetic	decreased
Diabetic given insulin	increased

When the effects of dietary and hormonal conditions were studied in detail it turned out that two types of control mechanisms have to be distinguished. One is short term: it can take as little as 30 to 90 minutes to come into effect. The other is, by comparison, long term. In the long-term control the level of several enzymes participating in fatty acid synthesis, and including acetyl CoA carboxylase, was found to be the decisive factor. On the other hand, most of the experimental evidence available suggests that the short-term control of fatty acid synthesis is mediated through the modulation of acetyl CoA carboxylase activity by allosteric effectors.

LONG–TERM CONTROL

The reason for studying the level of this particular enzyme under various conditions is related to the fact that the first step in the biosynthetic sequence leading specifically to fatty acids is the carboxylation of acetyl CoA. Accordingly, it would be important, teleologically speaking, to regulate fatty acid synthesis at the carboxylation step. In experiments with rat liver extracts it was found that fasting leads to a strong depression of acetyl CoA carboxylase and to a smaller depression of fatty acid synthetase (Numa, Matsuhashi and Lynen, 1961). Similar findings were also obtained with diabetic rats (Wieland et al., 1963) as well as with rats fed on a fatty diet (Bortz, Abraham and Chaikoff, 1963). On the other hand, the specific activities of both enzymes rise to very high levels in liver during high-carbohydrate, fat-free refeeding of starved rats (Gibson, Hicks and Allmann, 1966). The concentrations of other liver (and adipose tissue) enzymes contributing to the net conversion of glucose to triglyceride were also influenced by the dietary regime. Glucokinase, glucose-6-phosphate dehydrogenase, long-chain fatty acyl CoA desaturase and citrate cleavage enzyme are depressed in starvation and become elevated in response to refeeding (Gibson, Hicks and Allmann, 1966).

Citrate cleavage enzyme catalyses reaction 3:

(3) $ATP + CoA + citrate \rightarrow acetyl\ CoA + oxaloacetate + ADP + P$

This reaction is of importance in the total picture of fatty acid synthesis from carbohydrates, since citrate is considered to act as a vehicle for the transfer of the acetyl group from the site of its formation to the site of its utilization for lipogenesis (Lowenstein, 1968). The enzyme system for fatty acid synthesis was proved to be located in the cytoplasm together with the enzymes transforming glucose and other carbohydrates into pyruvate. Pyruvate, formed in the cytoplasm, then moves into the mitochondrion and is there oxidized to acetyl CoA, but as such cannot pass into the cytoplasm. However, acetyl CoA by condensing with oxaloacetate forms citrate which can be released from the mitochondrion. In the

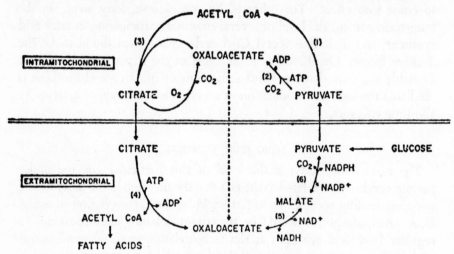

FIG. 1. A postulated role for citrate cleavage and malic enzyme in lipogenesis (Ball, 1966).

cytoplasm acetyl CoA is then regenerated from citrate by the action of the citrate cleavage enzyme (Fig. 1). The fact that this enzyme in liver or mammary gland goes through an adaptive increase during refeeding of the starved animal or the onset of lactation, respectively, provides a basis for believing that it is instrumental in the flow of glucose to triglycerides, as suggested originally by Lowenstein (1968) and by Bhaduri and Srere (1963).

Malic enzyme, catalysing reaction 4, is another enzyme which should be considered in this context.

(4) $Malate + NADP^+ \rightarrow pyruvate + CO_2 + NADPH + H^+$

Adipose tissue was found to be rich in both malic enzyme and citrate cleavage enzyme (Ball, 1966). Alterations in dietary regime which bring about enhanced lipogenesis in adipose tissue produce concomitant and nearly equal increases in the activity of both these enzymes, which may reach values five times normal. Ball (1966) has therefore suggested that oxaloacetate formed in the citrate cleavage reaction may be involved in the hydrogen transfer from NADH to NADP. According to this concept oxaloacetate is reduced by NADH, formed during the conversion of glucose to acetyl CoA, and the malate produced is then oxidized by malic enzyme to pyruvate and carbon dioxide with the generation of NADPH. This NADPH, along with TPNH produced in the pentose cycle, is used to reduce malonyl CoA to fatty acids (cf. equation 2). The pyruvate formed in the malic enzyme reaction (equation 4) in the cytoplasm enters the mitochondria where it is converted to oxaloacetate by pyruvate carboxylase (Fig. 1).

In other words the pyruvate carboxylase–malate dehydrogenase–malic enzyme system represents an ATP-requiring transhydrogenation, and can account for the transformation of reducing equivalents (NADH formed during the conversion of glucose to pyruvate via the Embden-Meyerhof pathway) into NADPH suitable for lipogenesis.

It is thought that the depression of enzyme activities during starvation is due to the repression of enzyme synthesis. Thus, the rise in activity attending high-carbohydrate, fat-free refeeding could be explained as adaptive synthesis of new enzyme. Evidence supporting this view has emerged in studies of Gibson, Hicks and Allmann (1966). As they found, specific inhibitors of protein synthesis, such as puromycin and actinomycin D, injected into rats at the beginning of or during the refeeding period block the anticipated rise in activity of acetyl CoA carboxylase, fatty acid synthetase and citrate cleavage enzyme.

The assumption that the changes in enzyme concentration are due to repression or induction of enzyme synthesis is further supported by experiments of Gellhorn and Benjamin (1966) with adipose tissue, who found that the rates of fatty acid production and their desaturation are reflected in the rates of RNA synthesis. That there is a functional relationship between these events is strongly suggested by the fact that the restoration of RNA synthesis with refeeding precedes the repair of the failure in fatty acid synthesis.

SHORT-TERM CONTROL

In addition to the repression of the *de novo* synthesis of various enzymes involved in fatty acid synthesis, as a result of starvation or diabetes, another regulatory mechanism was found to be responsible for the loss of synthetic capacity. As an example, when Korchak and Masoro (1962) compared the rates of fatty acid synthesis by liver slices and by liver homogenate fractions after a shorter fasting period (24 hours), they found that the synthetic activity of extracts had fallen only by about 50 per cent, whereas in liver slices at the same time fatty acid synthesis was depressed by 99 per cent. Similar findings were also obtained in experiments with diabetic rats or with rats fed a fatty diet. As Hill and his co-workers (1960) demonstrated, a 50 per cent decrease in the incorporation of acetate carbon into fatty acids by liver slices occurs as early as 1 hour after rats are fed 2 ml of corn oil by stomach tube. After 3 hours the capacity for fatty acid synthesis had fallen to less than 20 per cent. At the same time no apparent changes in the level of enzymes could be observed (Bortz, Abraham and Chaikoff, 1963). Such observations suggested that further factors, in addition to eventual variations in the concentration of acetyl CoA carboxylase and other enzymes, are involved in the regulation of fatty acid synthesis in intact liver cells. An obvious possibility seemed to be the modulation of enzyme activities. As was soon found in our laboratory, acetyl CoA carboxylase was the preferential point of control (Numa, Bortz and Lynen, 1965).

One of the unique features of this enzyme from animal sources is its activation by tri- and dicarboxylic acids, of which citrate and isocitrate are by far the most effective. It was discovered by Vagelos, Alberts and Martin (1963) and confirmed by the studies of Numa in our laboratory (Numa, Ringelmann and Lynen, 1965) and of Lane in New York (Gregolin et al., 1966a, b) that this activation by citrate or isocitrate, affecting the V_{max} of the enzyme, is accompanied by the aggregation of the inactive protomeric form of the enzyme to an active polymeric form, probably as a result of an allosteric transition. This aggregation could be followed by studies of the pure enzyme in the analytical ultracentrifuge (Gregolin et al., 1966a, b; Numa, Ringelmann and Riedel, 1966; Henniger, 1969), by sucrose density gradient centrifugation (Numa, Ringelmann and Lynen, 1965; Numa, Ringelmann and Riedel, 1966; Gregolin et al., 1966a, b; Gregolin et al., 1968), by electron microscopy (Gregolin et al., 1966a) or by light-scattering measurements (Henniger, 1969). The aggregation phenomenon was found with enzyme preparations from rat adipose tissue and from rat or chicken liver. The most careful studies were carried out with the chicken liver enzyme, which could be isolated in pure form (Gregolin, Ryder and Lane, 1968) and even be crystallized (Goto et al., 1967). The molecular dimensions of the two forms determined by the various techniques were found to be in good agreement.

TABLE II

PHYSICOCHEMICAL PROPERTIES OF THE TWO FORMS OF ACETYL CoA CARBOXYLASE FROM
CHICKEN LIVER

		Molecular weight measured by:				
Preparation	Sedimentation coefficient (a, b, c, d)	Ultra-centrifugation (a, b, c, d)	Light scattering (d)	Biotin content per molecule	Enzymic activity	Ratio of axes (d)
"Small" form	13·1–14·5S	410 000–440 000	—	1	inactive	1:10
"Large" form	42·4–62·5S	4–8·8 × 10^6	11–12 × 10^6	~10–20	active	1:35

(a) Numa, Ringelmann and Riedel (1966)
(b) Gregolin et al. (1966a)
(c) Gregolin et al. (1966b)
(d) Henniger (1969)

They are summarized in Table II. Studies with the electron microscope and by light scattering revealed a rod-like structure of both forms of the enzyme. The average dimensions, as calculated by Henniger, are shown in Fig. 2.

The comparison of the molecular weights indicated that the "large" form is built by aggregation of between 10 and 20 molecules of the "small" form. It became apparent that, within limits, a continuous array

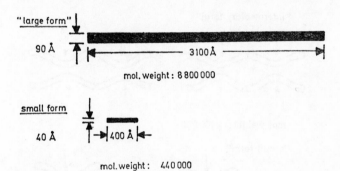

FIG. 2. Average dimensions of the two forms of acetyl CoA carboxylase
(Henniger, 1969).

of aggregational levels of carboxylase protomers is possible, the degree of aggregation being dependent upon a variety of factors, including the nature of the anion, ionic strength, and pH. Assuming that during the aggregation process which leads to the "large" form the structure of the "small" form is not changed significantly, Henniger (1969) suggested a double-helix-like structure of the "large" forms, as shown in Fig. 3. This hypothesis could also explain the existence of an "intermediate" form which was observed by Numa and co-workers (1967) in sucrose gradient centrifugation studies with the purified chicken liver enzyme. It should be briefly mentioned that the "small" form itself is built from subunits. Dissociation of chicken liver acetyl CoA carboxylase with sodium dodecyl sulphate gave rise to subunits of molecular weight about 100 000 (Gregolin *et al.*, 1968). The presence of non-identical subunits

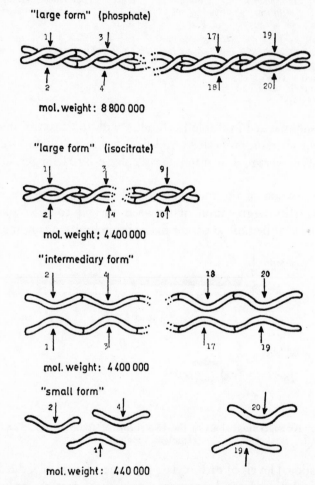

"large form" (phosphate)

mol. weight: 8 800 000

"large form" (isocitrate)

mol. weight: 4 400 000

"intermediary form"

mol. weight: 4 400 000

"small form"

mol. weight: 440 000

FIG. 3. Hypothetical structures of the various forms of acetyl CoA carboxylase (Henniger, 1969).

is indicated by the fact that there is a single biotinyl prosthetic group on the 410 000–440 000 molecular weight protomer. By treating the enzyme with 6M-guanidine·HCl in phosphate buffer and in the presence of mercaptoethanol Henniger (1969) detected even subunits possessing a molecular weight of about 60 000. The results of these dissociation experiments seem to indicate that the "small" form of acetyl CoA carboxylase itself must have a rather complex quaternary structure.

With respect to the control of fatty acid synthesis, only the reversible transformation of the "small" form of the enzyme into the "large" form seems to be involved. All available evidence suggests that only the aggregate is catalytically active. In this context experiments of Numa and Ringelmann (1965) with the rat liver enzyme were very instructive. Unlike the chicken liver enzyme the rat liver acetyl CoA carboxylase requires activation by prior incubation with citrate before the full enzyme activity can be elicited (Numa and Ringelmann, 1965; Chang et al., 1967; Fang and Lowenstein, 1967). Within limits the rate of activation during the preincubation period depends on the concentration of enzyme (as expected for an aggregation process) and citrate respectively and on the temperature (Fang and Lowenstein, 1967; R. Okazaki and S. Numa, personal communication, 1968). When crude rat liver extracts were preincubated at 37°c and in presence of 10 mM-citrate, maximum activity for fatty acid synthesis was found only after 30 minutes.

In previous measurements of acetyl CoA carboxylase and fatty acid synthetase activity of rat liver extracts this unique feature was not taken into consideration (Numa, Matsuhashi and Lynen, 1961) with the result that the acetyl CoA carboxylase activity found was too low. More recent studies with fully activated enzyme preparations showed that carboxylase and synthetase are present in about equal activity (Chang et al., 1967; Okazaki and Numa, personal communication, 1968). However, it is rather doubtful whether in vivo acetyl CoA carboxylase in rat liver is maximally activated. The citrate concentration necessary for full activation of the enzyme was found to be 10 mM* whereas the citrate concentration of liver measured in vivo amounted only to about 0·3 mM (Regen, 1967; Spencer and Lowenstein, 1967). Therefore it is still valid to assume that in vivo the carboxylation of acetyl CoA is the rate-limiting step of fatty acid synthesis, as originally proposed by Ganguly (1960) and by Numa, Matsuhashi and Lynen (1961).

The rather slow activation of rat liver acetyl CoA carboxylase, which seems to be related to the kinetics of protomer aggregation, enabled Numa and Ringelmann (1965) to prove the close correlation between aggregation and enzyme activity. In Figs. 4 and 5 their crucial experiments can be seen. Acetyl CoA carboxylase, partially purified from rat liver, was incubated with 10 mM-citrate alternately at 0° and 25°c. At various

---

* The citrate concentration required for half maximal activation of rat liver acetyl CoA carboxylase was found to be 3·9 mM (Numa, Ringelmann and Lynen, 1965).

time-intervals enzyme activity was measured under standard conditions (Fig. 4). In parallel, the sedimentation behaviour of the enzyme in a sucrose density gradient at 20°c and at 4°c was studied (Fig. 5). In the sedimentation experiments, fatty acid synthetase from yeast as a standard enzyme with a known and temperature-independent sedimentation coefficient of 43s was centrifuged together with the carboxylase in the same tubes, which additionally contained 10 mM citrate. As shown in Fig. 4, the enzyme, which had been activated by citrate at 25°c, was reversibly inactivated by exposure to cold. This reversible activation and inactivation of acetyl CoA carboxylase was accompanied by profound

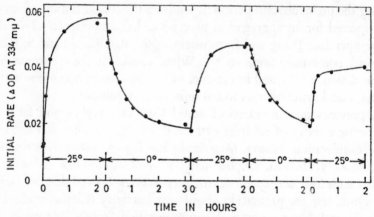

FIG. 4. Reversible inactivation of citrate-activated acetyl CoA carboxylase at 0°c (Numa and Ringelmann, 1965).

changes in its sedimentation behaviour. As can be seen from Fig. 5, the carboxylase sedimented faster at 25°c than at 4°c. By comparison with synthetase the sedimentation coefficient of the carboxylase at 25°c was estimated to be about 50s and that at 4°c between 17 and 20s. Within the limits of error both values are in close agreement with the sedimentation coefficients of the "large" and "small" forms measured for the polymer and protomer of acetyl CoA carboxylase (Table II). These experimental results are very instructive because they may be interpreted to indicate that the catalytically active conformation of the carboxylase protein cannot be induced by citrate alone but in addition requires the protein–protein interactions in the polymeric state, which occur only at elevated temperatures.

The crucial role of aggregation for the appearance of enzyme activity is also indicated by experiments with the purified chicken liver carboxylase. Gregolin and co-workers (1966a, b) as well as Numa, Ringelmann and Riedel (1966) found that the enzyme in phosphate buffer is present as the "large" form and when assayed in the absence of citrate or isocitrate

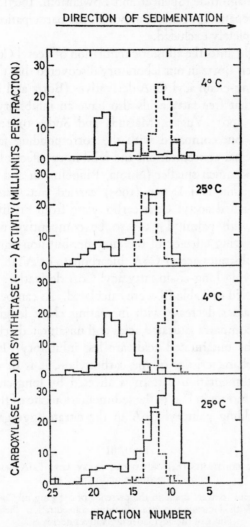

FIG. 5. Density gradient centrifugation studies of acetyl CoA carboxylase at 4°C and at 25°C (Numa and Ringelmann, 1965). The solid lines represent the carboxylase activity and the dashed lines the fatty acid synthetase activity.

is initially active but decays within several minutes, as a result of the transition into the "small" form. The dissociation and the concomitant loss of enzymic activity could be prevented by the addition of citrate or isocitrate to the assay mixture. The physiological significance of the stimulatory effect of citrate alone remains obscure. Decreases in the citrate level in the liver caused by fasting, diabetes or fat feeding were not of significant magnitude (Spencer and Lowenstein, 1967) to account for the depressed hepatic lipogenesis, although the participation of this factor cannot be completely excluded.

Further insight into the allosteric regulation of acetyl CoA carboxylase was gained when Bortz in our laboratory discovered that the enzyme was inhibited by long-chain acyl CoA derivatives (Bortz and Lynen, 1963a). It was shown that free fatty acids also have an inhibitory effect on the enzyme (Levy, 1963; Yugari, Matsuda and Suda, 1964; Korchak and Masoro, 1964), but compared with the corresponding fatty acyl CoA derivatives at equal concentrations the inhibitory effect of the free acids was found to be much smaller (Numa, Ringelmann and Lynen, 1965). Numa, Ringelmann and Lynen (1965) carried out systematic kinetic studies with purified acetyl CoA carboxylase from rat liver and found the inhibition with palmityl CoA to be competitive with regard to citrate, the activating effector of the enzyme, but non-competitive with regard to the substrates acetyl CoA, bicarbonate or ATP. From experiments with varying long-chain fatty acyl CoA derivatives the inhibitory constants, K_i, listed in Table III were calculated. As can be seen from the table, the K_i values decrease with increasing chain length of the acyl radical. If one compares saturated acids and unsaturated acids of the same chain length, the unsaturated acids are less inhibitory. In view of the sedimentation characteristics of the carboxylase, it is of interest to see whether the sedimentation pattern is affected by long-chain acyl CoA derivatives. As shown in Fig. 6, the sedimentation rate of the enzyme was decreased by adding palmityl CoA to the citrate-aggregated form and

TABLE III

INHIBITORY CONSTANTS (K_i) OF VARIOUS FATTY ACYL COA DERIVATIVES
(Numa, Ringelmann and Lynen, 1965)

The measurements were made in the presence of 0·75 mg/ml bovine serum albumin. With increasing serum albumin concentration the inhibitory effect of the acyl CoA derivatives decreased.

Fatty acyl CoA	K_i (µM)
stearyl CoA	0·33
oleyl CoA	0·67
margaryl CoA	0·36
palmityl CoA	0·81
palmitoleyl CoA	2·5
myristyl CoA	26
decanoyl CoA	35

corresponded to the "small" form found in the absence of citrate. There is again a correlation between enzyme activity and the sedimentation behaviour of the carboxylase.

Our studies have been restricted to the properties of the enzyme *in vitro* and must be interpreted with some reservations with regard to the

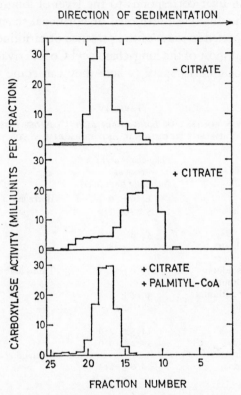

FIG. 6. Density gradient centrifugation studies of acetyl CoA carboxylase in the presence and absence of the effectors citrate and palmityl CoA (Numa, Ringelmann and Lynen, 1965).

physiological regulation of fatty acid synthesis. It is, nevertheless, conceivable that our findings might in fact represent part of the cellular control mechanism. Elevated concentrations of fatty acids in the blood as a result of increased lipolysis in adipose tissue are associated with starvation, diabetes or alimentary fat-loading, conditions in which fatty acid synthesis is known to be almost fully blocked (cf. Table I). An elevated concentration of fatty acids in the blood may lead to an influx of fatty acids into the tissues where they become esterified with coenzyme A. In agreement with this it was found in several laboratories that the concentration of the long-chain acyl CoA compounds in rat liver is markedly increased under all conditions of depressed fatty acid synthesis (Table IV). As can also be seen in the table, the level decreases when sugar is fed, a condition which leads to increased fatty acid synthesis. As Bøhmer, Norum and Bremer (1966) have shown, the level of long-chain acyl-carnitines in liver corresponds to the level of long-chain acyl CoA derivatives in the different nutritional or hormonal states of the animal. This is to be expected in view of the presence of carnitine acyltransferase and indicates that most of the long-chain acyl CoA derivatives are present in the extramitochondrial space, where they can come in contact with

TABLE IV

RAT-LIVER CONTENT OF LONG-CHAIN ACYL CoA DERIVATIVES IN
DIFFERENT NUTRITIONAL AND HORMONAL STATES

State	Long-chain acyl CoA content (nmoles/g fresh liver)	References
Balanced diet	14·9± 0·5	Bortz and Lynen (1963b)
Starved for 24 hours	57·7± 7·9	
Balanced diet	52·8± 8·6	
Starved for 48 hours	110·0±28	
Starved for 48 hours, fed fat for 48 hours	135·0±23	Tubbs and Garland (1964)
Starved for 48 hours, fed sugar for 48 hours	30·4± 9·5	
Alloxan diabetic	92·0±20	
Balanced diet	14·5± 2·6	
Alloxan diabetic	58·2±10·6	
Alloxan diabetic, insulin treated	16·0± 2·3	Wieland et al. (1965)

acetyl CoA carboxylase. Therefore it is tempting to speculate that fatty acid synthesis, at least in liver, is under negative feedback control. As can be seen from the metabolic map (Fig. 7), the inhibitory long-chain acyl CoA derivatives represent the last molecules in the synthetic sequence before subsequent incorporation into the "complex lipids". Further, the inhibition would affect the enzyme acetyl CoA carboxylase, the point at which fatty acid synthesis branches off from the many other reaction paths of acetyl CoA.

It is not yet known to what extent the increase in long-chain acyl CoA derivatives may be related to a fall in the concentration of α-glycerol phosphate, experimentally observed (Bortz and Lynen, 1963b) and related to the restrictions in carbohydrate metabolism in fasting or diabetic animals. α-Glycerol phosphate is required for the formation of triglycerides and phospholipids, and in this way can trap the long-chain fatty acids from their coenzyme A derivatives. In a secondary manner its concentration may affect the rate by which the long-chain acyl CoA derivatives are consumed. In experiments with the soluble supernatant fraction of rat liver extracts, the addition of microsomes stimulates fatty acid synthesis from acetate and acetyl CoA but not when malonyl CoA serves as substrate. The products of synthesis in the supernatant system

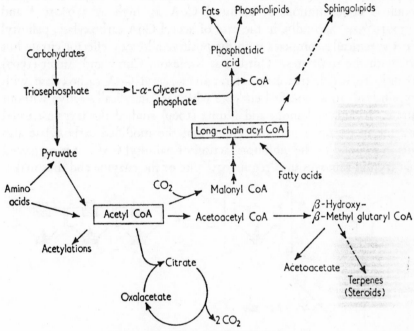

FIG. 7. Metabolic pathways of acetyl CoA.

were free fatty acids. In the system composed of the supernatant fraction plus microsomes, however, complex lipids were formed. Lorch, Abraham and Chaikoff (1963) concluded that one of the reaction products of fatty acid synthesis is an inhibitor of acetyl CoA carboxylase and that the further conversion of this intermediate by microsomal enzymes results in a non-inhibitory complex lipid. The long-chain acyl CoA derivatives are logical candidates for this role.

The assumption that long-chain acyl CoA derivatives are physiologically active as negative effectors of acetyl CoA carboxylase has been criticized by Taketa and Pogell (1966). In view of their finding that palmityl CoA was a rather general inhibitor of a number of enzymes they suggested that the inhibitory effects were rather unspecific and were due to the detergent-like properties of the long-chain acyl CoA esters. According to their view, the acyl CoA esters bind to many proteins and can thus produce conformational changes which may or may not alter enzyme activity. It seems to us that the interaction of palmityl CoA with acetyl CoA carboxylase is of a more specific nature. Firstly, the K_i value for palmityl CoA was determined to be $8 \cdot 1 \times 10^{-7}$ M, whereas 50 per cent inhibition of other enzymes, as studied by Taketa and Pogell, required concentrations of palmityl CoA as high as $3 \cdot 0 \times 10^{-4}$ and $2 \cdot 3 \times 10^{-2}$ M. Secondly, in the case of acetyl CoA carboxylase, palmityl CoA specifically competes with the positive allosteric effector citrate but not with the substrates. Thirdly, as Swanson, Curry and Anker (1967) recently reported, treatment of rat liver acetyl CoA carboxylase with trypsin leads to a modified enzyme which is catalytically active without citrate. Iritani, Nakanishi and Numa (1969) studied the trypsin-treated enzyme in greater detail and found that the modified carboxylase also fails to respond to the inhibitory action of palmityl CoA. It is suggested that trypsin removes the "regulatory" site of the enzyme and thus makes

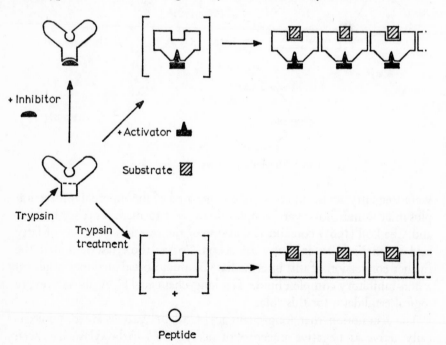

FIG. 8. Hypothetical model of acetyl CoA carboxylase regulation.

it insensitive towards the allosteric effectors citrate and palmityl CoA. The Japanese authors studied the trypsin-treated enzyme by sucrose density gradient centrifugation. They found, in contrast to the results of Swanson, Curry, and Anker (1968), that the "large" form sedimented. Ostensibly removal of the "regulatory" site of acetyl CoA carboxylase by trypsin leads to a modified protein which does not require citrate for aggregation and thereby assumes the conformation required for enzymic activity. This hypothesis is schematically illustrated in Fig. 8.

In view of the experimental evidence we feel confident that the control mechanism we have studied *in vitro* is of physiological significance.

SUMMARY

With respect to the regulation of fatty acid synthesis in animals a differentiation can be made between "long-term" and "short-term" control. In the long-term control the level of several enzymes participating in fatty acid synthesis was found to be a decisive factor. The short-term control is mediated through the modulation of the biotin enzyme acetyl CoA carboxylase. Effectors are citrate and long-chain acyl CoA derivatives, which possess antagonistic action as the result of direct competition for a common allosteric site. The activation of acetyl CoA carboxylase by citrate is accompanied by aggregation of the inactive protomeric protein (mol. wt. 420000) to an active polymer (mol. wt. 4–10 million). Activity and aggregation state are intimately associated. As Anker has found, liver acetyl CoA carboxylase is also activated by treatment with trypsin. Investigation of the trypsin-treated enzyme in Numa's laboratory revealed its polymeric form. The activity of the trypsin-treated enzyme is very little affected by citrate or by palmityl CoA, which seems to indicate that the allosteric site of the enzyme was removed by trypsin treatment. The subunit structure of this enzyme is discussed in detail.

REFERENCES

ATKINSON, D. E. (1966). *A. Rev. Biochem.*, **35**, 85.

BALL, E. G. (1966). *Adv. Enzyme Regulation*, **4**, 3.

BHADURI, A., and SRERE, P. A. (1963). *Biochim. biophys. Acta*, **70**, 221.

BØHMER, T., NORUM, K. R., and BREMER, J. (1966). *Biochim. biophys. Acta*, **125**, 244.

BORTZ, W. M., ABRAHAM, S., and CHAIKOFF, I. L. (1963). *J. biol. Chem.*, **238**, 1266.

BORTZ, W. M., and LYNEN, F. (1963*a*). *Biochem. Z.*, **337**, 505.

BORTZ, W. M., and LYNEN, F. (1963*b*). *Biochem. Z.*, **339**, 77.

CHANG, H. C., SEIDMAN, I., TEEBOR, G., and LANE, M. D. (1967). *Biochem. biophys. Res. Commun.*, **28**, 682.

CHANGEUX, J. P. (1961). *Cold Spring Harb. Symp. quant. Biol.*, **26**, 313.

CORI, C. F., and CORI, G. T. (1945). *J. biol. Chem.*, **158**, 341.

FANG, M., and LOWENSTEIN, J. M. (1967). *Biochem. J.*, **105**, 803.

Fischer, E. (1898). *Hoppe-Seyler's Z. physiol. Chem.*, **26**, 60.

Fritz, I. B. (1961). *Physiol. Rev.*, **41**, 52.

Ganguly, J. (1960). *Biochim. biophys. Acta*, **40**, 110.

Gellhorn, A., and Benjamin, W. (1966). *Adv. Enzyme Regulation*, **4**, 19.

Gerhard, J. C., and Pardee, A. B. (1961). *Fedn Proc. Fedn Am. Socs exp. Biol.*, **20**, 224.

Gerhard, J. C., and Pardee, A. B. (1962). *J. biol. Chem.*, **237**, 891.

Gibson, D. M., Hicks, S. W., and Allmann, D. W. (1966). *Adv. Enzyme Regulation*, **4**, 239.

Goto, T., Ringelmann, E., Riedel, B., and Numa, S. (1967). *Life Sciences*, **6**, 785.

Gregolin, C., Ryder, E., Kleinschmidt, A. K., Warner, R. C., and Lane, M. D. (1966a). *Proc. natn. Acad. Sci. U.S.A.*, **56**, 148.

Gregolin, C., Ryder, E., Kleinschmidt, A. K., Warner, R. C., and Lane, M. D. (1966b). *Proc. natn. Acad. Sci. U.S.A.*, **56**, 1751.

Gregolin, C., Ryder, E., and Lane, M. D. (1968). *J. biol. Chem.*, **243**, 4227.

Gregolin, C., Ryder, E., Warner, R. C., Kleinschmidt, A. K., Chang, H. C., and Lane, M. D. (1968). *J. biol. Chem.*, **243**, 4239.

Henniger, G. (1969). Thesis, University of Munich.

Hill, R., Webster, W., Linazasoro, J. M., and Chaikoff, I. L. (1960). *J. Lipid Res.*, **1**, 150.

Hsu, R. Y., Wasson, G., and Porter, J. W. (1965). *J. biol. Chem.*, **240**, 3736.

Iritani, N., Nakanishi, S., and Numa, S. (1969). *Life Sciences*, in press.

Korchak, H. M., and Masoro, E. J. (1962). *Biochim. biophys. Acta*, **58**, 354.

Korchak, H. M., and Masoro, E. J. (1964). *Biochim. biophys. Acta*, **84**, 750.

Koshland, D. E., Jr. (1958). *Proc. natn. Acad. Sci. U.S.A.*, **44**, 98.

Koshland, D. E., Jr. (1968). *Adv. Enzyme Regulation*, **6**, 291.

Levy, H. R. (1963). *Biochem. biophys. Res. Commun.*, **13**, 267.

Lorch, E., Abraham, S., and Chaikoff, I. L. (1963). *Biochim. biophys. Acta*, **70**, 627.

Lowenstein, J. M. (1968). *Biochem. Soc. Symp.*, **27**, 61.

Lynen, F., Hopper-Kessel, I., and Eggerer, H. (1964). *Biochem. Z.*, **340**, 95.

Masoro, E. J. (1962). *J. Lipid Res.*, **3**, 149.

Monod, J., and Jacob, F. (1961). *Cold Spring Harb. Symp. quant. Biol.*, **26**, 389.

Monod, J., Wyman, J., and Changeux, J. P. (1965). *J. molec. Biol.*, **12**, 88.

Numa, S., Bortz, W. M., and Lynen, F. (1965). *Adv. Enzyme Regulation*, **3**, 407.

Numa, S., Goto, T., Ringelmann, E., and Riedel, B. (1967). *Eur. J. Biochem.*, **3**, 124.

Numa, S., Matsuhashi, M., and Lynen, F. (1961). *Biochem. Z.*, **334**, 203.

Numa, S., and Ringelmann, E. (1965). *Biochem. Z.*, **343**, 258.

Numa, S., Ringelmann, E., and Lynen, F. (1965). *Biochem. Z.*, **343**, 243.

Numa, S., Ringelmann, E., and Riedel, B. (1966). *Biochem. biophys. Res. Commun.*, **24**, 750.

Perutz, M. F. (1969). *Eur. J. Biochem.*, **8**, 455.

Regen, D. cf. Lynen, F. (1967). In *Progress in Biochemical Pharmacology*, **3**, p. 29, ed. Kritchevsky, D., Paoletti, R., and Steinberg, D. Basel and New York: Karger.

Roberts, R. B., Abelson, P. H., Cowie, D. B., Bolton, E. T., and Britton, R. J. (1955). *Studies on Biosynthesis in* Escherichia coli. Washington, D.C.: Carnegie Institute. Publication No. 607.

Spencer, A. F., and Lowenstein, J. M. (1967). *Biochem. J.*, **103**, 342.

Swanson, R. F., Curry, W. M., and Anker, H. S. (1967). *Proc. natn. Acad. Aci. U.S.A.*, **58**, 1243.

Swanson, R. F., Curry, W. M., and Anker, H. S. (1968). *Biochim. biophys. Acta*, **159**, 390.

Taketa, K., and Pogell, B. M. (1966). *J. biol. Chem.*, **241**, 720.

Tubbs, P. K., and Garland, P. B. (1964). *Biochem. J.*, **93**, 550.

UMBARGER, H. W. (1956). *Science*, **123**, 848.

UMBARGER, H. W. (1961). *Cold Spring Harb. Symp. quant. Biol.*, **26**, 301.

VAGELOS, P. R., ALBERTS, A. W., and MARTIN, D. B. (1963). *J. biol. Chem.*, **238**, 535.

WARBURG, O. (1948). *Wasserstoffübertragende Fermente*, p. 37. Berlin: Arbeitsgemeinschaft Medizinischer Verlage, Dr. Werner Saenger.

WIELAND, O., NEUFELDT, I., NUMA, S., and LYNEN, F. (1963). *Biochem. Z.*, **336**, 455.

WIELAND, O., WEISS, L., EGER-NEUFELDT, I., TEINZER, A., and WESTERMAN, B. (1965). *Klin. Wschr.*, **43**, 645.

YATES, R. A., and PARDEE, A. B. (1956). *J. biol. Chem.*, **221**, 757.

YUGARI, Y., MATSUDA, I., and SUDA, M. (1964). *VI Int. Congr. Biochem.*, New York, Abstracts, p. 602.

DISCUSSION

Gopalan: I would like to raise the question of the possible physiological significance of the activation of acetyl CoA carboxylase by citrate in the development of fatty livers in malnutrition. There are two clinical types of protein-calorie deficiency in children. In one type, called marasmus, there is an extreme degree of muscular wasting but the liver is normal and there is no fatty change. In the other type, kwashiorkor, there is considerable oedema and the liver is fatty. Pineda (1968) demonstrated that whereas in marasmus isocitric dehydrogenase (ICDH) levels in white blood cells were increased, in cases of kwashiorkor these levels were considerably diminished. McLean (1966) has also found reduced activity of ICDH in the liver of kwashiorkor children. It could be argued that reduction of ICDH activity may bring about an increased citrate concentration which in turn may facilitate fatty acid synthesis through activation of acetyl CoA carboxylase.

Lynen: Might not the fatty deposition in the liver be due to a defect of lipoprotein synthesis whereby newly synthesized lipid cannot be released from the liver into the circulation?

Gopalan: This is what is generally believed and it is possibly the correct explanation. However, in view of your observations, the question is whether citrate activation of acetyl CoA carboxylase plays at least a contributory role in the development of the fatty livers in kwashiorkor.

Lynen: It would be interesting to measure the citrate concentration in tissues of kwashiorkor patients.

Gopalan: I am not aware of any studies on citrate concentration in the liver in cases of kwashiorkor, or, for that matter, of acetyl CoA carboxylase activity. Metcoff and co-workers (1960) have, however, found high levels of citrate in the muscle tissue in some but not all cases of kwashiorkor.

Ramalingaswami: Is there any information, Professor Lynen, on the rate of degradation of acetyl CoA carboxylase in starvation? Conceivably, you could measure a reduction in net enzyme activity if the enzyme was disappearing rapidly. I say this because one of the responses of the organism to starvation or acute protein deficiency is a rapid loss of nitrogen from the liver, and this is probably the rapidly turning over nitrogen of enzymic nature. Do you know the rate of catabolism of the enzyme in starvation or protein deficiency?

Lynen: The finding that the activity of acetyl CoA carboxylase after 24 hours' starvation has dropped by about 50 per cent seems to indicate that the half-life of acetyl CoA carboxylase is about 24 hours.

Jagannathan: It is interesting that citrate, ATP and coenzyme A are all very strong inhibitors, especially the three together, of phosphofructokinase. Might this provide a mechanism for regulating the relationship between glycolysis and fatty acid synthesis, both of which occur in the cytoplasm?

Lynen: This question is difficult to answer. Citrate activates acetyl CoA carboxylase, but ATP, at least in the absence of acetyl CoA, inhibits the enzyme, as a result of carboxylation of the enzyme (Numa *et al.*, 1967; Gregolin *et al.*, 1968).

White: Since there is some practical interest in over-nutrition, I wonder if one could influence the channelling of carbon from carbohydrate into fatty acids by use of analogues of citric acid. It is well known that fluoroacetate blocks the citric acid cycle because of the accumulation of fluorocitrate, probably because the citrate cleavage enzyme does not attack fluorocitrate. Have you any experience with fluorocitrate or other citric acid analogues with respect to their effects either on acetyl CoA carboxylase or on the citrate cleavage enzyme which might limit the ultimate channelling of carbon into fat?

Secondly, since the specificity of trypsin is well known, and citrate is a very strong acid, is there a possible interaction between the peptide which is liberated by trypsin from acetyl CoA carboxylase, and possibly basic in character, and citrate?

Lynen: The use of analogues is an interesting suggestion but we haven't done any experiments. But thinking of the practical applications of such a treatment, I wonder whether anyone would like to eat fluoroacetate?

White: I was not suggesting fluoroacetate itself, but the organic chemists could undoubtedly devise less noxious analogues of citrate, unless the blocking of the cycle is dependent upon accumulation of a toxic metabolite. Should it not be possible to study a number of synthetic analogues with respect to their allosteric effects on the enzyme?

Lynen: We have studied tricarballilic acid, the hydroxy-free analogue of citric acid, and found it to be completely inactive as an allosteric effector. With respect to the trypsin effect, in Numa's experiments a rather crude extract from rat liver was used and therefore the hypothetical peptide released from the enzyme cannot yet be studied.

Salganik: The local concentration of citrate may be much greater than the general concentration, and this would be important for the physiological effects.

Lynen: You are right. However, as already mentioned (see p. 40) we and others did not see any changes in the total concentration of citrate during starvation or during diabetes. It is not yet possible to determine local concentrations of these metabolites in the cells. This is one of the big handicaps in considering allosteric effects.

Harris: Professor Lynen, were the molecular weights of the small and the large form of acetyl CoA carboxylase extrapolated from sucrose gradient runs or were they analytical measurements?

Lynen: They were analytical measurements.

Harris: So we have to think in terms of about 20-25 subunits in the polymerized form. That is a huge concatenation for activity. Is your intermediate form enzymically active?

Lynen: The "intermediate" form is active, according to the experiments of Numa and his colleagues (1967).

Harris: The problem is that it becomes extremely difficult to make any serious model if 20 subunits are required to associate before you get any activity. So one would like to find a little activity before then, otherwise you are really not in business.

Lynen: They are long molecules. Gregolin and co-workers (1966) found filaments up to 4000 Å long in the electron microscope. But you raise the question of whether interaction of say two protein subunits already satisfies the requirements for generating an active centre. I cannot answer this question.

Monod: Do you know whether the small form of acetyl CoA carboxylase is itself made up of several subunits? This seems likely, since a molecular weight of 430 000 is very big for a single polypeptide chain. If the small form is made up of four or eight subunits, then it is probably what we call a closed crystal—a system which obeys the symmetry of the rotational point. It can be shown formally that once a structure of that kind has formed a closed crystal with more than a single axis of symmetry it cannot aggregate with itself into a further closed structure. It can aggregate only in the form of an ordinary crystal or, and this is most probable here, into an elongated fibre with a helical structure. This predicts that the large form would have to be polydispersed while the small form would be monodispersed.

Lynen: That is exactly the case. The "small form" (molecular weight 400 000) is composed of eight subunits. By treatment with urea or guanidine hydrochloride Henniger (1969) was able to split the small form successively into half, into quarters, and into eighth parts. Because the small form contains only one biotin molecule we must assume that one subunit contains biotin and the seven other subunits are biotin-free. The "large" form is a polydisperse one, as you suggest.

Monod: That distribution of biotin is difficult to accommodate; if the packing is symmetrical there should be as many binding sites as there are subunits.

Sarma: Would you expect a change in the polymeric form as a result of different types of fatty livers, for instance from a low protein diet? You can also have a fatty liver on high protein with orotic acid. Would you expect a change in polymeric form there?

Lynen: I wouldn't expect any change.

Sarma: The accumulation of fat in fatty livers can be either periportal or centrilobular under different dietary conditions. Can such differences influence the synthesis of acetyl CoA carboxylase?

Lynen: We did most of our experiments with palmityl CoA, but stearyl CoA and oleyl CoA also inhibit the enzyme (cf. Table III, p. 40). The kind of fatty acid bound to the CoA doesn't seem to matter too much.

Edelman: Have you measured the stoicheiometry and binding constants of citrate?

Lynen: No.

Monod: Has citrate any measurable affinity for the small form? Presumably it must have a much higher affinity for the aggregated form.

Lynen: We have not measured it.

Edelman: Is there any metal dependency of this enzyme, on divalent cations for instance?

Lynen: Magnesium is required.

Edelman: Does calcium substitute for magnesium?

Lynen: We don't know; calcium was not checked. As in many ATP-dependent reactions, manganese is active too.

Bhargava: Is there any evidence that the subunits of the small form are identical?

Lynen: In the ultracentrifuge they sediment as a homogeneous peak

(Henniger, 1969). Gregolin and co-workers (1968) cleaved the enzyme with sodium dodecyl sulphate and using gel electrophoresis they found no separation into two bands, even though one of the subunits must contain biotin.

Bhargava: Couldn't you separate the biotin-bound subunit from the subunits not bound to biotin?

Lynen: This has not been tried.

Burma: I wonder whether the low temperature gives you the simple protomeric form. One of the possible models in terms of citrate binding would be change of conformation due to change of temperature.

Lynen: I agree. We assume that at low temperature the complex disaggregates as a result of weakening hydrophobic interactions between the protomers. Breaking the protein–protein interaction should lead to a change in conformation.

REFERENCES

GREGOLIN, C., RYDER, E., KLEINSCHMIDT, A. K., WARNER, R. C., and LANE, M. D. (1966). *Proc. natn. Acad. Sci. U.S.A.*, **56**, 148.
GREGOLIN, C., RYDER, E., WARNER, R. C., KLEINSCHMIDT, A. K., CHANG, H. C., and LANE, M. D. (1968). *J. biol. Chem.*, **243**, 4239.
HENNIGER, G. (1969). Thesis, University of Munich.
MCLEAN, A. E. (1966). *Clin. Sci.*, **30**, 129–137.
METCOFF, J., FRENK, S., ANTONOWICZ, I., GORDILLO, G., and LOPEZ, E. (1960). *Pediatrics, Springfield*, **26**, 960–972.
NUMA, S., GOTO, T., RINGELMANN, E., and RIEDEL, B. (1967). *Eur. J. Biochem.*, **3**, 124.
PINEDA, O. (1968). In *Calorie Deficiencies and Protein Deficiencies*, pp. 75–87, ed. McCance, R. A., and Widdowson, E. M. London: Churchill.

THE EXPRESSION OF GENETIC INFORMATION: A STUDY WITH HYBRID ANIMAL CELLS

HENRY HARRIS

Sir William Dunn School of Pathology, University of Oxford

I DO NOT propose to review here the now very considerable body of work on interspecific cellular hybrids that has been done in many laboratories since J. F. Watkins and I showed that an inactivated virus could be used to fuse together cells from different animal species (Harris and Watkins, 1965). Instead, I shall limit myself to a consideration of one particular form of hybrid cell which is of special interest in the present context because the investigation of its properties has shed some new light on three questions which are central to our understanding of regulation in animal cells: how transcription of the genes is regulated in these cells; how information flows from the nucleus to the cytoplasm; and what the relationship is between transcription of the genes and translation of the corresponding RNA templates into protein. The particular hybrids that I shall discuss are those in which one of the parent cells is the nucleated erythrocyte of the hen. These hybrids differ from all others in two important respects. Hybrid cells are normally produced simply by the fusion of the two parent cells and thus contain both nucleus and cytoplasm from the two parents (Harris et al., 1966). When one of the parent cells is an erythrocyte, however, this is not, in general, the case. The virus that is used to fuse the cells together (Sendai) is a haemolytic virus, and, at the concentrations required to induce fusion, it lyses the erythrocytes before the cells coalesce. The erythrocytes thus lose their cytoplasmic contents, which in mature red cells are very largely haemoglobin, before the heterokaryons are formed. Fusion, in this case, takes place between the other cells in the combination and erythrocyte ghosts (Schneeberger and Harris, 1966). For practical purposes, then, the resultant heterokaryons contain the nuclei of both parents, but no appreciable cytoplasmic contribution from the erythrocytes. The second important special feature of these hybrids is the initially complete dormancy of the erythrocyte nucleus. During the course of differentiation, the erythrocytes of many vertebrates eliminate their nuclei altogether. In birds, reptiles, amphibians and certain other orders, however, the erythrocytes retain their nuclei; but these nuclei are completely inert: when the cells are mature, they synthesize neither RNA nor DNA, nor do they undergo mitosis.

When the dormant erythrocyte nucleus is introduced into the cytoplasm of a growing cell from the same or a different species of vertebrate, it resumes the synthesis of RNA and DNA. Both hen and frog erythrocyte nuclei can be reactivated in the cytoplasm of either human or mouse cells (Harris, 1966). This reactivation involves not only the resumption of RNA synthesis, but also, as I shall show presently, the ordered synthesis of specific proteins determined by the erythrocyte nucleus. We can therefore conclude at once that the signals emanating from the human or mouse cytoplasm are understood perfectly well by the hen or frog nuclei. In other words, these cytoplasmic signals are not, in this sense, species-specific.

The outstanding morphological event associated with the reactivation of the erythrocyte nucleus is a massive increase in its volume (Harris, 1966 and 1967). While accurate measurement of nuclear volume is difficult, it is likely that there is at least a twenty- to thirty-fold increase (Bolund, Ringertz and Harris, 1969). This expansion of the nucleus is associated with dispersion of its highly condensed chromatin, a process which is easily demonstrated with appropriate cytological stains. When the ability of the enlarged nuclei to synthesize RNA is measured, it is found that there is a direct relationship between the volume of the nucleus and the amount of RNA which it synthesizes (Harris, 1967). In other words, the bigger the nucleus, the more RNA it makes. Certain physical and cytochemical measurements that have been made on these reactivated nuclei are of particular interest. The increase in the volume of the nucleus is not simply due to the ingress of water. There is at least a 4–6-fold increase in dry mass, which is accounted for very largely by an increase in protein content (Bolund, Ringertz and Harris, 1969). Since erythrocyte nuclei that have been inactivated by ultraviolet light also show this increase in dry mass, the increase must be due to the movement of proteins into the nuclei from the cytoplasm of the cell. These proteins are, of course, human or mouse proteins, as the case may be, but they are nonetheless able to do whatever they need to do in the hen or frog nuclei. The dispersion of the chromatin is associated with structural changes which are reflected by a marked increase in the affinity of the chromatin for intercalating dyes, such as acridine orange and ethidium bromide, and by alterations in its melting profile indicative of diminished stability to heat denaturation (Bolund, Ringertz and Harris, 1969). It appears that, during the process of reactivation, the structure of the chromatin is loosened and the DNA becomes more accessible not only to macromolecules, but even to smaller molecules such as acridine dyes. It is reasonable to suppose that the DNA also becomes more accessible to the molecules involved in its transcription, so that, as the nucleus enlarges, more of the chromatin opens up and more of the DNA is transcribed.

Many of the cytochemical changes which the erythrocyte nuclei undergo when they are reactivated in heterokaryons can be mimicked in the nuclei of erythrocyte ghosts by removal of divalent cations from the condensed chromatin (Ringertz and Bolund, 1969). It is therefore not improbable that the mechanism by which transcription of the genes is regulated in these nuclei involves the interaction of the chromatin with divalent cations, and possibly also other electrolytes. How specificity—that is, the transcription of some genetic areas and not others—can be achieved by an interaction of this sort has been discussed in detail elsewhere (Harris, 1968). In any case, if the reactivation of the dormant erythrocyte nucleus is achieved by a change in the immediate electrolyte environment of the chromatin, it is not surprising that this reactivation takes place in cytoplasm from widely different animal species. On this view, the signals which pass to the hen or frog nucleus from the human or mouse cytoplasm are of a quite general kind which one would expect to be common to all vertebrate cells.

While these initial experiments showed that the erythrocyte nuclei could be reactivated in the sense that they could be induced to resume the synthesis of RNA, it remained to be shown whether a hen nucleus operating in human or mouse cytoplasm could induce the synthesis of specific hen proteins. The first proteins which I chose to examine were species-specific surface antigens. These antigens can be detected on the cell surface with great precision by immunological methods involving the adsorption of suitably sensitized red cells (haemadsorption) (Watkins and Grace, 1967). With appropriate antisera, these techniques are quite specific and are capable of revealing minute quantities of these antigens on the surface of the cell. Since the formation of the heterokaryon involves the fusion of the membrane of the erythrocyte ghost with the membrane of the other cell, the heterokaryon initially shows the presence of hen-specific antigens on its surface. However, during the first few days of cultivation, the hen-specific antigens introduced into the surface of the heterokaryon by the process of cell fusion are eliminated; they appear to be displaced by the continued production of mouse or human antigens, as the case may be (Harris *et al.*, 1969). The hen-specific antigens thus disappear within the first few days, despite the fact that the hen nucleus is reactivated and continuously synthesizes large amounts of RNA. But the reactivation of the erythrocyte nucleus during this initial period remains incomplete in one respect. While autoradiographic studies show that, *grosso modo*, the nucleus synthesizes RNA all over its chromosomes, and sedimentation studies show that this RNA is of high molecular weight, the nucleus does not, at this stage, develop a visible nucleolus or synthesize normal mature 28s or 16s (ribosomal) RNA. All the RNA made in the erythrocyte nucleus before development of the nucleolus is of the type which shows "polydisperse" sedimentation in conventional sucrose gradients (Harris *et al.*, 1969).

On the 4th or 5th day after cell fusion, under appropriate conditions, the erythrocyte nuclei begin to develop nucleoli and to synthesize mature 28s and 16s RNA; and, at this point, hen-specific antigens begin to reappear on the surface of the heterokaryon. The development of nucleoli in the erythrocyte nuclei and the appearance of hen-specific antigens on the surface of the cells run *pari passu*. If similar heterokaryons are made with erythrocytes from a chick embryo, reactivation of the erythrocyte nucleus and development of the nucleolus take place more rapidly than in adult hen erythrocyte nuclei; and, in this case, the hen-specific antigens appear on the surface of the heterokaryons sooner. Again, there is a close relationship between the development of the nucleolus in the erythrocyte nucleus and the appearance of the hen-specific surface antigens (Harris *et al.*, 1969). This difference between adult hen and embryonic chick erythrocyte nuclei indicates that the same sets of genes in the same cytoplasm may be switched on at very different rates, and these rates are determined, not by any differences in the signals which the cytoplasm transmits to these genes, but by differences in the physiological states of the nuclei which contain them. In both cases, nonetheless, the process of reactivation is a slow one, requiring some days for completion.

Since surface antigens are perhaps rather special proteins, a similar investigation has been made on the synthesis of a soluble enzyme. The enzyme chosen for study was inosinic acid pyrophosphorylase (E.C. 2.4.2.8.), which catalyses the condensation of hypoxanthine with ribosyl phosphate. This enzyme was chosen because a mouse cell line which lacked inosinic acid pyrophosphorylase activity was available (Littlefield, 1964). The enzyme is essential for the incorporation of hypoxanthine into RNA. The ability of the reactivated erythrocyte nucleus to determine

the synthesis of this enzyme in the defective mouse cells was investigated
by direct enzyme assay and by autoradiographic procedures designed to
detect the incorporation of tritiated hypoxanthine into RNA. Again it
was found that no enzyme was synthesized in the heterokaryon while
the erythrocyte nucleus lacked a nucleolus; but when the nucleolus
appeared, the enzyme began to be synthesized and the cell began to
incorporate hypoxanthine into RNA (Harris and Cook, 1969). Since the
antigen and the enzyme are neither structurally nor functionally related,
these observations make it very likely that similar kinetics will be observed
for other proteins.

It thus appears that the erythrocyte nucleus does not begin to determine the synthesis of specific proteins until it develops a nucleolus. If
the "polydisperse" RNA which the erythrocyte nucleus makes is the
RNA which carries the information for the synthesis of specific proteins,
the failure of the nucleus to determine the synthesis of these proteins
before the development of a nucleolus might be explained in two ways.
It is possible that the RNA made on the hen chromosomes may pass to
the cytoplasm of the cell, but may be unable to programme human or
mouse ribosomes; or it may be that the passage of this RNA to the
cytoplasm requires the participation of the nucleolus.

This latter possibility was examined with the aid of a microbeam of
ultraviolet light, which could be used to eliminate the activity of individual
nuclei or nucleoli within the cell. When RNA labelling patterns were
examined in heterokaryons in which the mouse nuclei were inactivated
by ultraviolet light, it was found that, until they developed nucleoli, the
erythrocyte nuclei made no detectable contribution to cytoplasmic RNA
labelling. Although these heterokaryons contained up to four enlarged
erythrocyte nuclei, which collectively synthesized very large amounts of
RNA, the level of cytoplasmic labelling in these heterokaryons, after
elimination of the mouse nuclei, was indistinguishable from that found
in single normal mouse cells in which the nucleus was inactivated. It
thus appeared that the RNA made on the hen chromosomes was not
transferred to the cytoplasm of the heterokaryon so long as the hen
nucleus lacked a functional nucleolus. On the other hand, when the
mouse nuclei in the heterokaryons were eliminated after the erythrocyte
nuclei had developed nucleoli, it could be shown that the erythrocyte
nuclei did contribute to cytoplasmic RNA labelling. The failure of the
reactivated erythrocyte nuclei to determine the synthesis of specific
proteins before the development of nucleoli was thus at least in part
attributable to the fact that the RNA which these nuclei were making

at this stage failed to reach the cytoplasm of the cell (Harris *et al.*, 1969).

Experiments on single HeLa cells in which the nucleolus only was inactivated by the microbeam confirm this interpretation of the observations made on erythrocyte nuclei in heterokaryons. In cells in which nucleolar activity had been effectively eliminated, synthesis of RNA in other parts of the nucleus continued, but this RNA was not transferred in detectable amounts to the cytoplasm of the cell. The level of cytoplasmic RNA labelling in such cells was virtually indistinguishable from that found in cells in which the whole nucleus had been irradiated (Sidebottom and Harris, 1969). This indicates once again that the activity of the nucleolus is essential, not only for the transfer to the cytoplasm of RNA synthesized at the nucleolar site, but also for the transfer of the RNA synthesized elsewhere in the nucleus.

These findings argue against any model for the transfer of information from nucleus to cytoplasm which postulates that the RNA carrying the information for protein synthesis diffuses from the nucleus into the cytoplasm and there attaches to pre-existing ribosomes; and they also argue against any model which postulates the passage of cytoplasmic ribosomes into the nucleus in order to release this RNA from the genes or serve as a vehicle for its transport to the cytoplasm. The hen erythrocyte nucleus in the heterokaryon does not begin to transfer detectable amounts of RNA to the cytoplasm of the cell and does not initiate the synthesis of specific proteins until it develops its own nucleolus. The pre-existing ribosomes in the cytoplasm of the hybrid cell cannot achieve the transfer of genetic information from the anucleolate erythrocyte nucleus.

CONCLUSIONS AND SUMMARY

The experiments which I have described permit the following conclusions: (1) Regulation of the transcription of DNA in animal cell nuclei is associated with condensation and dispersion of the chromatin. Inactivation and reactivation of nuclei is linked to changes in nuclear volume. (2) The signals which govern transcription of the DNA operate across wide species differences. (3) There is no mandatory, or even close, coupling between transcription and translation. In reactivated erythrocyte nuclei, many genes may be transcribed for several days without determining the synthesis of any proteins. (4) The process of switching genes on in a dormant nucleus is a relatively slow one extending over days rather than hours. Genetically identical nuclei in the same cytoplasm may be switched on at very different rates, and these rates are determined not by differences in the nature of the cytoplasmic signals, but by differences in the physiological state of the nuclei. (5) The RNA made in the nucleolus and the RNA made elsewhere in the nucleus are not transferred to the cytoplasm independently of each other. A nucleus does not transfer genetic information to the cytoplasm of the cell until it develops its own nucleolus and begins to synthesize its own ribosomal RNA.

REFERENCES

BOLUND, L., RINGERTZ, N. R., and HARRIS, H. (1969). *J. Cell Sci.*, **4**, 71–88.
HARRIS, H. (1966). *Proc. R. Soc. B*, **166**, 358–368.
HARRIS, H. (1967). *J. Cell Sci.*, **2**, 23–32.
HARRIS, H. (1968). *Nucleus and Cytoplasm*. Oxford: Clarendon Press.
HARRIS, H., and COOK, P. R. (1969). *J. Cell Sci.*, **5**, 121–134.
HARRIS, H., SIDEBOTTOM, E., GRACE, D. M., and BRAMWELL, M. E. (1969). *J. Cell Sci.*, **4**, 499–526.
HARRIS, H., and WATKINS, J. F. (1965). *Nature, Lond.*, **205**, 640–646.
HARRIS, H., WATKINS, J. F., FORD, C. E., and SCHOEFL, G. I. (1966). *J. Cell Sci.*, **1**, 1–30.
LITTLEFIELD, J. W. (1964). *Nature, Lond.*, **203**, 1142–1144.
RINGERTZ, N. R., and BOLUND, L. (1969). *Expl Cell Res.*, **55**, 205–214.
SCHNEEBERGER, E. E., and HARRIS, H. (1966). *J. Cell Sci.*, **1**, 401–406.
SIDEBOTTOM, E., and HARRIS, H. (1969). *J. Cell Sci.*, **5**, in press.
WATKINS, J. F., and GRACE, D. M. (1967). *J. Cell Sci.*, **2**, 193–204.

DISCUSSION

Monod: Is there any protein synthesis within the nucleolus itself?

Harris: This is a vexed question. I think that the case has not been decisively made that nuclei make any proteins at all. The evidence that they do is largely based on autoradiographic findings; but this type of experiment only demonstrates synthesis of protein at a particular site if one can exclude translocation of labelled protein occurring within the time of exposure to the radioactive precursor. My reasons for thinking that there may be little or no protein synthesis in the cell nucleus rest to some extent on the studies which indicate that there are few, if any, complete ribosomes in the nucleus. The particles inside the nucleus (and nucleolus) don't look like complete ribosomes in the electron microscope. They have a diameter of about 150Å, whereas a ribosome has a diameter of about 230Å or more. Nobody has really isolated from nuclei particles that behave under a variety of physical conditions in the same way as cytoplasmic ribosomes. There may be incomplete ribosomes in the nucleus but I can't see much evidence for complete ribosomes. And if there are no complete ribosomes in the nucleus, it is reasonable to suppose that there probably isn't much actual synthesis of protein. If there were some protein synthesis in the nucleolus you might imagine that the nucleolus made a specific protein which protected the polydisperse RNA; but the evidence for this idea is not good.

Cohn: Have you seen division in your hybrid cells? Can one isolate a clone of the heterokaryon?

Harris: Cell division does occur. We have clones of cells which have been derived from the fusion of mouse cells with hen erythrocytes.

Talwar: Professor Harris's observations have some similarity to the experience in the hormonal regulation of RNA synthesis. In most situations where messenger RNA is induced, there is also a simultaneous induction of ribosomal RNA synthesis (Talwar, Gupta and Gros, 1964; Korner, 1964; Tata, 1968).

I also have a question. There is evidence that mammalian nuclei synthesize in the first instance aggregated RNA molecules (Houssais and Attardi, 1966; Scherrer *et al.*, 1966; Penman *et al.*, 1963); while you are considering this overall scheme for the transfer of nuclear RNA to the cytoplasmic compartments, can we fit into the scheme the fate of these large precursor RNA's before they are transferred to the cytoplasm?

Harris: For my sins, I can lay claim to having introduced that confusing notion about 10 years ago, when John Watts and I first described the phenomenon of intranuclear RNA turnover (Watts and Harris, 1959; Harris, 1959). We showed that a lot of the RNA made inside the nucleus didn't appear in the cytoplasm, but broke down inside the nucleus. We had no explanation of the biological significance of this; and I do not believe anybody yet has a satisfactory explanation for this very large-scale process. I think I am beginning to see a glimmer of what intra-nuclear RNA turnover might mean, in the light of the experiments I have just described. The glimmer is this. If the nucleolus has to play some kind of regulatory role, that is, if the templates have to be "engaged" by some process operating at the nucleolus, one would expect, under conditions of constant metabolic shift, that there would be a constant fluctuation of "engagement" and "disengagement", and that cells in which a diminution of metabolic activity takes place ("step-down") would show more marked intranuclear turnover of RNA as the templates which fail to be "engaged" are eliminated inside the nucleus. On the other hand, as the metabolic level of the cell "steps up", I would expect intranuclear RNA turnover to be reduced, and transfer of RNA to the cytoplasm to be increased. And this is precisely what one finds. Intra-nuclear RNA turnover seems to be the elimination, within the nucleus, of those templates which fail to be engaged by the nucleolar mechanism. We have a lot of data on this. M. E. Bramwell has shown that high temperatures and low doses of actinomycin, both of which inhibit nucleolar activity, increase intranuclear RNA turnover at the expense of RNA transport to the cytoplasm (Bramwell, 1969); and E. Sidebottom has shown the same thing in cells in which the nucleolus has been irradiated with a microbeam (Sidebottom and Harris, 1969). The erythrocyte nucleus, reactivated in a heterokaryon, does nothing but turn over its RNA until it develops a nucleolus. Once this occurs the amount of RNA broken down inside the nucleus, relative to the amount transferred to the cytoplasm, falls. The "engagement" and "disengagement" of RNA templates is my present working hypothesis.

Bhargava: Is there nuclear fusion in these hybrid clones?

Harris: For practical purposes, the multinucleate state is not a form of life compatible with multiplication in mammalian cells. I don't know why this should be. There are, of course, many forms of life which operate perfectly well in the multinucleate state. But, in order to achieve multiplication, heterokaryons must first make a single nucleus, and this they do by way of mitosis. When two nuclei in a binucleate cell are synchronized they enter mitosis together. It commonly happens that a single metaphase plate and a single spindle are made; all the chromosomes line up along this one metaphase plate. The cell then divides to give two mononuclear daughters each of which contains, within a single nucleus, both the chromosomal sets of the parent cells. There are many variant forms of mitosis which result in nuclear fusion.

Bhargava: In these hybrid clones, have you evidence of the functioning of the two chromosome sets in respect of, say, synthesis of specific proteins?

Harris: We have evidence that both chromosomal sets are active. For example, both sets of surface antigens are expressed.

Bhargava: Have you tried to induce RNA synthesis in the nuclei by extracts of mouse or HeLa cells?

Harris: Thompson and McCarthy (1968) have studied the effects of cytoplasmic extracts on RNA and DNA synthesis in isolated hen erythrocyte nuclei. Their results parallel the results which I get in heterokaryons very closely, but I wish they didn't. The behaviour of isolated nuclei *in vitro* is exquisitely sensitive to physical conditions—ionic strength, cation concentration, and so on—and it is very easy to change the amount of RNA made, its molecular weight or its base composition, simply by changing the ionic strength or magnesium concentration. I am delighted that Thompson and McCarthy can mimic *in vitro* what I find *in vivo*, but I am a little worried by this.

Salganik: Are you sure that ribosome-like particles are involved in the transfer of messenger RNA into the cytoplasm? Perhaps the synthesis of ribosomal RNA merely coincides with the synthesis of special transfer proteins. It was established by G. P. Georgiev and his co-workers that messenger RNA molecules are associated in the nuclei with special proteins distinct from ribosomal ones into particles which they call "informomeres", and it was supposed that m-RNA is transported to the cytoplasm in this form (Samarina *et al.*, 1968). In the cytoplasm m-RNA molecules are also combined with non-ribosomal proteins in the form of so-called "informosomes" (Spirin, Belitsina and Aitkhozhin, 1964). Considering these results, the role of ribosomal particles in the transfer of m-RNA seems doubtful.

Harris: I am, of course, very familiar with this work. The difficulty is in identifying the "messenger". The number of sins committed in biochemistry by people saying that an RNA fraction is the "messenger", simply because it sediments at a particular place in a sucrose gradient, is vast.

Monod: One of the important results that comes out of your experiments is that there is no strong coupling between transcription and translation. It is extremely interesting that experiments by François Gros and his group indicate the same thing in bacteria, in the Lac messenger in *E. coli.* Instead of looking at pulses of the order of 2 minutes they go down to pulses of a few seconds and find that there is no strong coupling and that transcription may go on at much the same rate whether or not there is translation. The only thing that happens is that the messenger is broken down much faster when it is not translated.

Harris: That may be analogous to what I was just saying. It is difficult to get good measurements of rates of RNA turnover, because you can't "chase" RNA precursors either in bacteria or in animal cells. You can only infer a minimum rate of breakdown because you have no idea of the extent of reutilization of RNA breakdown products.

White: You mentioned two phenomena which appear to be rate limiting: the rate at which the nucleus enlarges, and the time at which the nucleoli appear. Have you considered looking at a system in which one can influence the rate of synthesis of a particular protein by the addition *in vitro* of a hormone? I am thinking for example of the synthesis of collagen by fibroblasts. Have you tried to either speed up or delay nuclear enlargement or nucleoli development by addition of a hormone?

Harris: No, I haven't tried any hormones.

Salganik: When we investigated the induction of transcription in animal cells caused by hormones, amino acids and carcinogens we found

3*

in every case that enhancement of messenger RNA synthesis is accompanied by an increase in ribosomal and transfer RNA synthesis (Salganik *et al.*, 1967). What do you feel about the synthesis of transfer RNA in heterokaryons when the red cell nuclei begin to synthesize ribosomal RNA?

Harris: This is an interesting question. I should have thought that if transfer RNA had been passing from the erythrocyte nucleus to the cytoplasm before the erythrocyte nucleus developed a nucleolus, we would have detected it. As far as we can tell from statistical measurements, no radioactive RNA is transferred from the erythrocyte nucleus before it develops a nucleolus. Whether the erythrocyte nucleus makes transfer RNA before it develops a nucleolus cannot be established easily, because the technique which I use for isolating nuclei is an aqueous technique; if you isolate mammalian nuclei in this way you don't have much 4s RNA left in them. To settle this point I should have to do a Behrens-type isolation of the nuclei, which doesn't lend itself to separation of the two sorts of nuclei in the heterokaryon. It may be that transfer RNA begins to be made at the same time as ribosomal RNA. There was a story some years ago that transfer RNA was synthesized at the nucleolar site, but I think most people disbelieve this now.

Salganik: In collaboration with Mosolov, Ganin and Reznik (Mosolov *et al.*, 1967) we have investigated the ability of tissue culture cells to seize nuclei previously isolated from homologous and heterologous cells. We observed that human tissue culture cells capture nuclei isolated from the same tissue much more intensively than those from L cells or mouse liver cells. In all cases, 7–10 days after the uptake, the captured nuclei were completely lysed and destroyed. When nuclei isolated from mouse liver were added to L cells the latter acquired the morphological features of epithelium-like cells, and these changes persisted through several generations of L cells.

Harris: A phagocytic cell may take small nuclei in. We see this phenomenon occasionally with erythrocyte nuclei. But if you make electron micrographic sections of such cells you can see that they are not true heterokaryons. Whereas in heterokaryons you see that the nuclei have normal double membranes and normal perinuclear organization, nuclei that have been *ingested* have a third membrane, showing that they are, in fact, inside a vacuole. If you wait you will find that they are digested. From the cell's point of view they are still outside.

Ramalingaswami: What proportion of the transplanted nuclei survive within the cells?

Harris: After about three days, 95 per cent or so have enlarged.

Tata: Have you ever attempted to introduce mitochondria rather than nuclei into cells?

Harris: No, I haven't.

Monod: In some old experiments on homologous transplantation of nuclei in amoebae the implanted nucleus very rapidly reached the same stage of the interdivision cycle that the host nucleus happened to be in. In your heterospecific fusions did you observe this?

Harris: R. T. Johnson and I (1969) have made a very detailed study of just this problem. The signals, not only for RNA synthesis, but also for DNA synthesis, are recognized perfectly well by the foreign nucleus, and the replication of the heterospecific chromosomes is synchronized with that of the homospecific ones. There is generally a very rapid imposition of synchrony of DNA synthesis and mitosis across species, but not always. There are some situations, even in intraspecific combinations, where physiological differences between the organization of the cell cycles of the two parent cells give rise to gross asynchrony.

REFERENCES

BRAMWELL, M. E. (1969). *J. Cell Sci.*, in press.

HARRIS, H. (1959). *Biochem. J.*, **73**, 362.

HOUSSAIS, J. F., and ATTARDI, G. (1966). *Proc. natn. Acad. Sci. U.S.A.*, **56**, 616.

JOHNSON, R. T., and HARRIS, H. (1969). *J. Cell Sci.*, in press.

KORNER, A. (1964). *Biochem. J.*, **92**, 449.

MOSOLOV, A. N., GANIN, A. F., REZNIK, L. G., and SALGANIK, R. I. (1967). *Genetika*, No. 11, 116.

PENMAN, S., SCHERRER, K., BECKER, T., and DARNELL, J. E. (1963). *Proc. natn. Acad. Sci. U.S.A.*, **49**, 654.

SALGANIK, R. I., CHRISTOLUBOVA, N. B., KIKNADZE, I. I., MOROZOVA, T. M., GRYASNOVA, I. M., and VALEEVA, F. S. (1967). In *Structure and Functions of Cell Nuclei*, p. 20. Moscow: Nauka.

SAMARINA, O. P., LUKANIDIN, E. M., MOLNAR, J., and GEORGIEV, G. P. (1968). *J. molec. Biol.*, **33**, 251.

SCHERRER, K., MARCAUD, L., ZAJDELA, F., LONDON, I. M., and GROS, F. (1966). *Proc. natn. Acad. Sci. U.S.A.*, **56**, 1571.

SIDEBOTTOM, E., and HARRIS, H. (1969). *J. Cell Sci.*, **5**, in press.

SPIRIN, A. S., BELITSINA, N. V., and AITKHOZHIN, M. A. (1964). *Zh. Obshch. Biol.*, **25**, 312.

TALWAR, G. P., GUPTA, S. L., and GROS, F. (1964). *Biochem. J.*, **91**, 565.

TATA, J. R. (1968). *Nature, Lond.*, **219**, 331.

THOMPSON, L. R., and McCARTHY, B. J. (1968). *Biochem. biophys. Res. Commun.*, **30**, 166.

WATTS, J. W., and HARRIS, H. (1959). *Biochem. J.*, **72**, 147.

THE ROLE OF CYCLIC AMP IN CERTAIN BIOLOGICAL CONTROL SYSTEMS

R. W. BUTCHER,* G. A. ROBISON AND E. W. SUTHERLAND

Departments of Physiology and Pharmacology, Vanderbilt University School of Medicine, Nashville, Tennessee

ADENOSINE $3',5'$-PHOSPHATE (cyclic AMP) occupies a central position in multicellular organism control systems as an intracellular mediator of the actions of a number of hormones, some of which are presented in Table I. The data implicating cyclic AMP in these actions have been extensively reviewed (Sutherland and Rall, 1960; Sutherland, Øye and Butcher, 1965; Robison, Butcher and Sutherland, 1968; Butcher *et al.*,

TABLE I

SOME HORMONE ACTIONS FOR WHICH THERE IS EVIDENCE OF THE INVOLVEMENT OF CYCLIC AMP

Tissue	Hormone(s)	Processes affected
Fat	catecholamines, ACTH, glucagon, TSH, LH	lipolysis
Adrenal cortex	ACTH	synthesis of glucocorticoids, others
Corpus luteum	LH	synthesis of progesterone
Testis	LH (ICSH)	synthesis of androgens
Ovary	LH	synthesis of progesterone
Thyroid	TSH	thyroxine release
Toad bladder	vasopressin	water and ion movement
Ventral frog skin	vasopressin	water and ion movement
Dorsal frog skin	MSH	melanophore dispersion
Liver	glucagon, catecholamines	phosphorylase activation, gluconeogenesis, urea production, K^+ release
Heart	catecholamines	force
Kidney, bone	parathyroid hormone	calcium and phosphate mobilization and excretion
Uterus (rat)	catecholamines	relaxation
Pancreas (β cells)	glucagon	insulin release

In the interest of brevity references have been omitted. The reader is asked to refer to Robison, Butcher and Sutherland (1968) which has an extensive bibliography, and, when available, to a monograph (Robison, Butcher and Sutherland, 1970).

Present address: Department of Physiology, University of Massachusetts School of Medicine, Worcester, Massachusetts.

1968), so it may be desirable to present a general view of the system as a control mechanism rather than to discuss the data again. Hopefully, this sort of presentation may be of greater interest to the reader, and in any event will be shorter.

It would be presumptuous of us to discuss the requirements imposed on biological control systems in detail. However, it is necessary to point out several fundamental characteristics of control mechanisms with special reference to the endocrine system, since it is as a mediator of hormone action that cyclic AMP has been most thoroughly studied. When considering endocrine systems,* it is obvious that without specificity between hormones and cells, chaos rather than control would be the result. Secondly, amplification is an obvious requirement. Hormones act at extremely low concentrations in physiological situations, and yet produce effects which are highly amplified when viewed at the molecular level. One obvious means of obtaining amplification in biological systems is by affecting multi-step processes, and so far this has always been found. A feature common to most control systems is that they are effective on a moment to moment basis. Thus, the system must be capable of increasing or decreasing its activity very rapidly. Finally, it is a time-honoured concept that many endocrine agents interact with a component of the cell membrane. It is our intention to examine the cyclic AMP mechanism in terms of these requirements.

THE CYCLIC AMP MECHANISM

The components of the cyclic AMP mechanism are illustrated in Fig. 1. After release from an endocrine gland, a hormone is transported to its target cell where it interacts with the adenyl cyclase system, a membrane-bound enzyme catalysing the cyclization of ATP to cyclic AMP, pyrophosphate being the other product (Sutherland, Rall and Menon, 1962; Rall and Sutherland, 1962). A second enzyme, the cyclic nucleotide phosphodiesterase, inactivates cyclic AMP by converting it to ordinary 5'-AMP (Butcher and Sutherland, 1962). To date, these two systems are the only ones clearly recognized as controlling components of the cyclic AMP mechanism.

Several mammalian adenyl cyclase systems have been extensively studied, but unfortunately, because they are particulate and very labile,

---

* For the purpose of this discussion we are using the definition of Huxley (1935) in which a hormone is a substance secreted by a cell or group of cells and is transported to cells at a distant site, wherein it causes an effect for the benefit of the entire organism.

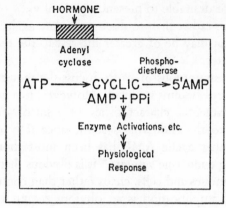

FIG. 1. The components of the cyclic AMP mechanism.

highly purified preparations have not been obtained (for example, see Sutherland, Rall and Menon, 1962). The subcellular distribution of adenyl cyclase has been studied in several tissues of higher species and it has so far been found in association with membranous fractions (Sutherland, Rall and Menon, 1962; Davoren and Sutherland, 1963b; Øye and Sutherland, 1966; De Robertis et al., 1967). Conversely, in *Brevibacterium liquefaciens* adenyl cyclase is freely soluble (Hirata and Hayaishi, 1965). Adenyl cyclase activity is widely distributed in nature, having been identified in all mammalian tissues studied so far with the exception of mature erythrocytes, and also in birds, amphibia, reptiles, several invertebrate phyla, and in unicellular organisms including cellular slime moulds and bacteria.

The distribution of the phosphodiesterase appears to parallel that of adenyl cyclase. While no hormonal control of the phosphodiesterase has been convincingly demonstrated, the enzyme is inhibited by several cellular constituents—for example, certain ribonucleotides, including ATP, pyrophosphate (a by-product of the adenyl cyclase reaction), and citrate (Cheung, 1967). In addition, several pharmacological agents including the methyl xanthines (Butcher and Sutherland, 1962) and puromycin are effective inhibitors (Appleman and Kemp, 1966).

In general, in the absence of stimulation by exogenous hormones, intracellular levels of cyclic AMP are similar in a variety of tissues (of the order of $0 \cdot 1$ to $0 \cdot 5$ nanomoles per gramme of tissue (wet weight), or about $0 \cdot 5$ to $2 \cdot 5$ picomoles per mg of protein). Assuming an even distribution throughout the intracellular water, the concentration would thus be somewhere between 10^{-7} and 10^{-6} M, two to four orders of magnitude lower than the other adenine nucleotides. The actual concentration of free cyclic AMP may of course be much less than 10^{-7} M, because of protein binding and the like, and it is tempting to speculate (even though the subcellular distribution of cyclic AMP has not been well studied) that in certain parts of some cells it may be much higher. Interestingly, the levels of cyclic AMP obtained with maximal stimulation by relevant hormones vary widely in different types of cells.

While the intracellular concentrations of cyclic AMP at any given instant must depend at least in part upon the relative activities of adenyl cyclase and phosphodiesterase, these activities may be influenced by a number of factors besides hormones. For example, under some conditions, the rate of efflux into the external environment may be a factor in determining the intracellular level of the nucleotide (Davoren and Sutherland, 1963a; Makman and Sutherland, 1965). In addition, other mechanisms for the breakdown or even the synthesis of cyclic AMP may exist.

Because our knowledge of the enzymic steps involved in the production of most physiological responses has been incomplete, it has been necessary to resort to criteria to determine whether cyclic AMP is or is not involved in a particular hormone response (Sutherland, Øye and Butcher, 1965). These criteria bear close resemblance to those proposed by Koch for pathogenic organisms and to those more recently suggested by Clark (1937) for hormone and/or drug actions. They entail looking for positive qualitative, quantitative and temporal correlations between effects of the hormone on cyclic AMP levels and the physiological response. In general, this has been done with cell-free systems and also with intact tissue preparations. Cell-free preparations of the adenyl cyclase system can be very useful, since soluble proteins (including much of the phosphodiesterase) and many small molecules can be washed away, thus minimizing the danger of being misled by indirect actions. Also, conditions can be more effectively controlled and more easily varied. However, there are disadvantages to cell-free preparations as well. In several systems, certain physiological effects of hormones (e.g. lipolysis in fat, steroidogenesis in endocrine tissues, the inotropic effect of the catecholamines on the heart, etc.) have not been established in the absence of intact cells. In addition, studies with intact cell preparations provide the best approximation to physiological mechanisms of hormone action, for the fully integrated intra- and extracellular control of cyclic AMP levels can at present be studied only in intact cells. Finally, there are certain qualitative differences between intact and cell-free preparations. For example, insulin and the prostaglandins antagonize the actions of lipolytic hormones on cyclic AMP levels in intact fat cells but have not been observed to do so in cell free preparations (Butcher *et al.*, 1966; Butcher, Baird and Sutherland, 1968).

No matter how extensive studies on the correlations between cyclic AMP levels and a physiological event might be, they cannot prove a role for cyclic AMP in the hormone action, for there is always the question of coincidence rather than cause and effect. Therefore, we have tested the ability of exogenous cyclic AMP or derivatives of cyclic AMP to mimic the hormone under study. Cyclic AMP does not enter certain cells well and, not surprisingly in view of its very rapid destruction by the phosphodiesterase, is ineffective in several systems (Robison *et al.*, 1965). However, as part of a collaborative project, Dr Theo Posternak of Geneva synthesized a series of mono- and di-acylated derivatives of cyclic AMP (Posternak, Sutherland and Henion, 1962; Falbriard, Posternak and Sutherland, 1967). The most successful of these compounds has been N^6-2'-O-dibutyryl cyclic AMP, which has been used in a number of systems (for example, Butcher and co-workers, 1965, and Henion, Sutherland and Posternak, 1967).

While studies with exogenous cyclic AMP or derivatives of the compound have been very useful for this purpose, there are limitations on how much may be deduced from them. It has been recognized for years that exogenous nucleotides are toxic compounds to cells (Green and Stoner, 1950). The concentrations of cyclic AMP or the derivatives required to mimic hormones are in most cases extremely high, often three to four orders of magnitude higher than intracellular levels, and, in the absence of any knowledge about what the relationships between extra-cellular and intracellular concentrations of cyclic AMP might be, one must be circumspect in trying to interpret any results obtained. In addition, it should be noted that the reason for the greater effectiveness of the derivatives is presently unknown. While they were initially designed to enter cells more easily by reducing the polar nature of cyclic AMP, this has not been proved. In addition, at least with purified enzyme preparations, dibutyryl cyclic AMP is not as well hydrolysed as free cyclic AMP, and this may contribute to its greater activity. Finally, at least in the phosphorylase activation system, the 2'-O-acyl group must be removed from the derivative for it to be active (Posternak, Sutherland and Henion, 1962). Whether this same situation pertains in other enzyme systems is at present unknown.

SPECIFICITY

In evaluating the contributions of the cyclic AMP mechanism to the specificity of hormone action, two points have to be considered: the interactions between hormones and adenyl cyclase systems, and secondly, the specificity of the action of cyclic AMP itself.

The actions of hormones on adenyl cyclase systems listed in Table I serve to illustrate how specific the mechanism is. Thus there is in an operational sense specificity between hormones and adenyl cyclase systems. Unfortunately, because of the difficulty in obtaining highly purified preparations of adenyl cyclase with the retention of hormonal sensitivity, it has been impossible to define the adenyl cyclase system, and a minor controversy over whether the "hormone receptor" is or is not a part of, or is close to, the adenyl cyclase system has occurred. To date, no evidence dissociating the receptor and adenyl cyclase has been presented. For example, in the adrenal cortex, cyclic AMP has mimicked effects of ACTH on increased steroidogenesis (Haynes, Sutherland and Rall, 1960; Grahame-Smith et al., 1967), glucose oxidation (Jones, Nicholson and Liddle, 1968), growth and the maintenance of adrenocortical function (Ney, 1969), ascorbic acid depletion (Earp, Watson and Ney, 1969), the hydrolysis of cholesterol esters (Davis, 1969), and glycogenolysis (Haynes, Sutherland and Rall, 1960). Thus, it may be that the only action of ACTH on the adrenal is to increase cyclic AMP levels via activation of adenyl cyclase. The same has so far been the case with those effects of the catecholamines which have the characteristics of β-adrenergic actions. Also, despite the lability of the system, particulate preparations of adenyl cyclase from a variety of tissues have been washed repeatedly in hypotonic and hypertonic solutions and some have been carried through several additional purification procedures, including lyophilization and extraction with dry ether, without losing their capacity to respond to hormones (Sutherland, Rall and Menon, 1962). These preparations retained the specificities exhibited by the "receptors" of intact tissues.

In any event, a possible model of the adenyl cyclase system which at least provides a useful working hypothesis is that it consists of at least two types of subunits, a regulatory subunit facing the extracellular fluid and a catalytic subunit with its active centre bordering on the interior of the cell. Interaction between part of the regulatory subunit and a relevant hormone leads to a conformational perturbation which is extended to the catalytic subunit, altering the activity of the latter. One regulatory enzyme known to be composed of subunits such as this is bacterial aspartate transcarbamylase. Gerhart and Schachman (1965) studied the enzyme from *Escherichia coli*, and found that one type of subunit possessed all of the catalytic activity, while the other type of subunit possessed all of the binding sites for CTP, the normal allosteric inhibitor in this species. Thus, the catalytic subunit of adenyl cyclase might be expected to be basically similar in all tissues, while the structure of the regulatory subunits (or receptors) would vary, thus accounting for hormonal specificity. This would be similar to aspartate transcarbamylase from bacteria, where the catalytic subunit seems to be similar in all species but the regulatory subunits of different species vary in their sensitivity to allosteric effectors.

In some tissues, such as fat, two or more structurally different hormones are known to be capable of stimulating adenyl cyclase, and the question arises as to whether the hormones stimulate the same adenyl cyclase system or whether there are separate ones. Most of the available evidence suggests that in a given tissue, different hormones ultimately stimulate the same catalytic function. Where studied, the effects of maximally effective concentrations of hormones were not additive; rather, it was found that the combinations of hormones reached only the activity produced by the more effective of the two (Makman and Sutherland, 1964; Butcher, Baird and Sutherland, 1968). In terms of the subunit model this could be explained by postulating separate regulatory subunits for each hormone which controls the same catalytic subunit. In those cases where an effect is produced by two or more hormones of markedly different size, such as adrenaline and glucagon in the liver, it is conceivable that the pattern of forces with which the smaller hormone interacts may be a part of the presumably larger pattern of forces with which the larger hormone interacts. It is also possible that the hormones interact with entirely separate systems, at least in some tissues, as suggested by Bitensky, Russell and Robertson (1968) for rat liver.

The second part of the specificity of the cyclic AMP mechanism is related to the action of cyclic AMP itself, once it is increased or decreased in the target cell. A variety of cellular processes are known to be influenced by cyclic AMP levels (shown in Table I), and it seems likely that more will be discovered. There is obviously no specificity to cyclic AMP itself. Rather, the specificity of the action of cyclic AMP is determined by the enzymic profile of the particular cell in which it occurs. A good example of this is provided by the steroid-producing endocrine tissues (Fig. 2), in which increased levels of cyclic AMP stimulate the conversion of cholesterol to pregnenolone. Corticosterone is the chief adrenocortical glucocorticoid synthesized and released by rat adrenals, while human, dog and beef adrenals produce primarily cortisol. However, this is not because ACTH (acting through cyclic AMP) is affecting a different process. Karaboyas and Koritz (1965) have convincingly demonstrated that both cyclic AMP and ACTH act on the cholesterol side-chain cleavage system in beef and rat adrenocortical tissue. Rather, the difference lies in the enzymic profile of the two tissues. In the rat adrenal the 17β-hydroxylase system present in man, etc. is deficient or vestigial and as a result the primary product is corticosterone (17-desoxy cortisol).

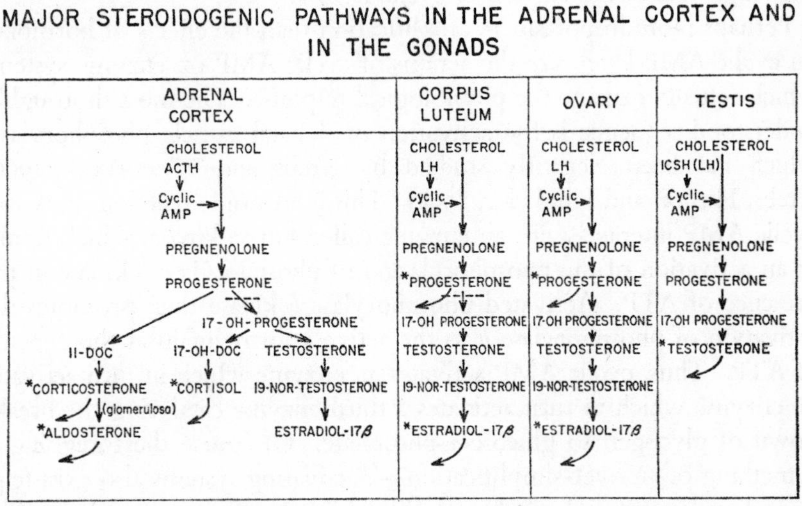

MAJOR STEROIDOGENIC PATHWAYS IN THE ADRENAL CORTEX AND IN THE GONADS

FIG. 2. Major steroidogenic pathways in the adrenal cortex and in the gonads. A dashed line across a step indicates that it is missing or in low activity in some species (e.g. the 17β-hydroxylase system in the rat). An asterisk next to a steroid indicates that it is a hormone secreted in significant quantities by the steroidogenic tissue.

The analogy could be carried further to the gonadal endocrine tissues, where, for example, increased cyclic AMP levels in the bovine corpus luteum, rabbit interstitial cells of the ovary, or the testis cause increased secretions of progesterone, oestrogens and 20-hydroxy-progesterone, or androgens respectively. In other words, a cell will respond to increased cyclic AMP levels with whatever mechanisms it has available.

AMPLIFICATION

In some cases, depending upon the system of measurement, there is clearly amplification at the level of adenyl cyclase; for example, the intravenous administration of ACTH into hypophysectomized rats resulted in increased adrenal cyclic AMP levels which were 180 times greater, on a molar basis, than the total amount of ACTH injected (Grahame-Smith et al., 1967). Conversely, in rat epididymal fat pads, the net increase in cyclic AMP concentrations in response to adrenaline was of the same order of magnitude as the concentration of the hormone (Butcher et al., 1965). However, in the presence of a phosphodiesterase inhibitor, amplification was apparent. In other words, the activation of adenyl cyclase was sufficient to produce amplification were it not for the very high rate of destruction of cyclic AMP.

Perhaps more important in amplification than the effects of hormones on cyclic AMP levels are the actions of cyclic AMP on enzyme systems which actually execute the physiological response. The most thoroughly understood sequence is the activation of skeletal muscle phosphorylase, which has been elegantly studied by Krebs and co-workers (1966; Krebs, Huston and Hunkeler, 1968). This is a cascade reaction, in which cyclic AMP interacts with an enzyme called kinase kinase, which results in an activation of the phosphorylation of phosphorylase b kinase in the presence of ATP. Activated phosphorylase b kinase then promotes the activation of phosphorylase b to the active a form, again at the expense of ATP. Thus, cyclic AMP activates an enzyme which in turn activates an enzyme which in turn activates a third enzyme catalysing the breakdown of glycogen to glucose-1-phosphate. Of course the foregoing is something of an over-simplification—inactivating systems also exist (e.g. phosphorylase phosphatase) and, in addition, effector molecules like ATP, 5'-AMP and Pi are involved in the regulation of phosphorylase activity. Nonetheless, the potential amplification obtainable by the activation of biological catalysts is obvious. Unfortunately, our knowledge of many of the other systems in which cyclic AMP plays a regulatory role is meagre. Although in the steroidogenic systems cyclic AMP is known to act upon the rate-limiting step in steroid hormone synthesis, cholesterol side-chain cleavage, the mechanism by which this system is activated is unclear and controversial. In other systems even less is known; even the sites of action of cyclic AMP are unknown, for example in processes such as the increased movement of water and ions across the toad bladder, the inotropic effect on the heart, or the dispersion of melanin granules in the dorsal frog skin. While it seems reasonable to expect that these processes will ultimately involve the activation or inactivation of enzymes, our ignorance is too profound to permit any more than speculation.

MODULATION

From the standpoint of minute to minute control, the cyclic AMP mechanism is quite dynamic. For example, within a few seconds after the addition of a relevant hormone, cyclic AMP levels in a number of tissues have been significantly increased (e.g the cyclic AMP levels were maximally increased within 3 seconds after the addition of adrenaline to isolated perfused rat hearts; Robison *et al.*, 1965). In addition, a variety of agents are capable of very rapidly decreasing cyclic AMP levels. Examples of these include insulin and the prostaglandins on adipose tissue (Butcher *et al.*, 1968; Butcher and Baird, 1968), insulin on liver (Jefferson *et al.*, 1968), α-adrenergic stimuli on the β-cells of the pancreas (Turtle and Kipnis, 1967) and α-adrenergic stimulation and melatonin on the frog skin *in vitro* (Abe *et al.*, 1969a, b).

ADDITIONAL COMMENTS

The identification of cyclic AMP in bacteria (Makman and Sutherland, 1965; Hirata and Hayaishi, 1965) and the recent demonstration of actions of the nucleotide on enzyme induction in *E. coli*, for example by Perlman and Pastan (1968), perhaps offer an explanation for the seeming ubiquitousness of cyclic AMP. If we can assume from the presence of the cyclic AMP mechanism in the unicellular organisms available today that it existed in the unicellular forms present before the evolution of multicellular organisms, then it becomes obvious that cyclic AMP is part of a very primitive mechanism. The adenyl cyclase system of *B. liquefaciens* is dependent upon the presence of pyruvate or certain other α-keto acids for activity (Hirata and Hayaishi, 1965). Thus in an evolutionary sense cyclic AMP may have pre-dated hormones, at least as defined by Huxley.

In general, cyclic AMP has been considered an intracellular compound. In the "second messenger" hypothesis, adenyl cyclase and the cyclic AMP mechanism were viewed as a means for transmitting extracellular information provided by the hormone (or first messenger) to the interior of the cell via changes in cyclic AMP levels (the intracellular second messenger). However, cyclic AMP may not in all cases act within the cell in which it is formed. For example, Bonner's group has provided evidence to show that cyclic AMP is an extracellular mediator in the aggregation of cellular slime moulds (Konijn *et al.*, 1968). The avian erythrocyte, a cell in which cyclic AMP levels are increased rapidly upon exposure to adrenaline, and which has an active pump for cyclic AMP which will transport the nucleotide to the medium even against a concentration gradient, is another case where extracellular effects of cyclic AMP may be important (Davoren and Sutherland, 1963a). Finally, the presence of cyclic AMP in urine and the demonstrations that it is excreted at higher rates in response to parathyroid hormone (Chase and Aurbach, 1967) and glucagon (Broadus et al., 1969) suggest that it can be released from the kidney and liver and perhaps other organs as well.

While cyclic AMP is at present unique as an intracellular mediator in the actions of hormones, it seems probable that others will be discovered. A likely candidate is guanosine $3',5'$-monophosphate (cyclic GMP), which is found in tissues and urine of several mammalian species by Hardman, Davis and Sutherland (1966). While no direct hormonal control on the levels of cyclic GMP has been demonstrated, it is clear that the excretion of the compound can be affected by procedures such as hypophysectomy and thyroidectomy. In addition, an enzyme system which catalyses the production of cyclic GMP from GTP, and which is distinct from adenyl cyclase, has been identified in rat lung and certain other tissues (Hardman and Sutherland, 1969).

SUMMARY

Adenosine $3',5'$-phosphate (cyclic AMP) holds a central position in control systems of multicellular organisms as an intracellular mediator of the actions of several hormones. Its mechanism of action has been discussed in terms of the basic characteristics of systems of control mediated by hormones: (1) specificity of hormone action; (2) amplification of effect of hormones; and (3) the possibility of rapid change in activity of the control mechanism.

Acknowledgements

This research was supported in part by grants HE 08 332, AM 07642 and MH 11468.

REFERENCES

ABE, K., BUTCHER, R. W., NICHOLSON, W. E., BAIRD, C. E., LIDDLE, R. A., and LIDDLE, G. W. (1969a). *Endocrinology*, **84**, 362–368.
ABE, K., ROBISON, G. A., LIDDLE, G. W., BUTCHER, R. W., NICHOLSON, W. E., and BAIRD, C. E. (1969b). *Endocrinology*, in press.
APPLEMAN, M. M., and KEMP, R. G. (1966). *Biochem. biophys. Res. Commun.*, **24**, 564–568.
BITENSKY, M. W., RUSSELL, V., and ROBERTSON, W. (1968). *Biochem. biophys. Res. Commun.*, **31**, 706–712.
BROADUS, A., NORTHCUTT, R. C., HARDMAN, J. G., KAMINSKY, N. I., SUTHERLAND, E. W., and LIDDLE, G. W. (1969). *Clin. Res.*, **17**, 65.
BUTCHER, R. W., and BAIRD, C. E. (1968). *J. biol. Chem.*, **243**, 1713–1717.
BUTCHER, R. W., BAIRD, C. E., and SUTHERLAND, E. W. (1968). *J. biol. Chem.*, **243**, 1705–1712.

BUTCHER, R. W., HO, R. J., MENG, H. C., and SUTHERLAND, E. W. (1965). *J. biol. Chem.*, **240**, 4515–4523.

BUTCHER, R. W., ROBISON, G. A., HARDMAN, J. G., and SUTHERLAND, E. W. (1968). *Adv. Enzyme Regulation*, **6**, 357–390.

BUTCHER, R. W., SNEYD, J. G. T., PARK, C. R., and SUTHERLAND, E. W. (1966). *J. biol. Chem.*, **241**, 1651–1653.

BUTCHER, R. W., and SUTHERLAND, E. W. (1962). *J. biol. Chem.*, **237**, 1244–1250.

CHASE, L. R., and AURBACH, G. D. (1967). *Proc. natn. Acad. Sci. U.S.A.*, **58**, 518–525.

CHEUNG, W. Y. (1967). *Biochemistry, N.Y.*, **6**, 1079–1087.

CLARK, A. J. (1937). In *Heffter's Handbuch der Experimentellen Pharmakologie*, vol. 4, ed. Heubner, W., and Schuller, J. Berlin: Springer Verlag.

DAVIS, W. W. (1969). *Fedn Proc. Fedn Am. Socs exp. Biol.*, **28**, 701.

DAVOREN, P. R., and SUTHERLAND, E. W. (1963a). *J. biol. Chem.*, **238**, 3009–3015.

DAVOREN, P. R., and SUTHERLAND, E. W. (1963b). *J. biol. Chem.*, **238**, 3016–3023.

DE ROBERTIS, E., RODRIGUEZ DE LORES ARNAIZ, G., ALBERICI, M., BUTCHER, R. W., and SUTHERLAND, E. W. (1967). *J. biol. Chem.*, **242**, 3487–3493.

EARP, H. S., WATSON, B. S., and NEY, R. L. (1969). *Clin. Res.*, **17**, 22.

FALBRIARD, J. G., POSTERNAK, TH., and SUTHERLAND, E. W. (1967). *Biochim. biophys. Acta*, **149**, 99–105.

GERHART, J. C., and SCHACHMAN, H. K. (1965). *Biochemistry, N.Y.*, **4**, 1054–1062.

GRAHAME-SMITH, D. G., BUTCHER, R. W., NEY, R. L., and SUTHERLAND, E. W. (1967). *J. biol. Chem.*, **242**, 5535–5541.

GREEN, H. N., and STONER, H. B. (1950). *Biological Actions of the Adenine Nucleotides.* London: Lewis.

HARDMAN, J. G., DAVIS, J. W., and SUTHERLAND, E. W. (1966). *J. biol. Chem.*, **240**, 3704–3705.

HARDMAN, J. G., and SUTHERLAND, E. W. (1969). *J. biol. Chem.*, submitted.

HAYNES, R. C., SUTHERLAND, E. W., and RALL, T. W. (1960). *Recent Prog. Horm. Res.*, **16**, 121–138.

HENION, W. F., SUTHERLAND, E. W., and POSTERNAK, TH. (1967). *Biochim. biophys. Acta*, **148**, 106–113.

HIRATA, M., and HAYAISHI, O. (1965). *Biochem. biophys. Res. Commun.*, **21**, 361–365.

HUXLEY, J. S. (1935). *Biol. Rev.*, **10**, 427.

JEFFERSON, L. S., EXTON, J. H., BUTCHER, R. W., SUTHERLAND, E. W., and PARK, C. R. (1968). *J. biol. Chem.*, **243**, 1031–1038.

JONES, D. J., NICHOLSON, W. E., and LIDDLE, G. W. (1968). *J. clin. Invest.*, **47**, Abstract No. 152, 51a.

KARABOYAS, G. C., and KORITZ, S. B. (1965). *Biochemistry, N.Y.*, **4**, 462.

KONIJN, T. M., BARKLEY, D. S., CHANG, Y. Y., and BONNER, J. T. (1968). *Am. Nat.*, **102**, 225.

KREBS, E. G., DELANGE, R. J., KEMP, R. G., and RILEY, W. D. (1966). *Pharmac. Rev.*, **18**, 163.

KREBS, E. G., HUSTON, R. B., and HUNKELER, F. L. (1968). *Adv. Enzyme Regulation*, **6**, 245–256.

MAKMAN, M. H., and SUTHERLAND, E. W. (1964). *Endocrinology*, **75**, 127–134.

MAKMAN, R. S., and SUTHERLAND, E. W. (1965). *J. biol. Chem.*, **240**, 1309–1314.

NEY, R. L. (1969). *Endocrinology*, **84**, 168–170.

ØYE, I., and SUTHERLAND, E. W. (1966). *Biochim. biophys. Acta*, **127**, 347–354.

PERLMAN, R. L., and PASTAN, F. (1968). *J. biol. Chem.*, **243**, 5420.

POSTERNAK, TH., SUTHERLAND, E. W., and HENION, W. F. (1962). *Biochem. biophys. Acta*, **65**, 558–560.

RALL, T. W., and SUTHERLAND, E. W. (1962). *J. biol. Chem.*, **237**, 1228–1232.

ROBISON, G. A., BUTCHER, R. W., ØYE, I., MORGAN, H. E., and SUTHERLAND, E. W. (1965). *Molec. Pharmac.*, **1**, 168.

ROBISON, G. A., BUTCHER, R. W., and SUTHERLAND, E. W. (1968). *A. Rev. Biochem.*, **37**, 149–174.

ROBISON, G. A., BUTCHER, R. W., and SUTHERLAND, E. W. (1970). *Cyclic AMP.* New York: Academic Press. In press.

SUTHERLAND, E. W., ØYE, I., and BUTCHER, R. W. (1965). *Recent Prog. Horm. Res.*, **21**, 623–646.

SUTHERLAND, E. W., and RALL, T. W. (1960). *Pharmac. Rev.*, **12**, 265–299.

SUTHERLAND, E. W., RALL, T. W., and MENON, T. (1962). *J. biol. Chem.*, **237**, 1220–1227.

TURTLE, J. R., and KIPNIS, D. M. (1967). *Biochem. biophys. Res. Commun.*, **28**, 797.

DISCUSSION

CYCLIC AMP IN BACTERIAL SYSTEMS

Monod: I would like to indicate what has been found on cyclic AMP in bacteria by Dr Ullman and ourselves on one hand and by Perlman and Pastan (1968) on the other. Both groups find essentially that cyclic AMP is an antagonist of the so-called catabolite repression effect. It has long been known that virtually all enzymes involved in the breakdown of carbohydrates are inducible by their specific substrates or analogues, and the rate of synthesis of these enzymes is greatly lowered in the presence of glucose or other rapidly assimilated carbohydrates. The observations of Sutherland and Martin suggested to us that there might be a relationship between this effect, whose mechanism is completely unknown, and cyclic AMP. Using the galactosidase system and setting up conditions where the catabolite repression effect is magnified, where we can obtain 99 per cent inhibition of galactosidase synthesis, we find that in such conditions cyclic AMP has an extremely strong effect. It may restore the synthesis of enzyme back to almost normal levels; that is to say, it causes 50 to 100-fold increases in the rate of synthesis. This effect is completely specific for cyclic AMP; it doesn't work with other nucleotides, cyclic or not. We haven't tried cyclic guanosine monophosphate, however.

We could also show that both the catabolite repression effect and its reversal by cyclic AMP are completely independent of the structure of the known control elements, the I gene and the operator. We can observe both the catabolite repression effect and its antagonizing by cyclic AMP in strains where the control system is completely deleted. So it is independent of specific regulation by induction and repression.

Burma: Professor Monod, in studies on the effect of cyclic AMP on catabolite repression there must be a permeability problem. In catabolite repression sufficient concentration of the repressor is required inside the cell. How much of the cyclic AMP will go inside the cell and what will be the internal concentration?

Monod: We need a high external concentration of cyclic AMP, of the order of 5 mM. We can't say what the inside concentration is, but some cyclic AMP does get in. There is a double difficulty that presumably the cells are highly impermeable to cyclic AMP, and furthermore it has been shown by Sutherland that there is active excretion of cyclic AMP in the presence of glucose, and our experiments had to be done with glucose. So it is not so strange that one has to use high concentrations. Perlman and Pastan used a different technique; they treated cells with EDTA so that they became more permeable. They obtain effects at concentrations 50 times less than in our conditions. What they gain in doing that, they lose because the magnitude of the effect they observe is much lower. Our effects are 50 or 100-fold while theirs is only about 3-fold. But the difference suggests that a high concentration is required because of the permeability barrier to cyclic AMP.

Butcher: The ratio of exogenous extracellular cyclic AMP to intracellular levels is probably very high. In addition, the phosphodiesterase in some strains of *E. coli* is apparently very active (Brana and Chytil, 1966), although Makman and Sutherland (1965) found no activity.

MODE OF ACTION OF CYCLIC AMP

Lynen: Referring to the effect of prostaglandins antagonizing lipolytic hormones, is it an effect on adenyl cyclase or on the phosphodiesterase?

Butcher: I cannot give a final answer because the prostaglandins do not lower cyclic AMP production in the absence of cell structure, and the turnover of cyclic AMP is so high in intact cells that it is almost impossible to determine the site of action using them. So although everything is compatible with an effect on adenyl cyclase, it hasn't been proved.

Lynen: You mentioned the effect of dibutyryl cyclic AMP. Does the level of cyclic AMP change in the presence of this compound, or is it just simulating the effect of cyclic AMP and that is why one has to use so much higher concentrations?

Butcher: We don't really know. The derivatives were designed to enter cells better but this has not been demonstrated. One thing that we do know is that they are not hydrolysed at such a high rate as free cyclic AMP by purified preparations of the phosphodiesterase (Posternak, Sutherland and Henion, 1962). In the case of phosphorylase activation the acyl group on the 2'-O- position must be removed before it is active, although the N^6-mono-acyl derivatives were active. In the case of lipolysis or steroidogenesis we do not know if the di-acyl derivatives are active as such or if they must be converted to different forms first.

Lynen: The butyryl group bound to N^6 is still there; it could be that it is just a poor substance.

Butcher: N^6-monobutyryl cyclic AMP was as active as free cyclic AMP on phosphorylase activation.

Monod: In bacteria the butyryl compound is inactive.

Jagannathan: You indicated that ATP is required for the action of cyclic AMP on phosphorylase kinase. Do you visualize any kind of covalent linkage between cyclic AMP and this enzyme?

Secondly, is the known adenyl cyclase activity able to account for the rather sudden increase in the total concentration of cyclic AMP within a few seconds?

And finally, is there any evidence for non-covalently bound cyclic AMP or cyclic AMP localized in the membrane or elsewhere? Some of the effects of the catecholamines, especially the increase in the concentration of AMP (several-fold in 15 seconds), might be due to liberation of bound cyclic AMP.

Butcher: ATP is required for the activation of phosphorylase b kinase but, as shown by the elegant studies of Krebs, Huston and Hunkeler (1968), cyclic AMP does not interact primarily with phosphorylase b kinase but rather with phosphorylase b kinase kinase, which is also a phosphate-donating enzyme. The nature of the interaction between phosphorylase b kinase kinase and cyclic AMP has not been reported. The energy of the $3',5'$-phosphate bond in cyclic AMP is extremely high. Greengard, Hayaishi and Colowick (1969), working with a soluble, partially purified adenyl cyclase from *Brevibacterium liquefaciens*, found they could reverse the cyclization reaction very well, i.e. cyclic AMP + pyrophosphate→ATP. They estimated the free energy of hydrolysis of the $3'$ bond to be 11–12 Kcal/mole, which is higher than the α–β and β–γ bonds of ATP. Therefore, it is not inconceivable that the cyclic phosphate bond might be involved in a covalent reaction. On the other hand, it might be an allosteric effect.

Monod: From looking simply at the kinetic curves of activation of kinase kinase, which are not too precise of course, because the whole system is complicated, they don't look like absorption isotherms. Moreover, the activation is obtained at such fantastically low concentrations of the nucleotide that one is tempted to believe that the activation of kinase kinase is in fact due to a covalent reaction.

Butcher: To answer Dr Jagannathan's other point, I feel that adenyl cyclase can account for the increase in cyclic AMP, and the very rapid increases in cyclic AMP which we see in tissues are probably not due to a release of bound cyclic AMP, since we measure the total cyclic AMP concentration of the tissue. However, there is preliminary evidence that in the liver and in adipose tissue some cyclic AMP is rather firmly attached to high-speed particulate fractions (Corbin, Sneyd and Park, 1969), which might not be metabolically active in the sense that freely soluble cyclic AMP would be.

ROLE OF CYCLIC AMP IN HORMONE ACTION

Talwar: A large number of hormones evidently increase the concentration of cyclic AMP. One could perhaps imagine that the membrane-linked adenyl cyclase enzyme is a system closely associated with a variety of proteins which have an affinity for a large number of hormones. The associated proteins act as receptors, and their combination with the appropriate hormone activates also the adenyl cyclase. There is thus an increase in the product of the enzyme reaction (cyclic AMP) in every case even though the inducing hormone is different.

I want to turn now to another issue. Exogenous cyclic AMP has been used to mimic the effects of oestradiol, with the assumption that cyclic AMP may be involved as a mediator in the action of this hormone. Hechter and his colleagues (1967) showed that the uterus would respond to cyclic AMP *in vitro* by an increased synthesis of RNA and proteins, but the experimental conditions were rather unphysiological, incubation had to be continued for several hours and the medium had to be deprived of glucose. Dr S. K. Sharma in our laboratory has tried to evolve a system in which, by employing an enriched medium and stretching of the uterus, an effect of exogenous cyclic AMP is obtained on the uptake and incorporation of uridine and radioactive amino acids into RNA and proteins. This effect appears to be related to the ability of the tissue to pick up these metabolites, and may have something to do with the action of cyclic AMP on the general permeability of the tissue to these metabolites. In the initial period the specific activity changes essentially in proportion to the expansion of the pool size. Here then one can have an apparent stimulatory effect of cyclic AMP on the incorporation of a precursor, but it does not fully simulate the true effects of oestrogens. If the uteri are maintained in organ culture for 24 or 48 hours in the presence of cyclic AMP, none of the histological effects usually caused by oestrogens *in vivo* are obtained. Finally, another issue on which I would seek your comments pertains to a report in which it has been shown that low levels of cyclic AMP are lipogenic in adipose cells whereas high levels are lipolytic (Blecher, 1967).

Butcher: I don't recall a report on cyclic AMP as lipogenic, although the anti-lipolytic effects of adenine nucleotides are recognized (Vaughan, 1960). This is a property not only of cyclic AMP but of all exogenous adenine nucleotides, probably because they are toxic to cells. To return to the uterus, Dr Hechter and his co-workers also showed that several nucleotides besides cyclic AMP had the same effect, which in my mind causes doubts about the specificity of cyclic AMP in this particular mechanism.

Tata: I want to mention some recent results from our laboratory on the extent of mimicry by cyclic AMP of the action of TSH on thyroid hormone production. We asked ourselves whether dibutyryl cyclic AMP mimics all actions of TSH including its biosynthetic actions or whether its activity was restricted to the very rapid adjustment of the levels of metabolites or breakdown of thyroglobulin. Our conclusion (Kerkof and Tata, 1969) is that the mimicry is restricted to breakdown of preformed thyroglobulin or the uptake of phosphate ions or iodine. The latter effects could give a false impression of mimicry by AMP of the biosynthesis of phospholipid or thyroglobulin. Thus, for example, the effect of cyclic AMP on the phospholipid effect of TSH can be explained by an increase in the uptake of radioactive phosphate and not by an effect of cyclic AMP on net phospholipid synthesis.

Butcher: Since I have not worked with thyroid at all I wonder what the concentration of dibutyryl cyclic AMP was after such a long incubation. I would like to mention also a result obtained by Ney (1969) which is interesting in this context. He asked the question: could cyclic AMP be involved in the trophic effect of ACTH on the adrenal gland? Previous experiments had shown that two weeks after hypophysectomy, when the adrenal was atrophic and non-responsive to ACTH in terms of steroid hormone production, the cyclic AMP mechanism was still responsive (Grahame-Smith *et al.*, 1967). Ney hypophysectomized rats and tried to maintain them (in terms of adrenal weight and function) with ACTH or with dibutyryl cyclic AMP, which he gave subcutaneously in gelatin. He found that dibutyryl cyclic AMP did maintain adrenal weight and function. This is certainly an extremely complicated system especially since it was *in vivo*, but it might suggest that under certain circumstances cyclic AMP can be involved in mammalian protein synthesis in some way. I think the bacterial work also supports this, although the effect of cyclic AMP is perhaps not direct.

Fortier: Professor Butcher, you have raised the issue of the target-gland specificity of the pituitary trophic hormones, which may be contrasted with the relative non-specificity or polyvalence of the lipolytic response of fat cells. This target-gland specificity is particularly interesting with respect to ACTH and LH, in view of the similarity of the steroid synthetic pathways which they influence. Doesn't it suggest that the site of the organ specificity of these trophic hormones is beyond cyclic AMP and might involve the guanyl cyclase or adenyl cyclase systems?

Butcher: This is exactly what I think. Where cyclic AMP is involved, the specificity between hormones and cells must be at the level of the adenyl cyclase system. There is no specificity to cyclic AMP—what happens once it is increased is determined entirely by what the cell can do. The same might be true with guanyl cyclase, but so far no clear effects of hormones on this system have been found.

White: May I make two brief comments about specificity? It is rather interesting that if one classifies the hormones as they are often classified, into those that are protein, polypeptide or derived from amino acids on the one hand, and steroids on the other, then as far as I am aware no steroid hormone effect has been shown as yet to be mediated by cyclic AMP. The fact that steroidogenesis appeared later in evolution may or may not be pertinent. Secondly, does not the direct addition of adrenal cortical steroids to the epididymal fat pad (Fain, Kovacev and Scow, 1965) also cause fatty acid release without any change in cyclic AMP concentration? Is this an instance where there may be another mechanism operative, and thus all lipolysis may not be mediated by cyclic AMP?

Butcher: I agree that there is little convincing evidence that steroid hormones act directly by raising cyclic AMP levels. The effects of glucocorticoids on epididymal fat pads are best seen in adrenalectomized animals, in which lipolysis is defective. It is a very different effect from the rapid effects of hormones acting through cyclic AMP, which take place in minutes. The cyclic AMP mechanism is in no way defective in adipose tissue of adrenalectomized rats (Corbin, Sneyd and Park, 1969). However, the lipolytic mechanism is defective, not only to other lipolytic hormones, but also to exogenous cyclic AMP. This has been tentatively localized at the level of lipase activation. This sort of phenomenon has also been observed by Friedman, Exton and Park (1967) in the perfused rat liver in terms of gluconeogenesis and glycogenolysis.

Lynen: Is the slow effect of adrenal cortical hormones in some way

related to carbohydrate metabolism in the adipose tissue? Glycero-
phosphate is required for the re-esterification of the fatty acids released
by specific lipases.

Butcher: You have made a very good point, Professor Lynen. I have
certainly oversimplified things, for these are interlocking systems, and
carbohydrate metabolism, as you suggest, is surely involved. For example,
Jeanrenaud (1967) has presented data showing that glucocorticoids caused
lipase activation only at very high concentrations, and that the apparent
lipolysis seen with lower concentrations of steroids was in fact due to
decreased re-esterification.

Gopalan: Is there any evidence that impairment of the second messenger
system may be concerned in the pathogenesis of human disease?

We can recognize two phases in cases of malnutrition: a phase of
adaptation when the organism responds by muscular wasting and essential
organs are protected, and then a stage of dysadaptation where the organism
manifests effects of malnutrition in the form of damage to the liver and
other essential organs. The children suffering from marasmus may be
considered to represent the first phase. The hyperglycaemic response to
adrenaline is greatly exaggerated in cases of marasmus, while in children
suffering from kwashiorkor, who may be considered to represent the
second phase, this response is greatly retarded (Jayarao, 1965). The increase in plasma cortisol levels brought about by corticotropin is very
much greater in subjects suffering from marasmus than those suffering
from kwashiorkor. In these latter cases, after treatment with a high
protein diet, the response increased (Jayarao, Srikantia and Gopalan, 1968).
These observations would seem to justify the speculation that in the stage
of dysadaptation there is an exhaustion of adenyl cyclase systems leading
to an impairment in the response of tissues to certain hormones; or,
alternatively, it may be postulated that adenyl cyclase activity is actually
exaggerated in the stage of adaptation and that this may, in fact, be a
regulatory mechanism involved in adaptation to nutritional stress. I
would like to hear your views on this.

Secondly, what importance do you attach to the estimation of urinary
levels of cyclic AMP as a clinical tool? If, as you say, the increase in
adenyl cyclase in different tissues is dependent upon a specific response to
particular hormones, urinary levels will be a result of complex effects
and may not lend themselves to interpretation.

Butcher: Malnutrition has not been investigated in human beings at the
level of cyclic AMP and adenyl cyclase, to my knowledge. However,
some very interesting changes occur in the cyclic AMP responses of

adipose tissue in rats after prolonged fasting, where the response to catecholamines is diminished, or after fasting and refeeding, where it is enhanced (Corbin, Sneyd and Park, 1969). What this means in terms of malnutrition is of course unclear.

On the use of urinary cyclic AMP levels, one of the big problems is knowing the source of urinary cyclic AMP. It is clear from the work of Chase and Aurbach (1967) and also that of Broadus, Hardman and Sutherland (1969) that the kidney is one source, and certain clinical implications might arise from this. Also, Broadus, Hardman and Sutherland have quite clearly demonstrated that cyclic AMP is escaping from the liver, and this would make a contribution. However, cyclic AMP is almost ubiquitous, and we don't know how many other tissues it is escaping from.

CYCLIC AMP IN THE NERVOUS SYSTEM

Paintal: Have you any idea about the role of cyclic AMP in a system with a high metabolic rate? Some of the tissues which you have used have a rather low metabolism, like fat. The chemoreceptors of the carotid body and the aortic body have a very high metabolic rate, possibly the highest in mammalian tissues. Recently Joels and Neil (1968) have shown that in these tissues cyclic AMP plays a role. I wonder what it does here, because the reaction rates must be very rapid, of the order of milliseconds. Have you any light to throw on this?

Butcher: I have no data on the chemoreceptors themselves, but I don't think the fact that the reaction is so fast would exclude a role of AMP in such tissue, because the chief limitation we have in detecting fast changes in the cyclic AMP is our ability to fix the tissue properly at early times, and we haven't reached the millisecond level yet. Certainly the activities of adenyl cyclase and the phosphodiesterase in the central nervous system are high. The cerebral cortex is the most active, and although the other parts of the central nervous system are lower, they are higher than many other tissues (Sutherland, Rall and Menon, 1962; Butcher and Sutherland, 1962).

Whittaker: Would you elaborate further on the possible role of cyclic AMP in nervous tissue, and in particular in the mode of action of transmitters such as acetylcholine and noradrenaline? Transmitter action tends to be discussed in terms of changes in membrane conductance, but there is increasing evidence that some transmitters, particularly acetylcholine when interacting with muscarinic sites, may act primarily by controlling cell metabolism and only secondarily by affecting the excitability of cells. I wonder what has been done in that area.

Butcher: Almost nothing directly. With De Robertis and co-workers (1967) we did a preliminary study of the distribution of adenyl cyclase in cerebral cortex and found a rough correlation between the distribution of adenyl cyclase and that of nerve endings. It was not a clear correlation but simply an association or a higher specific activity. Catecholamines and also histamine increase cyclic AMP levels in brain slices; this has been shown by Kakivchi and Rall (1968). They also showed that the most potent stimulator of cyclic AMP accumulation in brain was adenosine.

Whittaker: I find the association of adenyl cyclase and synaptosomes a little difficult to understand, because these structures presumably primarily represent the presynaptic element and the amount of postsynaptic membranes isolated with the synaptosomes may be quite low. We would expect this system to be in the target cell which would be on the postsynaptic side, and presumably that is broken down by the fractionation procedure used.

Butcher: I have had reservations about our work because I don't know what contributions glia made, and because only about half of the adenyl cyclase activity was in the "mitochondrial" fraction, the rest being in the so-called microsomal fraction. Since we didn't fully characterize the microsomal fraction, one can only say that in the mitochondrial fractions, the highest specific activity of adenyl cyclase was in the fraction richest in nerve endings.

Whittaker: And those would also to some extent be contaminated with microsomal material.

REFERENCES

BLECHER, M. (1967). *Biochem. biophys. Res. Commun.*, **27**, 560.
BRANA, H., and CHYTIL, F. (1966). *Folia microbiol., Praha*, **11**, 43–46.
BROADUS, A. E., HARDMAN, J. G., and SUTHERLAND, E. W. (1969). In preparation.
BUTCHER, R. W., and SUTHERLAND, E. W. (1962). *J. biol. Chem.*, **237**, 1244–1250.
CHASE, L. R., and AURBACH, G. D. (1967). *Proc. natn. Acad. Sci. U.S.A.*, **58**, 518–525.
CORBIN, J. D., SNEYD, J. G. T., and PARK, C. R. (1969). In preparation.
DE ROBERTIS, E., RODRIGUEZ DE LORES ARNAIZ, G., ALBERICI, M., BUTCHER, R. W., and SUTHERLAND, E. W. (1967). *J. biol. Chem.*, **242**, 3487–3493.
FAIN, J. N., KOVACEV, V. P., and SCOW, R. O. (1965). *J. biol. Chem.*, **240**, 3522–3529.
FRIEDMAN, N., EXTON, J. H., and PARK, C. R. (1967). *Biochem. biophys. Res. Commun.*, **29**, 113–119.
GRAHAME-SMITH, D. G., BUTCHER, R. W., NEY, R. L., and SUTHERLAND, E. W. (1967). *J. biol. Chem.*, **242**, 5535–5541.
GREENGARD, P., HAYAISHI, O., and COLOWICK, S. P. (1969). *Fedn Proc. Fedn Am. Socs exp. Biol.*, **28**, 467.

HECHTER, O., YOSHINAGA, K., HALKERSTON, I. D. K., and BIRCHALL, K. (1967). *Archs Biochem. Biophys.*, **122**, 449.
JAYARAO, K. S. (1965). *Am. J. Dis. Child.*, **110**, 519.
JAYARAO, K. S., SRIKANTIA, S. G., and GOPALAN, C. (1968). *Archs Dis. Childh.*, **43**, 365.
JEANRENAUD, B. (1967). *Biochem. J.*, **103**, 627.
JOELS, N., and NEIL, E. (1968). In *Arterial Chemoreceptors*, p. 153, ed. Torrance, R. W. Oxford: Blackwell.
KAKIVCHI, S., and RALL, T. W. (1968). *Molec. Pharmac.*, **4**, 367–388.
KERKOF, P. R., and TATA, J. R. (1969). *Biochem. J.*, **112**, 729–739.
KREBS, E. G., HUSTON, R. B., and HUNKELER, F. L. (1968). *Adv. Enzyme Regulation*, **6**, 245–256.
MAKMAN, R. S., and SUTHERLAND, E. W. (1965). *J. biol. Chem.*, **240**, 1309–1314.
NEY, R. L. (1969). *Endocrinology*, **84**, 168–170.
PERLMAN, R. L., and PASTAN, F. (1968). *J. biol. Chem.*, **243**, 5420.
POSTERNAK, TH., SUTHERLAND, E. W., and HENION, W. F. (1962). *Biochim. biophys. Acta*, **65**, 558.
SUTHERLAND, E. W., RALL, T. W., and MENON, T. (1962). *J. biol. Chem.*, **237**, 1220–1227.
VAUGHAN, M. (1960). *J. biol. Chem.*, **235**, 3049–3053.

INSULIN AND GROWTH HORMONE CONTROL OF PROTEIN BIOSYNTHESIS

A. KORNER

School of Biological Sciences, University of Sussex, Brighton

THERE is abundant evidence that both growth hormone and insulin can stimulate protein biosynthesis but the mechanism by which they do so is still unknown. Many of the protein anabolic effects of the two hormones are similar but enough differences between them have been uncovered to suggest that they do not act in the same way. In this article, I consider the possible points in the protein biosynthetic process where hormones might act and then assess the available evidence on the question of whether insulin and growth hormone act at these points.

PROTEIN ANABOLIC ACTIVITY OF GROWTH HORMONE AND INSULIN

Hypophysectomy of a variety of animals arrests growth if they are young and results in a fall in body weight if they are adult. Injection of suitable growth hormone preparations causes resumption of growth accompanied by increased body weight, protein and ribonucleic acid (RNA) accumulation, loss of lipid, alterations in carbohydrate metabolism and other changes (see review by Knobil and Hotchkiss, 1964). Growth hormone stimulates the incorporation of labelled precursors into protein and RNA of a variety of tissues when it is injected into animals or when it is added to the medium surrounding various tissue preparations (Kostyo and Knobil, 1959; Talwar *et al.*, 1962; Korner, 1960, 1965).

After injection of growth hormone into rats for one hour or less, or after perfusing liver with growth hormone for 30 minutes or less, isolated ribosomes are more active in incorporating amino acids into protein in an *in vitro* system, are more responsive when synthetic polynucleotides are added to act as artificial messenger RNA templates, and contain less monomeric and dimeric ribosomes and more polysomes than similar preparations from control tissues (Korner, 1961, 1965; Jefferson and Korner, 1967).

Growth hormone has not yet been shown to exert any of these effects when added to cell-free systems and the reason for this failure is unknown. It is possible that whole cells are necessary because the hormone acts at the cell surface or that the cell-free systems so far prepared are deficient in other ways so that the hormone cannot act. No evidence of the involvement of 3',5'-AMP in the anabolic action of growth hormone has yet emerged.

It is difficult to demonstrate a protein anabolic effect of insulin in whole animals because its well-known hypoglycaemic action causes considerable metabolic and hormonal changes which obscure its action in stimulating protein synthesis. It can, however, rapidly stimulate amino acid incorporation into protein in isolated tissue preparations such as rat diaphragm and perfused rat heart by mechanisms which are independent of the known effect of insulin in stimulating glucose uptake by muscle (see review by Wool et al., 1968). Ribosomes isolated from muscle of diabetic rats are less efficient at incorporating amino acid into protein in vitro than normal ones and the sucrose gradient profiles show less polysomes than in normal preparations. Injection of insulin into the diabetic rats restores these impairments very rapidly. Insulin, too, has no effect when added to cell-free systems.

HORMONAL CONTROL OF PROTEIN BIOSYNTHESIS

The outlines of the protein biosynthetic mechanism are clear and generally accepted but a number of important details are still obscure, particularly for mammalian tissues. The rate-limiting step of the reactions is unknown.

Any increase in rate or amount of protein biosynthesis must come, in the most general terms, from one or more of the following changes:

(1) An increase in the availability or activity of messenger RNA (m-RNA).
(2) An increase in the availability or activity of amino acyl transfer RNA (t-RNA) complexes.
(3) An increase in the availability or activity of ribosomes or of the factors required for initiation, translation or termination of protein biosynthesis.

Again in most general terms, the increase in availability of a factor could arise from one or more of (a) increased biosynthesis, (b) increased transport to site of utilization or (c) decreased degradation.

Activity here is used in the widest sense: one could envisage ways in which ribosomes, for example, could be available in large numbers at the correct locus and yet be unable to take part in protein biosynthesis until after some alteration in their physico-chemical characteristics.

Clearly, hormones could affect any or all of these parameters. They might affect more than one and the ones affected might vary with other metabolic perturbations of the cells.

DO INSULIN AND GROWTH HORMONE STIMULATE m-RNA AVAILABILITY OR ACTIVITY?

Synthesis

The messenger hypothesis (Jacob and Monod, 1961) sought to explain how bacteria produce different kinds of proteins in response to alterations in environment. Some hormones seem to act by stimulating the synthesis of proteins not previously synthesized by the cells on which they act and may act by a classic derepression mechanism. Insulin and growth hormone, however, seem to cause increased synthesis of those proteins which are already synthesized by the organism. If they act through a derepression mechanism, they must either stimulate the synthesis of some new and crucial m-RNA which is the template for the synthesis of a new protein which, in turn, triggers off subsequent changes; or they must act in a more general fashion to speed up the synthesis of a large number of the m-RNA species already being synthesized by the animal.

Insulin and growth hormone do stimulate a rapid rise in the labelling of RNA by precursors and, on long-term treatment, increase the RNA content of a number of tissues. Although the RNA polymerase activity of isolated liver nuclei is enhanced by treatment of rats with growth hormone, this enhancement appears to be dependent on increased protein synthesis. Treatment of rats with cycloheximide, or other inhibitors of protein synthesis, a few minutes before killing the animals abolished the enhancement of activity brought about by growth hormone (Korner, 1968). Growth hormone stimulates the labelling of all kinds of RNA including ribosomal and t-RNA rather than exclusively RNA which might be messenger RNA. There is, unfortunately, no method available for assaying the m-RNA content of mammalian tissues: the only really satisfactory one would be to measure the synthesis *in vitro* of a protein, specific to the added, putative m-RNA, which is not synthesized by the system in the absence of the added m-RNA. Since such a system is not yet available for mammalian tissues, Nirenberg's S-30 system from *E. coli* has been used to assay the m-RNA content of rat liver (Korner, 1968). The results show no difference in the stimulating activity of RNA from normal and hypophysectomized rats and a fall in stimulatory activity after growth hormone treatment. If these results can be accepted at face value, and it must be emphasized that they may well reflect protection of endogenous m-RNA of the *E. coli* or some other artefact rather than a measure of liver m-RNA, they suggest that growth hormone does not stimulate the synthesis of m-RNA as much as that of other types of RNA.

It is true that Wool and Munro (1963) have reported that insulin causes the appearance of a new peak in RNA extracted from rat diaphragm and analysed by sucrose density gradient analysis. However, the peak was only obtained in some but not all experiments and several workers have pointed to the dangers of relying solely on sucrose gradients as a means of analysing RNA species.

The possibility that insulin and growth hormone act on protein synthesis through the synthesis of RNA has been considered by some workers to have been disproved by the demonstration that both hormones can exert anabolic activity after RNA synthesis had been inhibited by actinomycin. Actinomycin injection into rats failed to stop the growth hormone protein anabolic effect assayed in isolated liver ribosomes (Korner, 1964), although the effect was diminished by the drug. Actinomycin did not abolish the anabolic effects of growth hormone (Martin and Young, 1965) and of insulin (Eboué-Bonis *et al.*, 1963) when it was added to the medium bathing isolated rat diaphragm.

Objections to the interpretation of experiments with actinomycin have been raised. Actinomycin might, for example, have actions other than the inhibition of DNA-directed RNA synthesis. Until this is shown not to be true, the argument that a hormone must act through RNA synthesis because its action is inhibited by actinomycin cannot be used. But the fact that the rapid anabolic action of insulin and growth hormone can be obtained even when actinomycin has abolished RNA synthesis is strong evidence that these hormones do not act through RNA synthesis, at least in their initial effects. It is possible, and indeed probable, that the later, prolonged, manifestation of the action of the hormones requires RNA synthesis but this may result from increased protein synthesis or the two effects may be independent actions of the hormones. Increased protein synthesis brought about by insulin and growth hormone is, however, not dependent on DNA-directed RNA synthesis.

Degradation

Very little information is available about the degradation of m-RNA in mammalian tissues. What there is suggests that the average half-life is long compared with bacterial m-RNA, but this does not preclude the existence of a class of m-RNA molecules of short half-life which are influenced in their decay by hormones. Several enzymes capable of degrading RNA have been shown to exist in mammalian tissues. A simple experiment was carried out of measuring the decay in protein synthetic activity of cell-free systems prepared from liver homogenates preincubated at various stages in the preparative process. No differences were detected in the rate or extent of decay in systems from liver of normal, hypophysectomized or growth hormone-treated rats. These measurements are, however, crude; they might not reveal differential decay or destruction in m-RNA in differently treated animals. More sophisticated experiments are needed on this point.

Transport

The movement of m-RNA, and indeed other types of RNA, from nucleus to cytoplasm is an obvious point where control could be exerted but there is almost no evidence available on this. The fact that most of the RNA synthesized in the nucleus does not emerge into the cytoplasm (Harris, 1968) and the many specific alterations in the size of the precursors of ribosomal RNA (see Perry, 1967 for a review) suggest a complex system of enzymes and of transporting and selecting mechanisms. This topic requires more attention than it has hitherto received. Hormones may speed up the movement of m-RNA to the cytoplasm but if they do the effect must be a very rapid one to explain the almost instantaneous effect that insulin has on protein synthesis.

Activity

Instances are known where m-RNA is available with ribosomes and other components of the protein synthetic system but protein synthesis does not occur until some change has taken place. Fertilization of sea-urchin eggs, for example, is followed by increased protein synthesis on polysomes, present before fertilization, but inactive through masking (Monroy, Maggio and Rinaldi, 1965). The evidence that is beginning to emerge that m-RNA is transported from the nucleus and acts as a ribonucleoprotein complex (Parsons and McCarty, 1968) offers means of explaining how a hormone could trigger the activity of available m-RNA. One could imagine that m-RNA could be altered by hormone action so that it can now accept ribosomes, or that preformed but inactive polysomes could be activated.

The available evidence, then, suggests that neither insulin nor growth hormone act, at least in the first instance, by stimulating m-RNA synthesis nor by inhibiting its degradation. There is no evidence on whether they stimulate its transport from the nucleus or its activity.

AVAILABILITY OR ACTIVITY OF AMINO ACYL t-RNA

Little is known about any hormonally wrought changes in the amount of amino acylating enzymes or of t-RNA in cell-sap of tissues. It is true that it has been shown (Korner, 1961) that the amino acid incorporation obtained with liver ribosomes *in vitro* was not greatly altered by using cell-sap from normal, hypophysectomized or growth hormone-treated rats but these experiments were usually carried out in the presence of excess cell-sap. When the effect of different concentrations of cell-sap on protein synthesis was examined it was found that normal liver cell-sap was more efficient than cell-sap from hypophysectomized rats. It is not clear from these studies nor from similar ones with diabetic tissues (Wool *et al.*, 1968) where the defect lies: it is possible that the difference is not in the provision of amino acylated t-RNA molecules but in the different amounts of an inhibitor in the different cell-sap fractions.

It is of course well known that both insulin and growth hormone stimulate amino acid uptake by some tissues (Kostyo, Hotchkiss and Knobil, 1959; Wool, 1965; Hjalmarson and Ahren, 1965). It has been argued that the increase in protein synthesis brought about by these hormones is not a secondary result of the increased supply of amino acids. One of the experiments on this point consisted of pre-labelling rat diaphragm with radioactive amino acids and showing that incorporation of these precursors into protein was enhanced when the hemidiaphragm was incubated with insulin (Wool and Krahl, 1959). Another experiment was the demonstration that label from non-amino acid precursors could enter proteins to a greater extent when diaphragms were incubated with insulin (Manchester and Krahl, 1959). Since the precursors have to be changed to amino acids intracellularly it was argued that insulin must act on protein synthesis at some point in addition to amino acid entry to the cell. However, an objection can be raised to this conclusion as a result of experiments with perfused rat liver. Jefferson and Korner (1967) were able to demonstrate a direct effect of growth hormone on protein bio-synthesis in liver perfused *in situ*. It was noticed that high concentrations of amino acids (ten times the usual plasma concentration) enhanced the growth hormone effect and could stimulate protein synthetic activity in the absence of the hormone. Further study (Jefferson and Korner, 1969) indicated that eleven of the amino acids were needed to obtain good amino acid incorporation, high activity of isolated ribosomes and a high proportion of polysomes in this system. Omission of any one of these eleven amino acids resulted in low incorporation, less polysomes and relatively inactive isolated ribosomes, and a small growth hormone effect.

Now suppose insulin or growth hormone acts by speeding the entry of one particular amino acid the concentration of which is limiting. Increased provision of this amino acid, or prevention of leakage of it, would stimulate protein synthesis from non-amino acid precursors or from labelled amino acids already inside the cell. Supply of amino acids could, of course, act directly to stimulate protein biosynthesis if one of them is limiting. It is possible, however, that amino acids could also have a more indirect effect on protein biosynthesis. The results from the liver perfusion experiments show an interesting parallel between the effects of amino acids and the effects of growth hormone. Both stimulate protein synthesis from injected labelled amino acids; both result in greater activity of isolated ribosomes; both result in a greater proportion of polysomes and less monomeric ribosomes being present.

Recently we were able to show that amino acids, fed to rats, increased the activity of tyrosine transaminase and tryptophan pyrrolase of liver by a mechanism involving RNA synthesis (Labrie and Korner, 1968). Following this we have found that feeding amino acids to rats, like treatment of them with growth hormone, stimulates synthesis of RNA.

I am not suggesting that insulin or growth hormone act simply by increasing the supply of amino acids: such an explanation is too simple. I merely wish to indicate that more searching comparisons need to be made before we can identify the differences in mode of anabolic action of amino acids and of growth hormone and insulin.

ACTIVITY OR AVAILABILITY OF RIBOSOMES OR PROTEIN FACTORS

Availability of ribosomes

The protein anabolic effects of both growth hormone and insulin can be obtained earlier than can be explained by the synthesis of new ribosomes. The appearance of newly synthesized ribosomal subunits in the cytoplasm of mammalian cells does not occur for at least 20 minutes (Perry, 1967; Hogan and Korner, 1968) which is longer than is needed to obtain hormonal effects on protein synthesis. This fact and the failure of actinomycin to abolish the hormone effects show that the profound changes in ribosome content and in the amount of endoplasmic reticulum of tissues (Cardell, 1967) after hypophysectomy and growth hormone treatment cannot be the explanation of at least the rapid changes in protein synthetic activity brought about by insulin and growth hormone.

It is possible that these hormones might facilitate the transport of pre-formed ribosomes or subunits to the cytoplasm from the nucleus. I know of no evidence on this point. If such an effect exists it would have to be rapid and massive if it is to explain the very quick effects of insulin and growth hormone on the considerable pool of cytoplasmic ribosomes.

It is doubtful if a change in rate of degradation of ribosomes can account for the rapid effects of these hormones, although it may contribute to the longer-term changes in ribosomal content, for the half-life of liver ribosomes is reported to be 2–5 days (Loeb, Howell and Tomkins, 1965) so that little change could be expected in the short time it takes insulin and growth hormone to act.

There is no information on the synthesis and degradation of the protein factors involved in the translation processes.

Activity of ribosomes

There is, however, evidence that both growth hormone and insulin can alter the activity of ribosomes of liver and muscle assayed *in vitro*.

Hypophysectomy reduces and growth hormone treatment increases the rate and extent of labelling of liver ribosomes with nascent peptide after a pulse dose of injected radioactive amino acid (Korner, 1960). The amino acid incorporating activity of ribosomes isolated from liver, heart or skeletal muscle is less, per unit weight of ribosomal RNA, if they come from hypophysectomized rats rather than from normal ones (Korner, 1959; Earl and Korner, 1966; Florini and Brever, 1966). Sucrose gradient analyses reveal less polysomes and more monomeric and dimeric ribosomes in ribosome preparations from hypophysectomized rats than from normal rats (Korner, 1964; Staehelin, 1965). These differences in incorporating ability of isolated ribosomes are maintained even when large amounts of synthetic polynucleotides are present in the *in vitro* system to supply any want in m-RNA. Similarly the differences were maintained when incorporation was stimulated by addition of rat liver RNA: it should be emphasized, however, that this material may be acting through a protective rather than through a stimulatory action. All of these effects

4*

of hypophysectomy could be reversed by treatment of the hypophysectomized rats with growth hormone and the hormone could stimulate the activity of liver ribosomes to greater than normal levels when injected into normal rats.

Somewhat similar effects have been reported for insulin. Muscle ribosomes were less active *in vitro* and contained less polysomes when taken from diabetic rats than from normal rats. Treatment of diabetic rats with insulin rapidly restored the deficiency (Wool *et al.*, 1968). The response of ribosomes from skeletal muscle of diabetic rats to polyuridylic acid (poly U) is complex. Those from diabetic rats are less active than normal ones at low magnesium ion concentration and more active at high magnesium ion concentration.

Although poly U cannot be regarded as equivalent to natural m-RNA because it does not have the start and stop codons which might occur (but for which no evidence has yet been produced) on mammalian m-RNA, the fact that ribosomal ability to attach to and translate this polynucleotide is affected by hormones does show that the hormones alter the ribosomes in some way.

It is often implied that, if a cell-free preparation of ribosomes contains a high proportion of polysomes, then its incorporating activity when the results are expressed as counts/min incorporated per mg ribosomal RNA will be high. The implication is that the number of ribosomes attached to m-RNA determines the specific activity obtained *in vitro*. This is not so. Polysomes containing only one ribosome, free to travel to the end of the m-RNA, will yield the same specific activity *per ribosome* as polysomes containing any number of ribosomes, if they are equally distributed on the m-RNA. A related fallacy is that the extent of phenylalanine incorporation into polypeptide (per mg ribosomal RNA) in the presence of poly U is a measure of the number of free ribosomes: it is in fact a measure of the proportion of active ribosomes in the preparation. Clearly, if results are expressed in terms of mg ribosomal RNA, the number of ribosomes on the m-RNA has already been taken into account and will, in an ideal case, not affect the specific activity obtained (see appendix for proof of this contention, p. 99).

The specific activity obtained *in vitro* (i.e. counts/min per mg protein per mg ribosomal RNA) is related to the length of message which can be translated, the proportion of active ribosomes attached to the message, the total number of ribosomes in the systems, attached or unattached to the message, and the relative proportions of these which are able to initiate peptide synthesis.

The length of message could be changed by hormonal treatment directly (see above) or indirectly. For example, if the polysomes are less fragile, fewer 3' ends of the message not carrying termination codons would be expected to be generated. Normally when two ribosome populations are compared the total number of ribosomes is kept equal. If peptide chain initiation is stopped the specific activity is dependent on the proportion of active to inactive ribosomes attached to the m-RNA. Several classes of inactive ribosomes could exist: some have been identified. Monomers could be inactive because of lack of active m-RNA; because of damage or lack of a factor; because they were released from a polysome still carrying a nascent peptide or a piece of m-RNA. Polysomes could contain inactive monomers because the correct termination codon is missing, causing inactivity at the 3' end of the m-RNA; because the ribosome is non-specifically bound to the polysome; or because the monomer which is itself defective is attached to the polysome, or the polysomes themselves could be inactive because they are non-specifically aggregated together.

Several questions about the changes in activity of ribosomes caused by hormonal treatment can be posed. One is to ask which subunit of the ribosome has been altered by the treatment. Martin and Wool (1968) tackled this question by separating muscle ribosomes into their 60s and 40s subunits and joining the 60s subunit from normal muscle with the 40s subunit from diabetic rat muscle and *vice versa*. These reconstituted hybrid ribosomes were then assayed for protein synthetic activity *in vitro* in the presence of added polynucleotide. These experiments showed that diabetes had altered the larger of the ribosomal subunits.

We have used the same method to test the effect of hypophysectomy and growth hormone treatment on liver ribosomes. Our preliminary data (Barden and Korner, 1969) indicate that it is the smaller subunit which is affected by growth hormone.

The method of separating ribosomes into subunits has some possible technical objections: the preparations of subunits are not completely pure. This apart, however, these results reveal a major difference between the action of insulin and growth hormone.

A second question can be asked about the hormonally caused changes in ribosomal activity. Is the change one of degree, that affects all the ribosomes, or is it one of kind, that alters the percentage of ribosomes which are active? Wool and Kurihara (1967) have sought to answer this question by measuring the number of molecules of puromycin which react with a known number of ribosomes. Puromycin reacts with and releases nascent polypeptide chains which are being synthesized on ribo-somes, so this method gives an indication of the proportion of ribosomes which are synthesizing protein. The results showed that only about 25 per cent of the ribosomes from normal muscle were active by this defini-tion and that diabetic rat muscle had about 9 per cent active ribosomes. Treatment of diabetic rats with insulin for five minutes brought the percentage of active ribosomes back towards the normal figure. Insulin increases the number of ribosomes which are active rather than the activity of each ribosome.

One surprising thing about this result is the small percentage of ribo-somes which are active even in normal tissue. This low figure must be compared with the 70–75 per cent of muscle ribosomes which occur as polysomes. Wool suggests that some ribosomal aggregates might sedi-ment with the appearance of polysomes which are not in actuality active structures, or that ribosomes can become attached to m-RNA in a way that does not allow them to synthesize protein. Another possibility, not considered by Wool, is that some of the nascent peptide chains will be attached to t-RNA molecules held in sites on the ribosomes where they are unable to react with puromycin. In the absence of translocase, which seems to be a supernatant enzyme and was not present in Wool's experi-ments, the figure for active ribosomes will be an underestimate. If this is the case, differences may exist in the rate of translocation between ribo-somes from differently treated rats. Again more work is needed on this point but Wool's result remains an interesting and important one.

We have found that differences exist in the ability of ribosomes from liver to react with puromycin in a way which is dependent on the circu-lating growth hormone in the rat. We have also carried out the same experiment in a different way (at the suggestion of Dr R. J. Jackson) by labelling nascent peptide chains and assaying the release of them from isolated ribosomes by puromycin. Similar results were obtained with the two methods. However, we noticed that the use of detergent in pre-paring ribosomes reduces their ability to react with puromycin, so the question is raised: how far are the differences noted *in vitro* reflections of differences *in vivo* and how far are they changes produced during prepar-ative procedures? This is a difficult point to answer but it should be emphasized that even when prepared without detergent, ribosomes from livers of rats without circulating growth hormone were still much less reactive with puromycin than those from rats with circulating growth hormone.

Quite another question can be asked about the alteration in ribosome activity brought about by hormones. Is the effect only on the initiation processes where the ribosome is first attached to the 5' end of the m-RNA, or is there a change in the efficiency of the translation process or in the termination step?

The direct approach to this problem is to study the initiation process in mammalian tissues and we are attempting to do this. Meanwhile we have adopted an indirect method by inhibiting the initiation step and then examining the changes in the translation process brought about by hormones. Dextran sulphate appears to be able to inhibit initiation without inhibiting translation in an *E. coli* system (Munro, 1965) and I have shown that it can also do so in rat liver systems (Korner, 1969). Added before poly U it stops the stimulation of polyphenylalanine synthesis usually seen with the synthetic polynucleotide but it does not inhibit the incorporation of amino acids not coded for by the poly U. I found that, added to ribosomes before the other components of the cell-free system, it caused some inhibition of basal incorporation, probably by preventing initiation. However, the difference in protein synthetic activity of ribosomes from normal, hypophysectomized and growth hormone-treated rats was about the same in the presence as in the absence of dextran sulphate. This result does not prove that the initiation processes are not affected by hormones: it does show that this alone is not the cause of the differences in ribosome activity and that these differences persist after the ribosomes have become attached to m-RNA.

Wool found that injection of cycloheximide, which inhibits protein synthesis, prevented the insulin correction of the defect in ribosomes of diabetic rats. He argues (Wool *et al.*, 1968) that protein synthesis, possibly of a specific translation factor, is required for insulin action. It is possible, however, that insulin requires general but not specific protein biosynthesis for its action. Wool has gone on to try to identify this putative protein factor. It will be recalled that skeletal muscle ribosomes from diabetic rats respond to poly U better than normal ones at high magnesium ion concentration but worse at low magnesium ion concentration. Wool found that preincubation of normal ribosomes rendered them more like diabetic ones in this regard, suggesting that normal ribosomes contain a factor, removed by preincubation, that diabetic ones lack. The diabetic defect in polyphenylalanine synthesis could be corrected by preincubating the ribosomes with poly U and phenylalanyl t-RNA at high magnesium ion concentration. Wool suggests that the defect, which occurs on the 60s subunit of the ribosome, is in the factor which enables t-RNA to be bound to the ribosome and template.

Ribosomes from hypophysectomized rat liver are less able than normal ones to synthesize polyphenylalanine on poly U at all concentrations of magnesium ion. If Wool's explanation of his data is accepted, this difference in behaviour suggests that growth hormone and insulin differ in their effects on ribosomes. This is not surprising since the defect seems to be associated with the large subunit in the case of diabetes and the small subunit in the case of hypophysectomy. Although we are still ignorant of how they act, we can conclude that the mechanisms by which growth hormone and insulin stimulate protein biosynthesis differ.

SUMMARY

It seems unlikely that insulin or growth hormone stimulate protein biosynthesis by altering the rate of synthesis or degradation of messenger RNA or of ribosomes. No evidence is available about the possibility that they alter the rate of transport of messenger RNA or ribosomal subunits to the cytoplasm or change the activity of these substances.

Both hormones enhance the activity of isolated ribosomes, assayed *in vitro*. Diabetes appears to alter the activity of the larger ribosomal subunit, possibly by loss of a protein factor involved in binding of transfer RNA molecules. It has been suggested that diabetes results in a fall in the already small percentage of ribosomes which are active in protein biosynthesis.

Hypophysectomy seems to alter the activity of the smaller ribosomal subunit. The ribosomes are less responsive to polyuridylic acid at all concentrations of magnesium ion than ones from normal rat liver. Since they differ in this respect from ribosomes of diabetic rats, the effect of hypophysectomy cannot be attributed to lack of a factor for binding t-RNA molecules. This lesser activity of ribosomes persists when initiation of peptide chain synthesis has been inhibited.

Acknowledgements

The work discussed in this paper is part of the programme of a Group supported by the Medical Research Council, to whom I express my thanks.

REFERENCES

BARDEN, N., and KORNER, A. (1969). *Biochem. J.*, in press.
CARDELL, R. R., JR. (1967). *Biochim. biophys. Acta*, 148, 539–552.
EARL, D. C. N., and KORNER, A. (1966). *Archs Biochem. Biophys.*, 115, 445–449.
EBOUÉ-BONIS, D., CHAMBAUT, A. M., VOLFIN, P., and CLAUSER, H. (1963). *Nature, Lond.*, 199, 1183–1184.
FLORINI, J. R., and BREVER, C. B. (1966). *Biochemistry, N.Y.*, 5, 1870–1876.
HARRIS, H. (1968). *Nucleus and Cytoplasm*. Oxford: Clarendon Press.
HJALMARSON, A., and AHREN, K. (1965). *Life Sciences*, 4, 863–869.

HOGAN, B. L. M., and KORNER, A. (1968). *Biochim. biophys. Acta*, **169**, 139–149.
JACOB, F., and MONOD, J. (1961). *J. molec. Biol.*, **3**, 318–356.
JEFFERSON, L. S., and KORNER, A. (1967). *Biochem. J.*, **104**, 826–832.
JEFFERSON, L. S., and KORNER, A. (1969). *Biochem. J.*, **111**, 703–712.
KNOBIL, E., and HOTCHKISS, J. (1964). *A. Rev. Physiol.*, **26**, 47–74.
KORNER, A. (1959). *Biochem. J.*, **73**, 61–71.
KORNER, A. (1960). *Biochem. J.*, **74**, 462–471.
KORNER, A. (1961). *Biochem. J.*, **81**, 292–297.
KORNER, A. (1964). *Biochem. J.*, **92**, 449–456.
KORNER, A. (1965). *Recent Prog. Horm. Res.*, **21**, 205–240.
KORNER, A. (1968). *Ann. N.Y. Acad. Sci.*, **148**, 408–418.
KORNER, A. (1969). *Biochim. biophys. Acta*, **174**, 351–358.
KOSTYO, J. L., HOTCHKISS, J., and KNOBIL, E. (1959). *Science*, **130**, 1653–1654.
KOSTYO, J. L., and KNOBIL, E. (1959). *Endocrinology*, **65**, 525–528.
LABRIE, F., and KORNER, A. (1968). *J. biol. Chem.*, **243**, 1116–1119.
LOEB, J. N., HOWELL, R. R., and TOMKINS, G. M. (1965). *Science*, **149**, 1093–1095.
MANCHESTER, K. L., and KRAHL, M. E. (1959). *J. biol. Chem.*, **234**, 2938–2942.
MARTIN, T. E., and WOOL, I. G. (1968). *Proc. natn. Acad. Sci. U.S.A.*, **60**, 569–574.
MARTIN, T. E., and YOUNG, F. G. (1965). *Nature, Lond.*, **208**, 684–685.
MONROY, A., MAGGIO, R., and RINALDI, A. N. (1965). *Proc. natn. Acad. Sci. U.S.A.*, **54**, 107–111.
MUNRO, A. J. (1965). *Abstr. 2nd Meeting Fedn Europ. Biochem. Soc. Vienna*, p. 2.
PARSONS, J. T., and McCARTY, K. S. (1968). *J. biol. Chem.*, **243**, 5377–5384.
PERRY, R. P. (1967). *Prog. Nucl. Acid Res.*, **6**, 219–251.
STAEHELIN, M. (1965). *Biochem. Z.*, **342**, 459–468.
TALWAR, G. P., PANDA, N. C., SARIN, G. S., and TOLANI, A. J. (1962). *Biochem. J.*, **82**, 173–175.
WOOL, I. G. (1965). *Fedn Proc. Fedn Am. Socs exp. Biol.*, **24**, 1060–1070.
WOOL, I. G., and KRAHL, M. E. (1959). *Nature, Lond.*, **183**, 1399–1400.
WOOL, I. G., and KURIHARA, K. (1967). *Proc. natn. Acad. Sci. U.S.A.*, **58**, 2401–2407.
WOOL, I. G., and MUNRO, A. J. (1963). *Proc. natn. Acad. Sci. U.S.A.*, **50**, 918–920.
WOOL, I. G., STIREWALT, W. S., KURIHARA, K., LOW, R. B., BAILEY, P., and OYER, D. (1968). *Recent Prog. Horm. Res.*, **24**, 139–213.

APPENDIX

The proposition that specific activity per mg ribosome obtained *in vitro* is independent of the number of ribosomes on m-RNA, if reinitiation is ignored and if it is assumed that there are no inactive ribosomes on the m-RNA and no blocks to translation, seems to be self-evident. But it can be demonstrated as follows:

Let L = length of m-RNA,
  n = number of ribosomes attached to it
and s = counts/min per mg protein per ribosome (or per mg ribosomal RNA).

If the ribosomes are equally spaced on the m-RNA the distance between ribosomes will be:

(a) $\dfrac{L}{n-1}$ if there is a ribosome at the extreme $5'$ and $3'$ ends of the m-RNA,

or

(b) $\dfrac{L}{n+1}$ if neither extreme end of the m-RNA is occupied by a ribosome;

or some value between these extremes.

In case (a) total distance travelled by all the ribosomes on the m-RNA is:

$$\frac{L}{n-1} \sum [1 \text{ to } (n-1)] = \frac{L(n-1)n}{(n-1)2} = \frac{Ln}{2}$$

In case (b) total distance is:

$$\frac{L}{(n+1)} \sum [1 \text{ to } n] = \frac{L}{(n+1)} \cdot \frac{n(n+1)}{2} = \frac{Ln}{2}$$

In both extreme cases and, therefore, in any intermediate ones,

$$S = \frac{K.Ln}{2.n} = \frac{KL}{2}$$

where K is a constant depending on rate of translation, specific activity of precursor, etc.

S, the specific activity obtained *in vitro*, is then independent of n, the number of ribosomes on m-RNA. This applies for an ideal case where all ribosomes on the message are active and no unattached ribosomes are present. If x of the ribosomes on m-RNA are inactive and there are y active unattached and z inactive unattached ribosomes present in the system,

$$S = \frac{KL(n-x)}{2.(n+y+z)}$$

DISCUSSION

White: First I would like to make a comment on the inhibitory effect of cortisol on protein synthesis by lymphoid cells. One locus of action of the steroid appears to be the ribosomes; ribosomes prepared from thymic cells of rats injected with cortisol have an impaired capacity to incorporate amino acids in an *in vitro* system (Peña, Dvorkin and White, 1966). This action can be dissociated from a second effect of cortisol *in vitro* on amino acid transport by rat thymocytes (Makman, Nakagawa and White, 1967).

Secondly, have you had the opportunity to look for an action of growth hormone at the transport level?

Korner: Both insulin and growth hormone stimulate the entry of amino acids into cells. The question has been whether the protein synthetic effect of insulin or growth hormone is secondary to the increased entry of amino acids. The argument went back and forth for many years and then it appeared that for insulin it had been solved. Amino acids were injected into a rat and the diaphragm was removed and divided. One half was incubated with insulin and one half without. Insulin stimulated the incorporation into protein of amino acids that were already inside the diaphragm, showing that although insulin does stimulate amino acid uptake it must be having some effect on protein synthesis distal to transport (Wool and Krahl, 1959).

Manchester and Krahl (1959) used non-amino acid carbon-labelled materials such as α-ketoglutarate which would have to get into the cell to be transaminated before the label could enter protein. They found that insulin stimulated incorporation of this label into protein. So it looked as though an effect on transport alone could not explain the effect of insulin. However, our experiments on the perfused liver (Jefferson and Korner, 1969), showing that the supply of some single amino acids can be vitally important in protein synthesis, suggest the theoretical possibility that one particular amino acid might be rate limiting for protein synthesis and that insulin stimulates its entry.

Ramalingaswami: Drs V. R. Young and H. N. Munro (see McCance and Widdowson, 1968) have recently been looking at the concentration of ribosomes and their size distribution in the muscle of protein-deficient animals. It appears that the concentration of polysomes increases when insulin is added to skeletal muscle. In protein deficiency their number goes down, and it is suggested that this might be due to a diminution in the influx of amino acids into the muscle cell. At the same time there is a rise in the plasma cortisol level. I wonder what cortisone does to the aggregation of polysomes in muscle?

Korner: I have not worked with cortisone in muscle, but if you inject corticosteroids into the rat, as you know the liver actually gains weight, a situation similar to that resulting from treatment with growth hormone. The polysomes increase in number and protein synthetic activity goes up, probably because of the flood of amino acids to the liver from muscle protein. I don't think amino acids can be acting merely by supplying substrates. I think they are actually activating, in some fashion, the ability of ribosomes to synthesize protein.

Gopalan: Recently there have been reports showing that in extreme forms of protein-calorie malnutrition growth hormone levels in plasma are considerably raised (Pimstone *et al.*, 1966). There is also a marked increase in the levels of non-esterified fatty acids in the serum (Jayarao and Prasad, 1966). After treatment with high protein diets, these levels fall. Apparently, the increase in growth hormone concentration is not explainable on the basis of hypoglycaemia because the change in blood sugar concentration is not consistent. But there seems to be a relationship to the severity of the protein-calorie malnutrition. It is claimed that the lower the serum albumin falls, the higher the elevation of growth hormone concentration (Pimstone *et al.*, 1968). In view of your observation that the amino acid concentration in plasma may determine growth hormone levels, I wonder how these observations fit in. We do know that in malnutrition, the serum amino acid concentration is considerably reduced (Anasuya and Narasinga Rao, 1968).

Korner: I don't know if it fits in! One of the puzzling things about growth hormone is that, as you realize, the level in the blood varies with the glucose content. It also seems to vary for no apparent reason during the night in ways which cannot be related simply to its action on growth and protein synthesis. It could well be that growth hormone has two quite distinct actions, one determining whether one metabolizes fat or carbohydrate, and that its other action, growth, arises somehow as a result of this shift in metabolism. I can't believe that one grows in little fits and starts simply because one lowers one's blood sugar.

Tata: That growth hormone may rapidly affect the entry of amino acids and even nucleotides into cells is an important observation, but it is also very likely that ribosome production is an important part of the later events concerned with growth hormone action. To a large extent this late manifestation is masked, or complicated, by a change in the distribution of ribosomes on endoplasmic reticulum membranes (see Tata and Williams-Ashman, 1967).

Korner: I agree entirely with that. I hope I emphasized that it was the immediate effect of growth hormone which seemed not to be abolished by stopping RNA synthesis. I emphasized that the actual content of RNA does go up; you can actually see more ribosomes after growth hormone treatment. It seems reasonable that if you stimulate the protein-synthetic mechanism something is going to become rate limiting and you have to have more of it; as soon as you supply more of that you will need more of some other part of the system, and so on. Nevertheless, even though there are more ribosomes after growth hormone treatment the activity of the ribosome population is also increased and one can show anabolic effects of growth hormone in the absence of RNA or ribosome synthesis.

Monod: In microorganisms there is an extremely nice direct correlation between rates of growth, that is to say protein synthesis, and the ribosomal content of cells. Has any fairly strict relationship been established in cells of higher organisms, or could it be? Secondly, is there any analogue of the phenomenon known as stringency in bacteria? That is the fact that the rate of essentially ribosomal RNA synthesis depends on the availability of amino acids; that is, of each individual amino acid for its own sake.

Korner: There is not the clear correlation one gets in bacteria, but certainly those cells that are growing fast or are stimulated to grow do have more ribosomes and more endoplasmic reticulum than usual.

On the amino acid question, what I find most intriguing about the correlations observed between RNA synthesis and amino acids in mammalian tissues, which is somewhat similar to the situation in bacteria, is that if we simply feed amino acids to a rat, within an hour or two we find a great increase of RNA synthesis in the nucleus.

Monod: To prove the point you would need to have a cell suspension and to supply all amino acids except one essential one and test whether that blocks synthesis of ribosomal RNA.

Korner: We are trying to do this, but it is not as easy, technically, as it is with bacteria. Much work has been done on the need for particular amino acids in growing mammalian cells in culture and their relationship to RNA synthesis. The problem is complicated in mammalian cells because they degrade protein, which, on the whole, bacteria don't do.

Bhargava: Professor Korner, did you estimate the levels of free amino acids in your experiments?

Korner: We did try; it's not easy. They don't alter greatly in the liver after growth hormone treatment. Tryptophan is very low and may well be rather crucial.

Sarma: You said that eleven amino acids were necessary for stimulation of protein synthesis in perfused liver; were there any differences among the amino acids, in view of the report by Sidransky and co-workers (1968) that dietary tryptophan had a special role in the regulation of polyribosome aggregation and protein synthesis in the liver?

Korner: Tryptophan was one of the eleven, but it seemed no more or less essential than, say, methionine.

Salganik: I would like to comment on the stability of ribosomes in animal cells. The number of ribosomes in animal cells changes according to the intensity of protein synthesis. In hormone induction especially the number of ribosomes increases, and it decreases when induction ceases. It is tempting to suggest that not all ribosomes are stable in animal cells but only those which are almost permanently engaged in translation.

Korner: The average half-life of liver ribosomes is between $1\frac{1}{2}$ and 3 days. This half-life is long in the sense that one can get a growth hormone or insulin effect in 15 minutes, so that alterations in rate of ribosome turnover cannot be the cause of the initial effects. But, of course, the turnover of the ribosomes is going to be important in the way that Dr Tata has also suggested for the long-term action of the hormone.

Harris: I wonder how the idea ever began that the primary action of hormones is at the level of the genes? We have to assume that, for practical purposes, the genes in all our cells are the same; but hormones have specific target organs. How can one envisage any *specific* hormone effect as a primary action on genes?

Korner: The prime exponent of this hypothesis is Karlson (Karlson and Sekeris, 1966) who relied heavily on evidence obtained with ecdysone, the insect moulting hormone. At some stage in the development of the larva the cuticle is keratinized, probably by the formation of dopamine and dopa because of the production of the enzyme dopa carboxylase. One can see puffs on the chromosomes which are said to represent RNA synthesis, and Karlson used this, together with data of Sekeris and Lang (1964), to argue that the hormone was acting directly on the genome, to derepress a portion of it.

In a case where a hormone is inducing synthesis of an enzyme not previously synthesized in the cell it seems reasonable to argue that one is uncovering a portion of the genome. It doesn't seem so easy to argue this when hormones are merely stimulating the amount of protein already made by the cell. Unfortunately the experimental evidence on which Karlson relies is poor. The stimulation of incorporation of amino acids into protein by RNA extracted from treated larvae is minute, and calculation shows that it is hardly sufficient for one molecule of enzyme to have been synthesized.

Talwar: Another piece of evidence is the puffing pattern observed by Beerman and Clever in these larvae under the influence of hormones (Beerman, 1962; for other references see Beerman and Clever, 1964).

There was not only a physical increase in the size of the chromosome puffs but an actual demonstration that at the site of the puffs there is active synthesis of RNA. One should not assume that all hormones act at the gene locus, but nevertheless *some* hormones do, directly or indirectly.

Korner: Kroeger (1966) has argued that puffing is not a direct effect of hormones. Salts produce similar changes.

Monod: The idea that hormones act directly on genes is an absurd hypothesis anyhow, if by this is meant that, say, oestradiol sticks directly to some genetic segment. The hormone has to act on something which may act on something else, which may act on genes.

Salganik: There is some evidence that the state of the chromatin is altered under the action of the factors inducing transcription. We have established that hydrocortisone induction results in an increase of the amount of active chromatin in rat liver nuclei from 5–8 per cent to 11–17 per cent. The difference between these two fractions of chromatin, which were prepared according to the method of Frenster, Allfrey and Mirsky (1963), is rather clear. The incorporation of [^{14}C]adenine into active chromatin is $1 \cdot 6$–$3 \cdot 0$ times greater than that into repressed chromatin. However, the content of non-histone proteins in active chromatin is about 3–4 times greater than that in repressed chromatin (Salganik, Morozova and Zakharov, 1969).

White: I am still reserving final judgement on accepting the interesting hypothesis of Professor Karlson (1963). However, as Dr Salganik has just indicated, it has been reported from at least two laboratories (Dahmus and Bonner, 1965; Salganik, Morozova and Zakharov, 1969) and certain of our own data indicate (Nakagawa and White, 1969) that after the administration of a glucocorticoid there is a demonstrable alteration in chromatin of the target cell and this may be reflected in the "transcribing efficiency" of RNA polymerase. However, Gabourel and Fox (1969) reported recently that they were unable to detect differences in the transcription of thymocyte chromatin by *E. coli* RNA polymerase in comparisons of chromatin obtained from thymocytes of either control or glucocorticoid-injected rats.

Korner: We also find that RNA polymerase activity does go up after treatment with growth hormone (Pegg and Korner, 1965), but this is not a primary effect of growth hormone.

White: An unusual aspect of trying to focus on a single mechanism for explaining hormone action is the fact that the injection of cortisol into a rat produces opposite responses in two target cells. Thus, in liver, amino acid transport, and RNA and protein synthesis, are increased in rate. Yet in lymphoid cells, the steroid decreases the rates of these same processes. Either a single mechanism is operating in opposite ways, or two differing mechanisms are operating in each of the two target cells.

Monod: I wonder whether the endocrinologists feel that it is possible to classify hormones that have some effect on protein synthesis into two broad classes: those that affect the overall rate of protein synthesis, which seems to be the case for growth hormone, and those with a differential effect on the synthesis of certain specific proteins, which seems to be the case for oestrogens. These two mechanisms must be fundamentally different. If one of these mechanisms acts directly or indirectly at the transcriptional level, one would expect this to be the mechanism for differential effects. The mechanism working at the level of overall protein synthesis might act virtually anywhere.

Korner: Alternatively one could argue that even hormones which stimulate general protein synthesis act at some specific point, maybe at transcriptional level, to stimulate some RNA polymerase.

Monod: I am extremely suspicious of experiments where one tries to estimate the RNA polymerase activity of cells in higher organisms. RNA polymerase is such a dangerously tricky enzyme anyway, and especially as extracted from higher cells.

White: A striking example of a selective action of a hormone, apart from that of oestrogen in the uterus, is the action of glucocorticoids on the liver. Here there may result an acute, early increased synthesis of additional enzyme protein, for example tyrosine-α-ketoglutarate transaminase (Kenney, 1962), whereas some of the other liver enzymes do not change in activity.

Korner: I quite agree with Professor Monod about RNA polymerase, but I am not so sure that one can rigidly divide hormones into two classes. One knows that certain specific enzymes are stimulated by corticosteroids, but no one has done anything like a complete survey of all the enzymes in any cell to see which of them are affected.

White: A number of enzymes have been measured which do not change in activity after administration of a hormone *in vivo*.

Bhargava: Is any instance known where a specific effect of a hormone, as specific as the effect of an inducer like a galactoside in *E. coli*, has been shown to occur in a mammalian system without any concurrent non-specific effect? I suspect the answer would be no, but if so, should we not think in terms of just one class of hormones with a varying degree of non-specificity rather than two classes of hormones, one which is entirely specific, and the other which is entirely non-specific?

Korner: I think there is danger in either lumping or splitting; I agree that too rigid a division should be avoided.

REFERENCES

ANASUYA, A., and NARASINGA RAO, B. S. (1968). *Am. J. clin. Nutr.*, **21**, 723.
BEERMAN, W. (1962). *Protoplasmatologia*, **6D**, 1.
BEERMAN, W., and CLEVER, U. (1964). *Scient. Am.*, **210**, 50.
DAHMUS, M. E., and BONNER, J. (1965). *Proc. natn. Acad. Sci. U.S.A.*, **54**, 1370.
FRENSTER, J. H., ALLFREY, V. G., and MIRSKY, A. E. (1963). *Proc. natn. Acad. Sci. U.S.A.*, **50**, 1026.
GABOUREL, J. D., and FOX, K. E. (1969). *Fedn Proc. Fedn Am. Socs exp. Biol.*, **28**, 635.
JAYARAO, K. S., and PRASAD, K. P. S. (1966). *Am. J. clin. Nutr.*, **19**, 205.
JEFFERSON, L. S., and KORNER, A. (1969). *Biochem. J.*, **111**, 703–712.
KARLSON, P. (1963). *Perspect. Biol. Med.*, **6**, 203.
KARLSON, P., and SEKERIS, C. E. (1966). *Acta endocr., Copenh.*, **53**, 505.
KENNEY, F. T. (1962). *J. biol. Chem.*, **237**, 4395.
KROEGER, H. (1966). *Mem. Soc. Endocr.*, No. 15, 55–67.
MCCANCE, R. A., and WIDDOWSON, E. M. (ed.) (1968). *Calorie Deficiencies and Protein Deficiencies*, p. 73. London: Churchill.
MAKMAN, M. H., NAKAGAWA, S., and WHITE, A. (1967). *Recent Prog. Horm. Res.*, **23**, 195.
MANCHESTER, K. L., and KRAHL, M. E. (1959). *J. biol. Chem.*, **234**, 2938.
NAKAGAWA, S., and WHITE, A. (1969). Submitted for publication.
PEGG, A. E., and KORNER, A. (1965). *Nature, Lond.*, **205**, 904.
PEÑA, A., DVORKIN, B., and WHITE, A. (1966). *J. biol. Chem.*, **241**, 2144.
PIMSTONE, B. L., BARBEZAT, G., HANSEN, J. D. L., and MURRAY, P. (1968). *Am. J. clin. Nutr.*, **21**, 482.
PIMSTONE, B. L., WITTMANN, W., HANSEN, J. D. L., and MURRAY, P. (1966). *Lancet*, **2**, 779.
SALGANIK, R. I., MOROSOVA, T. M., and ZAKHAROV, M. A. (1969). *Biochim. biophys. Acta*, **174**, 755.
SEKERIS, C. E., and LANG, N. (1964). *Life Sciences*, **3**, 625.
SIDRANSKY, H., SARMA, D. S. R., BONGIORNO, M., and VERNEY, E. (1968). *J. biol. Chem.*, **243**, 1123.
TATA, J. R., and WILLIAMS-ASHMAN, H. G. (1967). *Eur. J. Biochem.*, **2**, 366–374.
WOOL, I. G., and KRAHL, M. E. (1959). *Am. J. Physiol.*, **196**, 961.

MECHANISM OF ACTION OF GROWTH HORMONE AND OESTROGENS

G. P. Talwar, B. L. Jailkhani, S. K. Sharma, M. L. Sopori,
M. R. Pandian, G. Sundharadas and K. N. Rao

Department of Biochemistry, All-India Institute of Medical Sciences, New Delhi

Most growth-promoting hormones stimulate at an earlier or later stage the synthesis of RNA and proteins in the organs influenced by them. The manner in which they bring about this effect may however not be the same, as will be evident from the discussion that follows on oestradiol-17β and pituitary growth hormone.

It is now apparent that the interaction of these hormones with the target tissues initiates a sequence of reactions, the products of the early events acting as inducers of subsequent reactions. There is a built-in chain of events. The metabolic picture differs according to the time at which it is seen after the administration of the hormone. It is clear that at any given time one is looking at only a part of the whole phenomenon. The coordination of events in the chain is a complex but necessary prerequisite to gaining complete understanding of the action of these hormones.

EARLY INTERACTIONS

It follows that the early interactions of the hormones with the components of the target tissues are important starting points to focus attention on and have relevance to subsequent developments. Oestrogens are rapidly and selectively taken up and retained by some organs which respond biologically to this hormone, suggesting the presence of specific "receptors" for this hormone in these tissues. Growth hormone on the other hand is localized in many organs of the body. The "receptors" for this hormone (if any) have a more ubiquitous distribution.

Uterine receptors for oestradiol

When $[6,7\text{-}^3H]$oestradiol-17β is injected into ovariectomized rats, it is rapidly taken up from the circulation by the uterus, vagina, pituitary and some other organs and is retained for several hours against concentration gradients (Jensen and Jacobson, 1962; Stone, 1963; Stone, Baggett and Donnelly, 1963; Eisenfeld and Axelrod, 1965; King and Gordon, 1966). Within the uterus, the steroid is not present in the free state but is found bound to a heavy particulate fraction (later characterized as nuclei; Talwar, 1969) and to a macromolecular component of the high-speed supernatant cytoplasmic fraction (Talwar *et al.*, 1964, 1965; Noteboom and Gorski, 1965).

Cytoplasmic fraction. The cytoplasmic uterine receptor for oestradiol has a sedimentation coefficient of $\sim 9 \cdot 5$s (Toft and Gorski, 1966; Jensen *et al.*, 1968; Kornman and Rao, 1968). The protein moiety is essential for the binding properties; the bound steroid is labilized by treatment with trypsin or pronase, while ribonuclease, deoxyribonuclease and neuraminidase are without action (Talwar, 1969; Talwar *et al.*, 1968; Jensen *et al.*, 1968).

The uterine cytoplasmic fraction has a high degree of stereo-specificity for binding oestradiol-17β. Cortisone and testosterone are not bound while progesterone is bound to a much lower degree. There is discrimination of binding for even a stereo-isomer, 17α-oestradiol (Talwar *et al.*, 1964; Noteboom and Gorski, 1965; Talwar *et al.*, 1968). The binding is non-covalent in nature. Weak bonds including hydrophobic interactions play an important part. It appears as if the receptor molecule has a glove-like compartment for fixing this steroid or other molecules resembling oestradiol in their molecular dimensions and steric configuration.

The binding of the steroid is optimal at pH values between 7 and $8 \cdot 5$, and is rapidly lost at a pH above $8 \cdot 5$ or below $6 \cdot 5$. The rate of binding is optimal at $37°$c. The temperature and the presence of the steroid are beneficial factors for the optimum binding of the receptor protein(s) (Talwar *et al.*, 1968).

Cytoplasmic receptors for oestradiol are present not only in the endometrium but also in the myometrium of the calf uterus (Talwar, 1969). The diaphragm has practically no such receptors. They are present in the vagina, pituitary, mammary gland and hypothalamus, but are either absent or present in small amounts in tissues which are not influenced by oestradiol. This observation prompts the belief that the presence or absence of receptor molecules for oestradiol in a tissue may be the molecular basis of the selective effect of this hormone on some but not all tissues of the body.

Nuclear "receptors". Radioactive oestradiol accumulates substantially in the nuclear fraction of the uterus. The nature of the receptor(s) present in the nucleus is not precisely known. Jensen and co-workers (1968) reported the extraction of an oestradiol-binding fraction from the nuclei with o·3 M-KCl. This fraction had a sedimentation coefficient of ~ 5s, and was believed to be different from the cytoplasmic "receptor" fraction (sedimentation coefficient ~ 9·5s). More recent observations of Kornman and Rao (1968) have however shown that the oestradiol-binding fraction extracted by o·3 M-KCl may not be a distinct entity; the ionic strength of the medium would disperse the 9·5s cytoplasmic fraction into a less aggregated component of 5s.

Relation of oestradiol binding to subsequent metabolic events in the uterus

The uptake of oestradiol by the uterus is by definition the earliest inter-action of the hormone with the tissue. The connexion of this event with the uterotropic action of oestradiol is not clearly known. The cyto-plasmic fraction from ovariectomized rat uteri was observed to inhibit the activity of purified *E. coli* RNA polymerase *in vitro* (Talwar *et al.*, 1964). In some of these preliminary experiments, less inhibition was obtained by addition of oestradiol. Reversal of the inhibition is however not repeat-edly obtained. The cytoplasmic fraction used in these experiments was not purified. The order of addition of reactants and other factors presumably influence the degree of inhibition obtained.

It has not been possible so far to obtain in an unambiguous manner the effects of oestradiol on the uterus *in vitro*. K. N. Rao in our laboratory has used a variety of media and incubation conditions with marginal success. A negative experiment may not necessarily be significant. It is possible that the "right" conditions have still to be discovered, but other workers have met with similar failures (Hechter *et al.*, 1966). It is not unlikely that additional factors (hitherto unrecognized) may be important in the action of oestradiol on the uterus.

Role of cyclic AMP

Szego and Davis (1967) have reported a doubling of the cyclic AMP content of the uterus within 15 seconds of injection of oestradiol-17β to rats. The increase is given by diethylstilboestrol but not by 17α-oestradiol which also has little biological action on the uterus. It is not clear whether oestradiol-17β acts directly on the adenyl cyclase system or whether the cyclic AMP is produced because of the initial release of biogenic amines. It is further not clear whether cyclic AMP is a true second-stage messenger of the hormone for the uterotropic action.

Dr S. K. Sharma in our laboratory has made extensive studies of the effect of cyclic AMP *in vitro* on uterine metabolic activities. Earlier studies of Hechter and co-workers (1966) have indicated that the action of cyclic AMP requires several hours for expression. It was also observed by these workers that the effect was better shown when the uterus was incubated for 2–6 hours in a medium devoid of glucose or any other energy-providing substrate. These experimental conditions appear to be

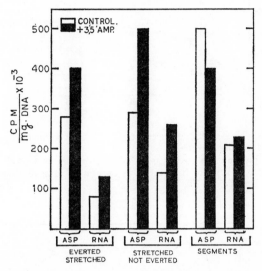

FIG. 1. *In vitro* system for the study of the effect of cyclic AMP on the rat uterus.

Uteri from ovariectomized rats were examined either as segments or lightly stretched and tied with surgical thread to the two hooks of a glass rod. They were preincubated for 15 minutes at 37° C in Earle's balanced salt solution with vitamins and amino acids (BME) in an atmosphere of 95 per cent oxygen and 5 per cent carbon dioxide. The tissues were then transferred to fresh incubation medium containing when indicated 3′5′-AMP, 0·1 mM; theophylline, 1·0 mM; and [5-³H]-uridine (10^{-7} M; $3·2 \times 10^6$ counts/min per 1·5 ml). The incorporation was studied for 2 hr. ASP: Cold-TCA-soluble pool.

non-physiological. Results to be reported in detail elsewhere show that with proper stretching and inversion of the tissue and by the use of enriched medium, the effect of cyclic AMP can be observed *in vitro* within 30 minutes (Figs. 1 and 2). The cyclic nucleotide increases the uptake and incorporation of [5-³H]uridine into RNA (Figs. 2 and 3). The initial rise in specific radioactivity of RNA is entirely due to the increased pool size of the radioactive precursor, suggesting an effect of cyclic AMP on the permeability of the membrane to some metabolites. Cyclic AMP also

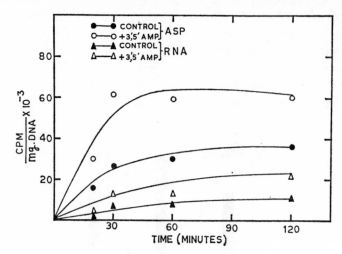

FIG. 2. Time-course of the effect of 3′,5′-AMP on the incorporation of [5-³H]uridine into RNA.

[5-³H]uridine was 10^{-7} M; 0.6×10^6 counts/min per 1·5 ml. Other conditions (except for the time of incorporation of uridine) were the same as described in Fig. 1.

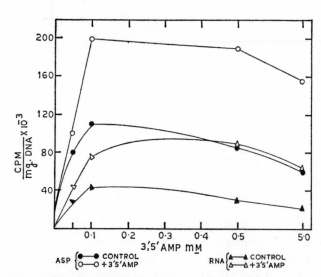

FIG. 3. Effect of different concentrations of 3′,5′-AMP on the incorporation of [5-³H]uridine into RNA.

[5-³H]uridine was 10^{-7} M; $1·5 \times 10^6$ counts/min per 1·5 ml. Other conditions (except the concentration of cyclic AMP) were the same as described in Fig. 1. The time of incorporation of the precursor was 2 hr.

influences the uptake of amino acids and their incorporation into proteins in the uterus. Cyclic AMP is not the only nucleotide to give these effects; AMP and GMP also induce similar changes, though to a smaller extent. It has not been possible so far to obtain the oestrogen-like effects of cyclic AMP on the endometrial cells, stroma or glandular structures of the uterus by incubating the organ for 24 or 48 hours in a medium containing cyclic AMP.

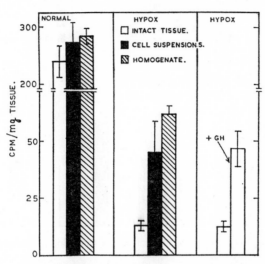

Fig. 4. Incorporation of [U-¹⁴C]glucose into fatty acids at various levels of tissue organization in normal and hypophysectomized rats.
Epididymal adipose tissue from normal and hypophysectomized (Hypox) rats was tested for its ability to metabolize [U-¹⁴C]glucose into fatty acids in the form of intact tissue, cell suspensions or homogenates. The incorporation time was 2 hr. Results are the mean of not less than 6 flasks ± standard error of the mean. GH: bovine growth hormone, 5 µg/ml.

Early interactions of growth hormone

Growth hormone stimulates the synthesis of RNA in rat liver, muscle and perhaps other tissues (Talwar et al., 1962; Talwar, Gupta and Gros, 1964; Korner, 1964; Breuer and Florini, 1966; Widnell and Tata, 1966; Gupta, 1967). The synthesis of both ribosomal and messenger RNA is increased. It is further observed that the hormone does not increase the template activity of the chromatin, nor does it introduce new species of RNA, detectable by hybridization techniques (Breuer and Florini, 1966; Drews and Brawerman, 1967; Gupta and Talwar, 1968). The action of the hormone is most marked in short-pulse experiments. The hormone influences essentially the rate of RNA synthesis.

The effect of this hormone on RNA synthesis is not shown earlier than 30–60 minutes after perfusion of the hormone *in vivo* (Jefferson and Korner, 1967). It is likely that the hormone does not act directly on RNA synthesis in this case, but influences other processes which affect the synthesis of RNA in course of time.

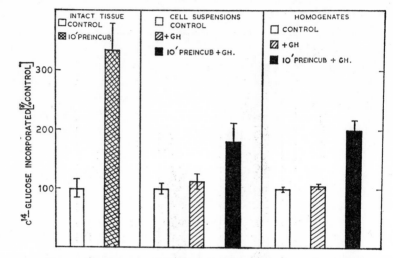

Fig. 5. Requirement of the integrity of tissue during preincubation period for response to growth hormone.

Intact epididymal adipose tissue was optimally stimulated on exposure of the tissue to bovine growth hormone (5 μg/ml) for 10 minutes at 37°C (preincubation period), at the end of which it was rinsed three times and transferred to Krebs Ringer bicarbonate buffer, pH 7·4, containing [U-¹⁴C]glucose (but not growth hormone unless indicated) and incubated for 2 hr. Cell suspensions and tissue homogenates were examined in parallel experiments. No stimulation was obtained when growth hormone was added directly to the medium containing the cell suspensions and homogenates. However, if the intact tissue was preincubated for 10 minutes with the hormone before being made into cell suspensions or homogenates, glucose incorporation was stimulated. The results are the mean of six flasks ± standard error of the mean.

A system which responds to growth hormone *in vitro* has been used to gain further insight into the early actions of growth hormone. Adipose tissue from hypophysectomized rats has a reduced capacity to metabolize [U-¹⁴C]glucose into fatty acids. Addition of growth hormone *in vitro* alleviates this deficiency to some extent and results in an increased conversion of glucose into fatty acids (Fig. 4). The presence of this hormone is not required continuously for this effect. Preincubation of the tissue with the hormone for 10 minutes *in vitro* fully programmes the stimulation, which is valid for a period of 2 hours' incubation of the tissue in a medium containing [^{14}C]glucose. The integrity of the tissue is essential for hormonal programming of the tissue during the preincubation period (Fig. 5).

The location of the hormone in the tissue has been investigated by the fluorescent antibody technique. The intact tissue was incubated at 37° c in the usual medium with growth hormone for 2 or 10 minutes. The tissue was removed and rinsed thoroughly through three changes of plain medium. Rabbit anti-growth hormone serum was used for the immunological reaction with the hormone bound to the tissue. These sites were then revealed with goat anti-rabbit globulin serum tagged with fluorescein isothiocyanate. Such experiments have shown the presence of the hormone on the membranous boundaries of the adipose cells. Radioautographic studies are in progress to confirm these observations. There is a fair indication that one locus of action of the growth hormone is on the membranous structures in adipose tissue.

The blood vessels and the endothelial cells lining blood vessels also show fluorescence. It is not unlikely that the hormone influences these structures also in the adipose tissue. There is also some suggestive evidence for intercellular communication between the different types of cells interwoven in an organized pattern in the adipose tissue. Data to be presented elsewhere in more detail show that the hormone when added *in vitro* to homogenates and cell suspensions prepared from the intact tissue preincubated with growth hormone for 10 minutes is effective in stimulating glucose incorporation (see also Fig. 5). The addition of the hormone to the homogenates or cell suspensions directly (without prior preincubation of the intact tissue with the hormone for 10 minutes) has no stimulatory effect. The hormone seems to require a synergistic event, taking place during the first 10 minutes of contact of the whole organized tissue with the hormone.

OESTROGENS AND THE ACTIVATION OF THE GENETIC LOCUS

In the experiments to be described below, two questions have been posed.

(1) Do oestrogens induce the synthesis of new messenger RNA? This question is pertinent since growth hormone affects essentially the *rate* of RNA synthesis in adult animals and does not appear to cause the appearance of new species of RNA. There is also evidence in many developing systems for the presence of a stable messenger RNA (m-RNA). The

regulation in such cases occurs at the level of its recognition and translation.

(2) Does the stimulation of RNA synthesis on giving oestrogens depend on the synthesis of protein(s)? There is controversy whether the growth-promoting hormones influence in the first instance the translation or the transcription processes. In course of time both would be affected as they are tightly coupled to each other in many systems.

The system employed for these investigations is the rooster liver. Administration of oestrogens to the roosters gives rise to the production of two proteins, namely phosvitin and lipovitellin (Schjeide and Urist, 1956). These egg yolk proteins are normally made by laying hens but not by roosters. Phosvitin has a sedimentation coefficient of 3·11s and is particularly rich in serine, which accounts for more than half of the total amino acids (Heald and McLachlan, 1963). Its synthesis in the rooster liver is prevented by actinomycin D if it is given up to 6 hours after the administration of the hormone (Greengard, Sentenac and Acs, 1965).

TABLE I

RATE OF INCORPORATION OF [^{32}P]ORTHOPHOSPHATE INTO CHICKEN LIVER RNA FOLLOWING OESTRADIOL ADMINISTRATION

Ten-day-old chicks were given 100 μg each of oestradiol-17β in 0·1 ml of propanediol. 100 μc of [^{32}P]orthophosphate was injected intraperitoneally to each chick and the incorporation allowed for 30 minutes before sacrifice.

Time after oestradiol administration	Relative specific radioactivity (counts/min per mg RNA/ counts/min per μg of phosphorus in acid-soluble pool)		
(*hours*)	*Expt 1*	*Expt 2*	*Average*
0	10·4	11·8	11·1
½	15·6	17·4	16·5
4	23·6	25·8	24·7
12	12·2	12·3	12·2
24	18·0	—	18·0
38	15·1	16·7	15·8

The effect of oestrogens on the rate of synthesis of RNA in the rooster liver is shown in Table I. The rate is distinctly elevated by 4 hours after the injection of the hormone. Some increase is discernible as early as 30 minutes. Similar results were obtained with [^{3}H]orotic acid as precursor. It was also observed that concomitant with the increased labelling of RNA, the specific radioactivity of the acid-soluble pool was increased in chicks treated with oestrogens. The rise in RNA synthesis is not prevented by cycloheximide (Table II), suggesting that the oestradiol-induced stimulation of RNA does not require the prior synthesis of proteins.

TABLE II

EFFECT OF CYCLOHEXIMIDE ON THE OESTROGEN-INDUCED SYNTHESIS
OF RNA IN THE CHICKEN LIVER

Cycloheximide (4 mg) was effective for 60 minutes in inhibiting the incorporation
of [^{14}C]leucine into proteins of the chick liver by 90 per cent.

Time after oestrogen (minutes)	Time of [5-H^3]uridine incorporation (minutes)	Cycloheximide action (minutes)	Specific activity (counts/min per mg RNA)*	Relative specific activity*
0 (Control)	30	—	1382	5·19
30	30	—	2473	11·20
30	30	60	2780	12·10
240	30	—	2235	10·0
240	30	60	2540	16·0

* Mean values of two experiments.

The RNA prepared from liver nuclei of control and oestrogen-treated chickens was analysed on sucrose density gradients to ascertain the types of RNA stimulated by the hormone. The oestrogen action *in vivo* was allowed for 3½ hours and a 30-minute pulse of radioactive precursors was given to control and hormone-treated animals. The nuclear RNA was chosen for examination for two reasons: (*i*) A pulse of 30 minutes labels primarily the nuclear fractions of RNA in animal cells and (*ii*) there is evidence to suggest that all species of RNA are not transferred from nuclei to the cytoplasm in animal cells (Penman, 1966; Shearer and McCarthy, 1967). The results show that the radioactivity following oestrogen treatment was increased mostly in a fraction sedimenting towards the top of the gradient (Fig. 6*a* and *b*). This could be an RNA fraction akin to transfer RNA (t-RNA) or it could be an m-RNA fraction. Experiments reported below have tested for both these possibilities.

The first series of experiments was designed to test for the presence of a new m-RNA from the point of view of its functional attributes. Administration of oestrogens leads to the formation of phosvitin which is rich in serine. Dr M. L. Sopori has investigated the incorporation of [U-^{14}C]-L-serine into proteins in an *in vitro* system consisting of microsomes and cell-sap from control and oestrogen-treated animals. The ability to incorporate this amino acid into proteins was distinctly elevated in the system derived from the hormone-treated chick liver (Fig. 7*a*). It was further observed that the stimulatory capacity resided mostly in the microsomal fraction. The cell-sap from control animals with microsomes

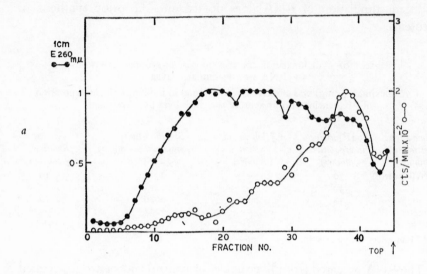

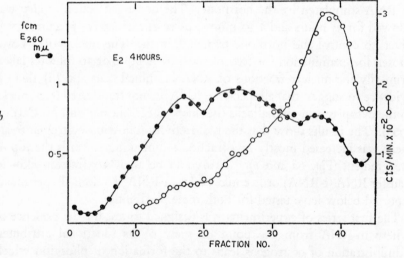

FIG. 6. Sedimentation profiles of nuclear RNA from (a) Control and (b) Oestrogen-primed rooster liver.

One mc of [³²P]orthophosphate was injected intraperitoneally to 26-day-old male chicks. The livers were removed after 30 minutes of incorporation of the radioactive precursor *in vivo*. RNA was extracted from nuclei with hot phenol and analysed on a 5–20 per cent gradient of sucrose. In the experiment of Fig. 6b the rooster received 2 mg oestradiol-17β in 0·4 ml propanediol intraperitoneally 4 hr before sacrifice.

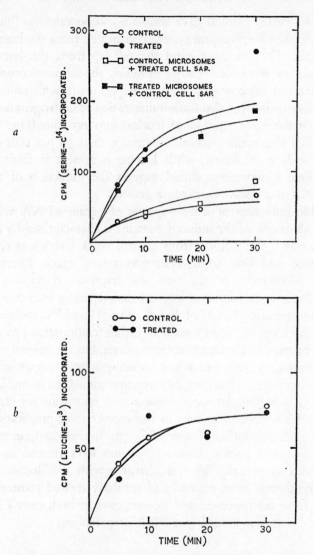

FIG. 7. Incorporation of (a) Serine and (b) Leucine in protein-synthesizing systems derived from control and oestrogen-treated roosters.

Total incubation mixture (0·2 ml) contained 1 μmole ATP, 3 μmoles
phosphoenol pyruvic acid, 10 μg pyruvate kinase, 0·1 μmole gluta-
thione (reduced), 1 μmole GTP, 1 mg cell-sap protein in 40 μl of
Medium B(Tris 50 mM), 0·8 mg protein of microsomes in Medium
A(Tris 35 mM, pH 7·8; MgCl₂ 10 mM, KCl 25 mM, sucrose 0·15 M),
0·2 μmoles each of lysine, alanine, aspartic acid, glutamic acid,
arginine, histidine, methionine, isoleucine, threonine, glycine and
phenylalanine, 2·5 n-moles (0·25 μc) of [14C]serine, or 0·33 moles of
[4,5-3H]leucine (5 μc). Ten mg of oestradiol was given intraperitoneally
24 hours before removal of the liver.

from hormone-treated animals gave almost the same values as (but may not be equal to) the homologous system constituted from the hormone-treated animals. On the other hand the cell-sap from the hormone-treated animals was not effective alone in raising the serine-incorporating capacity of control microsomes. Parallel experiments with radioactive leucine as precursor showed that the administration of oestrogens does not alter markedly the incorporation of leucine into proteins (Fig. 7b). It may be recalled that serine accounts for more than 50 per cent of the total amino acids in phosvitin, while leucine is present in fairly small amounts. These observations would suggest the induction of a new messenger RNA as a result of oestrogen action.

The possible induction of a new species of serine t-RNA was also investigated, since one of the induced proteins is characterized by a high content of serine. The t-RNA from control chick livers was charged with [^{3}H]serine and that from oestrogen-treated chick livers with [^{14}C]serine. A mixture of the two was fractionated on a DEAE-Sephadex column according to a method that separates into two peaks the three serine-specific t-RNAs of E. coli. The ^{3}H and ^{14}C radioactivity peaks were coincident in this experiment. Other fractionation procedures may have to be tried to confirm these indications, but the present suggestion is that oestrogen treatment is not inducing a new species of serine t-RNA in rooster liver. However, experiments conducted in our laboratories by Dr G. Sundharadas suggest that a new synthetase for charging serine on t-RNA is perhaps present in the rooster liver preparations 24 hours after the administration of oestradiol-17β. In these experiments the t-RNA from control rooster liver and rooster liver treated 24 hours previously with oestrogen (E$_2$) was charged with [^{14}C]serine, with amino acyl synthetase from control and similarly treated rooster liver. The enzyme from oestrogen-treated roosters gave in both cases a higher plateau value for serine charged on t-RNA preparations.

SUMMARY AND CONCLUSIONS

Growth-promoting hormones stimulate at an earlier or later stage the biosynthesis of RNA and proteins in tissues on which they act. Their mode of action may not however be the same. Growth hormone influences mainly the *rate* of RNA synthesis in livers of adult rats. Experiments discussed here suggest that one locus of action of this hormone is on the membranous structures of cells in adipose tissue of hypophysectomized rats.

To gain further understanding of the action of oestrogens on the genetic locus, the influence of oestradiol-17β on the transcription and translation processes in rooster liver has been studied. Oestrogens activate the incorporation of radioactive precursors into RNA, an effect which is not prevented by cycloheximide in doses sufficient to block leucine incorporation into chick liver proteins by 90 per cent. The microsomes prepared from oestrogen-treated rooster liver are more effective for the incorporation of serine in an *in vitro* system, suggesting the induction of a new m-RNA in rooster liver under the influence of oestrogens. There is also some suggestive evidence for the oestrogen-directed induction of a new amino acyl synthetase for charging of serine on t-RNA.

Acknowledgements

This work has received grants from the Department of Atomic Energy, Government of India; the Council of Scientific and Industrial Research, New Delhi; the Population Council Inc., New York and the World Health Organization, Geneva.

REFERENCES

BREUER, C. B., and FLORINI, J. R. (1966). *Biochemistry, N.Y.*, 5, 3857–3865.
DREWS, J., and BRAWERMAN, G. (1967). *Science*, 156, 1385–1386.
EISENFELD, A. J., and AXELROD, J. (1965). *J. Pharmac. exp. Ther.*, 150, 469–475.
GREENGARD, O., SENTENAC, A., and ACS, G. (1965). *J. biol. Chem.*, 240, 1687–1691.
GUPTA, S. L. (1967). Thesis, All-India Institute of Medical Sciences, New Delhi.
GUPTA, S. L., and TALWAR, G. P. (1968). *Biochem. J.*, 110, 401–406.
HEALD, P. J., and McLACHLAN, P. M. (1963). *Biochem. J.*, 87, 571–576.
HECHTER, O., YOSHINAGA, K., HALKERSTON, I. D. K., COHN, C., and DODD, P. (1966). In *Molecular Basis of Some Aspects of Mental Activity*, vol. 1, pp. 291–346, ed. Walaas, O. London: Academic Press.
JEFFERSON, L. S., and KORNER, A. (1967). *Biochem. J.*, 104, 826–832.
JENSEN, E. V., and JACOBSON, H. H. (1962). *Recent Prog. Horm. Res.*, 18, 387–414.
JENSEN, E. V., SUZUKI, T., KAWASHIMA, T., STUMPF, W. E., JUNGBLUT, P. W., and DeSOMBRE, E. R. (1968). *Proc. natn. Acad. Sci. U.S.A.*, 59, 632–638.
KING, R. J. B., and GORDON, J. (1966). *J. Endocr.*, 34, 431–437.
KORNER, A. (1964). *Biochem. J.*, 92, 449–456.
KORNMAN, S. G., and RAO, B. R. (1968). *Proc. natn. Acad. Sci. U.S.A.*, 61, 1028–1033.
NOTEBOOM, W., and GORSKI, J. (1965). *Archs Biochem. Biophys.*, 111, 559–569.
PENMAN, S. (1966). *J. molec. Biol.*, 17, 117–130.
SCHJEIDE, O. A., and URIST, M. R. (1956). *Science*, 124, 1242–1244.
SHEARER, R. W., and McCARTHY, B. J. (1967). *Biochemistry, N.Y.*, 6, 283–289.
STONE, G. M. (1963). *J. Endocr.*, 27, 281–288.
STONE, G. M., BAGGETT, B., and DONNELLY, R. B. (1963). *J. Endocr.*, 27, 271–280.
SZEGO, C. M., and DAVIS, J. S. (1967). *Proc. natn. Acad. Sci. U.S.A.*, 58, 1711–1718.
TALWAR, G. P. (1969). *Proc. 5th Int. Symp. Comp. Endocrinol.; Gen. comp. Endocrinology*, suppl. 2, 123–134.
TALWAR, G. P., GUPTA, S. L., and GROS, F. (1964). *Biochem. J.*, 91, 565–572.

TALWAR, G. P., PANDA, N. C., SARIN, G. S., and TOLANI, A. J. (1962). *Biochem. J.*, **82**, 173–179.
TALWAR, G. P., SEGAL, S. J., EVANS, A., and DAVIDSON, O. W. (1964). *Proc. natn. Acad. Sci. U.S.A.*, **52**, 1059–1065.
TALWAR, G. P., SEGAL, S. J., GUPTA, S. L., QUERIDO, A., KAKAR, S. N., NANDI, J. K., SRINIVASAN, C. N., and SOPORI, M. L. (1965). *Proc. II Int. Congr. Endocrinology.* Amsterdam: Excerpta Medica Foundation. International Congress Series No. 83, pp. 14–18.
TALWAR, G. P., SOPORI, M. L., BISWAS, D. K., and SEGAL, S. J. (1968). *Biochem. J.*, **107**, 765–774.
TOFT, D., and GORSKI, J. (1966). *Proc. natn. Acad. Sci. U.S.A.*, **55**, 1574–1581.
WIDNELL, C. C., and TATA, J. R. (1966). *Biochem. J.*, **98**, 621–629.

DISCUSSION

Lynen: I was interested to hear that growth hormone increased the incorporation of glucose into fatty acids. One wonders which step is activated at an increased rate. Did you also check the incorporation of acetate under the same conditions?

Talwar: Mr M. R. Pandian has checked acetate incorporation and finds that though it is a well-utilized precursor, the effect of growth hormone is not very marked. There is a requirement for glucose; with glucose plus acetate the incorporation under hormonal influence is highly stimulated.

Lynen: This may be due to the trapping of long-chain acyl CoA compounds by glucose metabolites. As I discussed in my paper (p. 40), the long-chain acyl CoA compounds (products of fatty acid synthetase) inhibit acetyl CoA carboxylase. Only when they can be removed does fatty acid synthesis from acetate occur.

Talwar: That is a very plausible explanation.

Butcher: When you were studying the effect of hypophysectomy and growth hormone on incorporation of glucose into fatty acids in adipose tissue you used wet weights of tissue as one parameter. We have lost much time trying to use wet weight as a measure of adipose tissue, because the amount of triglyceride changes so dramatically with many treatments, including hypophysectomy. I would suggest that tissue nitrogen, or better still tissue DNA, is a more reasonable measurement.

Secondly, with regard to the idea that intercellular effects are operating when you obtain stimulation of glucose uptake into fatty acids in whole fat pads but not in isolated fat cells, I would point out that collagenase treatment of fat pads may cause all sorts of artifacts. One can very rapidly lose the effects of hormones on cyclic AMP and other parameters if the collagenase preparation is not exactly right, or if the incubation is too long, etc. I don't think anyone yet understands this problem exactly. The effects of insulin are lost most rapidly in cells treated with collagenase (Kono, 1969).

Talwar: On the question of wet weight as a measure, I think you are possibly right. I may add that in our experiments the tissue weight taken for calculation is the weight before incubation and hence any changes in the content of triglycerides during incubation does not influence the results. We have assessed the data on the basis of proteins as well as DNA, and we get similar results. What we would like to avoid is expressing the results on the basis of fat content; that is very tricky with adipose tissue and gives false results.

We are very aware of the difficulties with collagenase and we have also tested the protease-free collagenase and find the same difference between intact tissue and isolated cells in the response to growth hormone. Of course, no enzyme treatment leaves the tissue totally intact in terms of its responding capacity, and these data have to be interpreted with that reservation in mind.

As far as we know Goodman has studied the lipogenic action of growth hormone only on intact adipose tissue and not on isolated fat cells; whereas Fain has studied in fat cells lipolysis and not lipogenesis caused by growth hormone and dexamethasone.

Butcher: In our hands the more purified commercial collagenases are the worst of all.

Prasad: Professor Talwar provided evidence to show that oestrogens stimulate synthesis of specific proteins in the liver. I would like to mention some of our recent data on the effect of oestrogens on liver of rats. We have been studying the mechanism of action of anti-oestrogens in inhibiting oestrogen-induced phenomena in order to explain the anti-fertility action of these compounds. We found that the oestrogen-induced increase in uterine glycogen, protein and sialic acid (Mohla and Prasad, 1968; Rajalakshmi, Prasad and Mohla, 1969) was inhibited by pretreatment with anti-oestrogens. Several anti-oestrogens like clomiphene, U11 100A and U11 555A which we have studied are also weakly oestrogenic. We studied liver glycogen and found that oestrogen induced an increase in liver glycogen. Pretreatment with anti-oestrogens blocked the effect of oestrogens on the liver. Our results suggest that the sites of action of oestrogen and of the weakly oestrogenic anti-oestrogens may be the same in the liver. We have postulated that anti-oestrogens may competitively bind the receptor sites in the liver, in a manner similar to that postulated for oestrogen-specific target sites in the uterus. Have any studies been made on the presence of oestrogen-specific receptor sites in the liver?

Talwar: One would expect specific receptors in the liver if the hormone is effective there, but nobody has looked for them so far. The situation is extremely complex for liver because in addition to the specific binding proteins we have also a variety of enzymes in liver metabolizing these steroids.

Jagannathan: Bornstein and his group have reported that incubation of growth hormone under mildly acid conditions produces two small fragments (Bornstein *et al.*, 1968; Bornstein, Armstrong and Jones, 1968). One is strongly inhibitory to fatty acid synthesis, both in cell-free systems and in slices. The other fragment inhibits the inhibition due to the first fragment. Could the effect of the preincubation of the hormone with the cells be partly due to what the hormone does to the cell, but also partly be caused by what the cells do to the hormone? There may be a labile enzyme in the cell which splits the hormone to give a more active fragment. Is anything known about whether the cells cause any alteration in the growth hormone?

Talwar: I don't know of any data on that, but it is an interesting possibility.

Korner: Dr Z. Laron in Israel has also done some experiments on splitting growth hormone into peptides and claims to get different effects of the different peptides on adipose tissue and on carbohydrate metabolism.

Sarma: I was intrigued by the differential effect on leucine incorporation and serine; the serine seems considerably higher in the same protein. Was any measurement made on the concentration of a possible protein phosphatase which might specifically influence the phosphoserine level?

Talwar: You mean that one should test for the protein by the phosphoprotein phosphatase assay method? An enzyme assay would need to be specific for this phosphoprotein, however.

Tata: In animal cells which synthesize hundreds of different proteins it is difficult to deduce either from hybridization or from sucrose gradient patterns the existence of message for a few new proteins. Using the technique of DNA-RNA hybridization to analyse hormone-induced RNA it was difficult, even in those cases like metamorphosis involving many new proteins, to demonstrate new messenger RNA (Wyatt and Tata, 1968). This is because relatively more label goes into ribosomal RNA so that the hybridization efficiency of labelled RNA may be lowered after hormone administration.

Talwar: As far as the functional characteristics are concerned, the capacity to incorporate serine has been measured with microsomal RNA, not nuclear RNA. Nobody can deduce the synthesis of a given messenger type of RNA from the sucrose density profiles of the nuclear RNA; they only show the sedimentation constants.

Tata: No, I was making a general comment that the type of experiments so often recorded with sucrose gradients or hybridization do not *prove* the existence of new messages, though in some cases these must be produced.

Korner: It seems to me that it is very important to have some criterion for recognizing messenger RNA in mammalian cells. None of those that have been used are in fact sufficient. For example, adding RNA to a Nirenberg S-30 system and getting stimulation is not proof, because this might merely be protecting endogenous RNA. One needs a system in which RNA can be added and the synthesis of a specific protein be shown, and to my knowledge this has not been done for mammalian RNA.

Edelman: Dr Tata, is there a differential rate of incorporation of nucleotides into ribosomal and messenger RNA when they are administered in short pulses to mammalian cells?

Tata: Yes. In HeLa cells the rate of transcription of ribosomal RNA is thought to be about 10 times that of other RNA.

Talwar: I may have wrongly given the impression that oestradiol in roosters induces the synthesis of m-RNA only. As I stated earlier, the hormones that we have investigated induce concomitantly the synthesis of ribosomal RNA. In this particular case the experiments were directed to seeing whether a new species of RNA capable of enhancing the synthesis of the protein (phosvitin) is present in the polysomes or microsomes from the hormone-treated rooster liver or whether the hormone acts at the translation level.

Studies on the development of sea urchins (see for review Gross, 1968) have shown that one can have a situation where stable messenger RNA's pre-exist, and the activation is only of reading or translational controls.

Jagannathan: Professor Harris has pointed out that the nucleolus has a definite role in the transfer of ribonucleic acid from the nucleus. A few years ago Neifakh and Repin (1964) reported on the existence of a lipoprotein in mitochondria which is reversibly released and bound, and has a very marked activating effect on pyruvate kinase. One experimental approach might be to see whether there are any specific substances or messengers which regulate the flow of metabolites from one part of the cell to the other, and whether one organelle can produce a substance which controls metabolism in another part of the cell. For instance, the level of citrate plays an important role in the regulation of metabolism,

5*

especially in the cytoplasm. One wonders whether there are specific factors which regulate the movement of citrate from mitochondria into the cytoplasm or whether this takes place entirely by passive diffusion.

Harris: While we are on the subject of messenger RNA, might I briefly summarize the position, as I see it, in animal cells? I don't know of any area in biochemistry in which confusion is greater. It is, of course, common ground that an RNA is made on the DNA, which goes out into the cytoplasm and carries information for the synthesis of specific proteins. If one wants to use the word "messenger" for this RNA, that is all right by me. The problem is how to identify it.

The great confusion arises because people attempt to apply to animal cells certain criteria which have been suggested for the identification of messenger RNA in bacteria. Rapidity of labelling, polydispersity of sedimentation in sucrose gradients, a high level of hybridizability with DNA, an "apparent" base composition resembling that of DNA by radioactive phosphorus pulse-labelling (in any case, a spurious method)— all these criteria, probably wrong for the most part even in bacteria, have been applied to identify "messenger" RNA in animal cells. If any of these criteria apply to an RNA in animal cells, they apply only to an RNA whose location is limited to the nucleus. The only "polydisperse" RNA which labels rapidly, which has an erratic "apparent" base composition as determined by the phosphorus-pulse technique and which hybridizes at a high level with DNA, is located in the nucleus. When you get out into the cytoplasm there is no real evidence for the existence of an RNA of this kind. Indeed, there is no evidence yet in animal cells that any RNA fraction sedimenting in a different position from the main ultraviolet-absorbing components actually carries information for protein synthesis. It is quite possible that the information is somewhere in the ultraviolet-absorbing peaks, and maybe we ought to look there.

Edelman: What would be your ultimate criterion?

Harris: We have no assay yet in an animal cell system that can identify "messenger" RNA.

Korner: Either that it will synthesize a specific protein, or one must identify a sequence of message and show it to code for a protein of known sequence.

Harris: Yes, for any particular template if you can get enough of it. But the real problem is how to identify "messenger" RNA as a class of molecules by some physical or chemical test. One can, with consummate ease, produce smears of radioactivity in sucrose gradients, but for the most part, they are meaningless.

Tata: There is a danger in over-interpreting the size of the polyribosome because of the possibility of a modification in the size of the nascent protein (for example, conversion of proinsulin to insulin).

Monod: Quite obviously a great technical breakthrough has to be made. I would say that the only identification of the messenger is by specific hybridization to known genetic structures. That has been done quite extensively in bacteria where the Tryp and the Lac messengers and several others are as well identified as possible even though neither tryptophan synthetase nor galactosidase has been synthesized *in vitro* by these messengers.

Harris: One has a much better chance of identifying a particular RNA template in a bacteriophage or bacterium, because DNA-RNA hybridization probably works in these systems, within limits. But, in animal cells, the latest evidence is that it doesn't work at all, except for reiterated DNA sequences (Melli and Bishop, 1969).

Monod: Is there any reason why it doesn't work apart from the fact that there are a thousand times more genes in an animal cell than in a bacterium?

Harris: There are two problems. One is the problem of heterogeneity, which is the obvious one. The other is a more primitive technical problem. In all these DNA-RNA hybridization experiments, and this applies also to some of the bacterial ones, far from enough care has been given to the secondary structure of the RNA that is applied during the hybridization experiment. For example, ribosomal RNA has a high degree of secondary structure, but even if you melt this right out at high temperature, it re-forms on cooling. At the 60° c temperature, or thereabouts, at which you do hybridization, depending, of course, on the ionic strength, you may be attempting to hybridize an RNA which already has a high degree of secondary structure and hence a low intrinsic hybridizability; and you might be comparing it with an RNA which is relatively unmethylated, which has a very low secondary structure and which is therefore intrinsically much more hybridizable. For animal cells, DNA-RNA hybridization remains an essentially uncharacterized reaction.

Edelman: In addition to the technical points, have you an alternative concept of how the ribosomes are organized—that is, their actual concatenation?

Harris: The "polydisperse" RNA in the nucleus of animal cells turns out to have very interesting properties. It is quite easy to make it sediment as a completely symmetrical 16s peak by lowering the ionic strength and removing the magnesium (Bramwell and Harris, 1967). This effect is reversible, and is thus not due to degradation in the ordinary sense. You can make this "polydisperse" RNA accumulate in the nucleus in "stepdown" conditions, or with low concentrations of actinomycin D. You can then get enough of it to analyse its base composition directly by acceptable methods. M. E. Bramwell (1969) has done this. Its base composition is then found to be identical to that of the 16s RNA (Tamaoki and Lane, 1967; Riley, 1969). But we know from autoradiographic studies that this RNA is made all over the chromosomes. I am therefore suggesting the possibility that the information may be somewhere in the 16s RNA. This may be nonsense. But it is possible that the "polydisperse" RNA made outside the nucleolus is an immature form of the 16s RNA, that the large ribosomal subunit only (28s RNA) is made in the nucleolus and that the two together constitute an informational unit. The DNA-RNA hybridization results are all against this; but then I don't believe the hybridization work with animal cells.

White: We have been studying the mechanism of action of steroids on lymphoid cells; some of our data may be of interest in view of those of Professor Talwar and also Professor Korner. Dr Maynard H. Makman and Miss Bernyce Dvorkin have developed an *in vitro* system for examining the effects of hormones on lymphoid cells; the latter can be either thymocytes or peripheral node cells or mesenteric lymph node cells of the rat or the rabbit, all of which respond with changes in a number of measurable parameters when the cells have been exposed to a thymolytic or lymphocytolytic steroid either *in vivo* or directly *in vitro* (Makman, Dvorkin and White, 1966, 1968). We are encouraged to think that the system has physiological significance because it responds only to steroids which are thymolytic or glycogenic *in vivo*. Moreover, the relative activity of these steroids *in vitro* is comparable to their relative biological activity in *in vivo* assays. Thus a steroid such as fluocinolone acetonide, a synthetic compound which is 200–300 times as active as cortisol *in vivo*, is active in the *in vitro* system at a concentration of 10^{-10}M, while for a similar degree of activity a concentration of 10^{-7}M cortisol is required. However, cortisone is inactive in the *in vitro* system; this can probably be attributed to the fact that cortisone must be converted to cortisol in order to show physiological activity and the lymphocyte apparently lacks the necessary enzymic mechanism for reducing the 11-oxygen to the 11-hydroxyl configuration.

In summary of our numerous data with this *in vitro* system, two distinct loci of action of thymolytic steroids on lymphocytes have been delineated. One is on transport: the entry of hexoses, amino acids, ribonucleotides, deoxyribonucleotides and rubidium is inhibited in the presence of an active steroid. Non-active steroids such as progesterone, testosterone and other non-thymolytic steroids do not affect transport. However, one can diminish the inhibitory effect of cortisol by adding high enough concentrations of progesterone.

In addition to these effects on transport, there is a direct intracellular effect of the steroids which appears to be on nuclear RNA synthesis. This is not a consequence of accelerated RNA degradation; the steroid inhibits incorporation of precursors into nuclear RNA and decreases the DNA-dependent RNA polymerase activity. This inhibition of RNA synthesis has at the present time not been reflected in inhibition of synthesis of any particular species of RNA. These inhibitory effects of cortisol on the metabolism of thymocytes *in vitro* are energy dependent: this may be derived from the glucose in the medium; although pyruvate or amino acids can serve as energy sources in the absence of glucose.

We do not know; however, some of the data obtained in our laboratory (Nakagawa and White, 1969) suggest that rather than altering the specificity of the polymerase, cortisol modifies, in some as yet unexplained manner, the nature of the template and/or the nature of the binding of the enzyme to the template.

Korner: Are the late effects dependent upon the early effects, or are they independent?

White: There are a number of reasons for believing that the effects on RNA synthesis are *not* dependent on or causally related to the earlier inhibition of transport phenomena. The polymerase effect is not completely blocked by puromycin, although the transport effect appears to be dependent upon continuing protein synthesis. The cortisol effect on RNA polymerase activity is also seen in an aggregate enzyme preparation from thymocytes exposed to cortisol. Hence, transport permeability factors and intranuclear pool sizes of RNA precursors are not variables in the action of cortisol on thymic RNA polymerase.

Burma: In bacterial systems the binding of the template DNA to the RNA polymerase varies with the nature of the template. In your case is the specificity of the enzyme with respect to the template modified?

White: I do not know!

REFERENCES

BORNSTEIN, J., ARMSTRONG, J. McD., and JONES, M. D. (1968). *Biochim. biophys. Acta*, 156, 38.
BORNSTEIN, J., KRAHL, M. E., MARSHALL, L. B., GOULD, M. K., and ARMSTRONG, J. McD. (1968). *Biochim. biophys. Acta*, 156, 31.

BRAMWELL, M. E. (1969). *J. Cell Sci.*, in press.
BRAMWELL, M. E., and HARRIS, H. (1967). *Biochem. J.*, **103**, 816.
GROSS, P. R. (1968). *A. Rev. Biochem.*, **37**, 631.
KONO, T. (1969). *J. biol. Chem.*, **244**, 1772–1778.
MAKMAN, M. H., DVORKIN, B., and WHITE, A. (1966). *J. biol. Chem.*, **241**, 1646.
MAKMAN, M. H., DVORKIN, B., and WHITE, A. (1968). *J. biol. Chem.*, **243**, 1485.
MELLI, M., and BISHOP, J. O. (1969). *J. molec. Biol.*, **40**, 117.
MOHLA, S., and PRASAD, M. R. N. (1968). *Steroids*, **11**, 571–583.
NAKAGAWA, S., and WHITE, A. (1969). Submitted for publication.
NEIFAKH, S. A., and REPIN, S. (1964). *Biochem. biophys. Res. Commun.* **14**, 86.
RAJALAKSHMI, M., PRASAD, M. R. N., and MOHLA, S. (1969). *Steroids*, **14**, 47–54.
RILEY, W. T. (1969). *Nature, Lond.*, **222**, 446.
TAMAOKI, T., and LANE, B. G. (1967). *Biochemistry, N.Y.*, **6**, 3583.
WYATT, G. R., and TATA, J. R. (1968). *Biochem. J.*, **109**, 253–258.

HORMONAL CONTROL OF METAMORPHOSIS

J. R. TATA

National Institute for Medical Research, London

METAMORPHOSIS is perhaps the most dramatic example of the action of developmental hormones. The initiation and continuation of this developmental process are obligatorily dependent on ecdysone in insects and on thyroid hormones in many amphibia. Metamorphosis can be distinguished from adaptational changes in that the process is begun, and often completed, in *anticipation* of a change in environment and therefore represents the triggering off of a pre-determined programme of partially differentiated cells.

ENDOCRINOLOGY OF METAMORPHOSIS

There is some analogy in the hormonal interactions which set off metamorphosis in insects and amphibia, as shown in Fig. 1. In both invertebrate and vertebrate metamorphosing species the initial stimulus may be an environmental one such as the length of daylight or sudden temperature changes. This environmental signal is transmitted via neural mechanisms to the hypothalamus or neurosecretory cells to be converted to a series of chemical signals or hormones. In the insect, the "brain hormone" produced by neurosecretory cells acts on the prothoracic gland to produce ecdysone which is the substance ultimately responsible for metamorphosis, or moulting (see Karlson, 1963; Schneiderman and Gilbert, 1964). Similarly in amphibia the production of thyrotropic hormone by the pituitary activates the dormant thyroid gland to produce the metamorphosis hormones, thyroxine and tri-iodothyronine (see Etkin, 1964). It is well known that the administration of thyroid hormones or ecdysone to immature larvae or tadpoles will cause a precocious induction of metamorphosis, and this is the basis for the design of much experimental work. However, in nature the process results from a delicate balance between the metamorphosing hormones and another group of endocrine secretions. The action of juvenile hormone in controlling larval

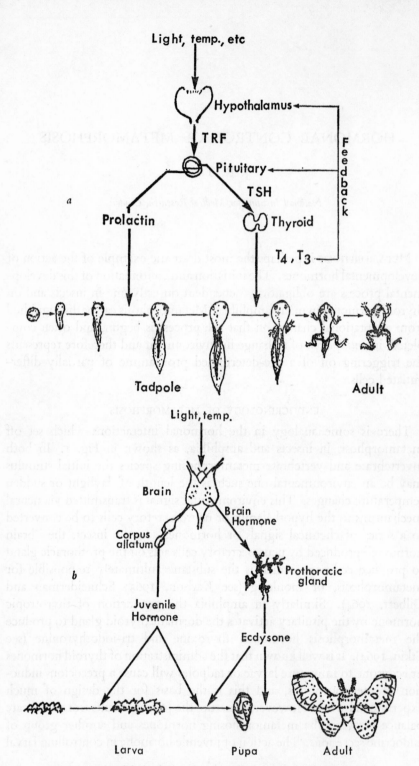

FIG. 1. A simplified scheme showing the hormonal transmission of external stimuli for metamorphosis in (*a*) amphibian and (*b*) insect larvae. See text for explanation.

and pupal moults in insects is well known (Williams, 1961), and more recently prolactin has been shown to control amphibian metamorphosis in an analogous fashion (Bern, Nicoll and Strohman, 1967; Etkin and Gona, 1967).

Behind the very obvious morphological manifestations of metamorphosis such as insect wing eruption or tadpole tail resorption are some very fundamental biochemical changes. These involve the induction *de novo* of new proteins or the preferential synthesis of proteins, some of which are summarized in Table I (see Karlson, 1963; Wyatt, 1962;

TABLE I

SOME MORPHOLOGICAL AND BIOCHEMICAL CHANGES CHARACTERISTIC OF META-
MORPHOSIS INDUCED BY THYROID HORMONE AND ECDYSONE IN AMPHIBIAN AND
INSECT LARVAE

Thyroid hormone, amphibia*		Ecdysone, insects†	
Tissue	Change induced	Tissue	Change induced
Liver	Maturation; induction of urea cycle enzymes, serum albumin	Fat body	Resorption and reorganization
Tail, gut	Resorption; synthesis of hydrolases (cathepsin, nucleases, collagenase)	Salivary gland	Regression; accumulation of hydrolases; chromosomal puffing
Skin	Hardening and pigmentation; collagen deposition	Epithelial cells	Cuticle and pigment formation; induction of dopa decarboxylase, polyphenol oxidase and cocoonase
Limb buds	Cell division; growth	Wing buds	Cell division; scale and pigment formation
Eye	Enzymes for conversion of vitamin A to rhodopsin		

* Data largely based on work on the bullfrog tadpole (*Rana catesbeiana*).
† Information compiled from work on various insects (*Calliphora, Chironomus, Cecropia*).

Schneiderman and Gilbert, 1964; Cohen, 1966; Frieden, 1967). Virtually no larval or pupal cell escapes the impact of thyroid hormones or ecdysone and some of the above biochemical changes, such as the artificial induction of urea cycle enzymes or the formation of polytenic chromosomal puffs, have contributed significantly to developmental biology. The adaptational advantages of the biochemical changes shown in Table I are of course quite obvious. For example, the formation of urea cycle enzymes will eliminate the toxic effects of ammonia and thus allow survival of the frog or toad on land and similarly the laying down of a cuticle will save the insect from dehydration.

I would now like to consider mainly the regulation of biosynthetic changes in thyroid hormone-induced amphibian metamorphosis, with which I am more familiar than the process in invertebrates. Many phenomena that are described below also occur in ecdysone-induced metamorphosis in insects, and some of them have already been reviewed (Wyatt, 1962; Karlson, 1963; Schneiderman and Gilbert, 1964). The following three facets of the regulation of protein synthesis will be considered.

(1) The biosynthetic and ultrastructural changes in cells, such as the hepatocytes, which undergo maturation during metamorphosis.

(2) The importance of additional RNA and protein synthesis in tissues, such as the tail, which are programmed for death or regression.

(3) The early acquisition of competence of *Xenopus* larvae to respond to thyroid hormones, in biochemical terms.

IMPORTANT FEATURES OF REGULATION OF PROTEIN SYNTHESIS

It is now becoming increasingly clear that in nucleated cells of higher organisms, regulation of protein synthesis is more complex than the simple mechanisms described for microorganisms. Before the above three features of metamorphosis are discussed it is worth noting the following aspects of RNA and protein synthesis that set apart nucleated animal cells from microorganisms and are relevant to this discussion:

(*a*) The nature and selective breakdown of nuclear RNA. A large amount of the RNA synthesized in the nucleus is ribosomal RNA with the result that rapidly labelled RNA is predominantly ribosomal precursor RNA or a high molecular weight RNA with no apparent function but not messenger RNA (m-RNA). Recent DNA–RNA hybridization studies have shown that only a fraction of the species of RNA present in the nucleus is present in the cytoplasm, but the mechanism of this selective restriction of RNA to the nucleus remains unknown (Georgiev, 1966; Shearer and McCarthy, 1967).

(*b*) The transfer of RNA from the nucleus to the cytoplasm is a process which requires the formation and maturation of ribonucleoprotein particles. Two mechanisms have been suggested. One is that m-RNA enters into the cytoplasm as a complex with sub-ribosomal particles or in combination with protein only, both of which may then be directly assembled into polysomes (McConkey and Hopkins, 1965). The other possibility is that free messenger does not complex with ribosomal precursors but that it is released into the cytoplasm as ribonucleoprotein particles which are somehow incorporated into polysomes. The role of the protein associated with m-RNA would be to modulate the activity of the template. Such particles have been detected both in the nucleus and in the cytoplasm (Samarina et al., 1968; Perry and Kelly, 1968; Henshaw, 1968).

(c) In intact cells it seems that there is some form of structural requirement in the synthesis of proteins, mainly involving binding of polysomes to membranes of the endoplasmic reticulum (Campbell, 1965). It is of some interest that the latter are themselves being continuously formed and broken down according to the demands for protein synthesis (Tata, 1968a).

REGULATION IN TISSUES UNDERGOING MATURATION:
THE HEPATOCYTE

The liver of the tadpole has been intensively studied by Frieden's (1967) and Cohen's (1966) groups with respect to the synthesis of proteins that characterize amphibian metamorphosis, such as urea cycle enzymes, serum albumin and adult haemoglobin. These and other workers had firmly established, especially in the bullfrog, Rana catesbeiana, that administration of thyroid hormones to pre-metamorphic tadpoles causes a de novo synthesis of these proteins. Because of this firm biochemical background, I undertook six years ago to investigate the formation and turnover of nuclear and cytoplasmic RNA in the pre-metamorphic bull-frog hepatocyte at different stages after the hormonal induction of metamorphosis (Tata, 1965, 1967a). Before these studies there was only one report showing that total tissue nucleic acid content was increased during metamorphosis (Finamore and Frieden, 1960), although Cohen's group has also been investigating the metabolism of RNA recently (Nakagawa and Cohen, 1967).

The most common experimental procedure in studying the regulation of protein synthesis has been to study sequential biochemical events that follow the administration of thyroid hormone to pre-metamorphic tad-poles. The lag period and the magnitude of response that will now follow depend on many factors such as temperature, the developmental stage of the larvae, the analogue of thyroid hormone used and the route of its administration (see Frieden, 1967). In Figs. 2 and 3 are summarized the responses of some of the major biosynthetic processes that are detected under our experimental conditions, during the lag period preceding the appearance of newly formed proteins as a result of administration of tri-iodothyronine to immature Rana catesbeiana tadpoles (Tata, 1965, 1967a). It can be seen that there occurred, well within the latent period (5–6 days) for new proteins to be detected, an acceleration of the rate of nuclear and cytoplasmic RNA synthesis. Somewhat similar results in the same species of tadpole have also been observed in Cohen's laboratory (Nakagawa and Cohen, 1967). In insects too a similar sequence is known to occur; a stimulation of nuclear RNA synthesis has been demonstrated in

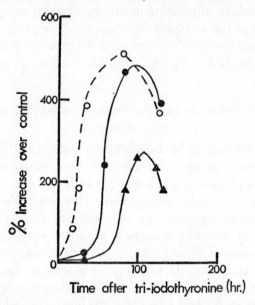

FIG. 2. Schematic representation of the stimulation of nuclear RNA synthesis and its turnover into the cytoplasm of liver of *Rana catesbeiana* tadpoles after induction of metamorphosis with tri-iodothyronine. O----O, specific activity of nuclear RNA labelled with [³H]uridine not corrected for changes in distribution of radioactivity in the acid-soluble fraction; ●——●, specific activity of nuclear RNA after correction for changes in acid-soluble radioactivity; ▲——▲, specific activity of RNA recovered in cytoplasmic ribosomes and polyribosomes. The earlier stimulation of labelling of nuclear RNA when not corrected for acid-soluble radioactivity is presumably due to a more rapid action of the hormone on the uptake or pool size of uridine. The abrupt downward trend of the curves is due to the dilution with nucleotides released by the regression of tail, gut, gills, etc. Data compiled from Tata (1965, 1967a).

anticipation of an increased rate of protein synthesis or appearance of specific enzymes in the wing and epidermal cells of *Calliphora* and *Cecropia* (see Karlson, 1963; Sekeris, 1967; Wyatt and Linzen, 1965). The curves for rates of RNA synthesis in Fig. 2 and turnover have been derived from values of RNA specific activity obtained after correction for changes in uptake of the radioactive precursor which occur much earlier after tri-iodothyronine, as has been found by Eaton and Frieden (1968). An early perturbation in the precursor pool size or uptake of precursors of RNA and protein in target tissues has been observed with hormones known to affect protein synthesis (Tata, 1968*b*). The question of pool sizes in radioactive labelling of constituents of non-regressing tissues is even more complicated when regression occurs in other tissues in the same organism during metamorphosis. Thus the abrupt downward trend in the specific activity of nuclear and cytoplasmic RNA is not due to a

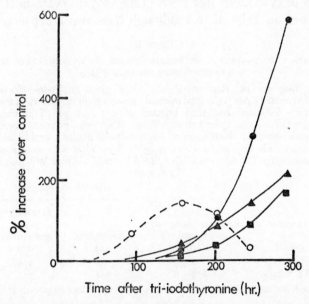

Time after tri-iodothyronine (hr.)

FIG. 3. Schematic representation of the lag period for the induction by tri-iodothyronine of *de novo* protein synthesis and the stimulation of amino acid incorporation into hepatic protein of *Rana catesbeiana* tadpoles. o----o, specific radioactivity of protein recovered in the liver microsomal fraction 40 minutes after the administration of a mixture of ^{14}C-labelled amino acids; ●——●, carbamyl phosphate synthetase; ▲——▲, specific activity of cytochrome oxidase in mitochondrial fraction; ■——■, serum albumin accumulation in blood. Data compiled from Tata (1965, 1967*a*).

sudden reversal of the accelerated rate of synthesis but merely reflects a dilution of radioactive precursors caused by the autolysis in regressing tissues, such as the tail, gut and gills. The same complication applies to measurement *in vivo* of rates of protein and phospholipid synthesis (see Fig. 4).

It would be tempting to suggest that the additional RNA made after hormone administration includes messengers for proteins like urea cycle enzymes and serum albumin, especially as a sustained RNA synthesis is important for metamorphosis to occur (Weber, 1965; Nakagawa, Kim and Cohen, 1967). Furthermore, a moderate increase in template activity of liver chromatin from thyroxine-treated bullfrog tadpoles has been observed (Kim and Cohen, 1966). In practical terms however, we failed to demonstrate, by base analysis, sucrose density gradient fractionation and DNA–RNA hybridization that a significant part of the additional nuclear RNA synthesized *in vivo* at the onset of metamorphosis was messenger- or even DNA-like RNA (Tata, 1967*a*; Wyatt and Tata, 1968). It can be seen in Table II that although there was a net increase in the

TABLE II

EFFECT OF INDUCTION OF METAMORPHOSIS ON HYBRIDIZATION OF TADPOLE LIVER NUCLEAR RNA

Tadpoles of *Rana catesbeiana* were given 3,3′,5-tri-iodo-L-thyronine (T_3) (1 μg/animal injected plus 10 μg/l in the water), at the times indicated and then injected with 5·2 μc of [³H]uridine (2·73 c/m-mole) 3 hr before killing. RNA was prepared from nuclei obtained from livers of 10–15 tadpoles pooled together and tested for hybridization with 6·6 μg RNA per filter with increasing amounts of DNA to reach DNA/RNA ratios of 20 (from Wyatt and Tata, 1968).

Hormone treatment	RNA specific activity (counts/min/μg)	Hybridization plateau* Counts/min	Percentage of input radioactivity
Control	30	14	6·9
T_3, 2 days	70	24	5·4
T_3, 6 days	155	32	3·2

* Each value is the mean from three tubes of 36 μg DNA and three tubes with 47 μg DNA.

amount of readily hybridizable nuclear RNA, there was a drop in the overall hybridization efficiency of the RNA formed during metamorphosis. This is interpreted to mean that very small increases in DNA-like RNA after hormone administration are accompanied by relatively massive increases in the rate of synthesis of ribosomal RNA. A consequence of the burst of nuclear RNA synthesis is the appearance of additional cytoplasmic RNA, mainly in the polyribosomes, which builds up to a maximum just before new proteins appear (Fig. 3). An increased rate of accumulation of newly synthesized ribosomes and polysomes coinciding with either a preferential or non-preferential synthesis of additional proteins seems to be a common feature of regulation of protein synthesis during growth and development (see Tata, 1967*b*, 1968*b*).

What is perhaps of utmost importance is to study the sequential events that occur between the initial burst of RNA synthesis following tri-iodothyronine administration and the appearance of newly synthesized proteins. Studies from our laboratory have shown that there occurs during this period a complex process of breakdown of "old" RNA, alterations in polysome profiles and a redistribution of ribosomes attached to membranes of the endoplasmic reticulum (Tata, 1967a). The latter process is accompanied by a simultaneous increase in the synthesis of ribosomes and membranes of the endoplasmic reticulum as judged by membrane phospholipid synthesis (Fig. 4a). There were no qualitative

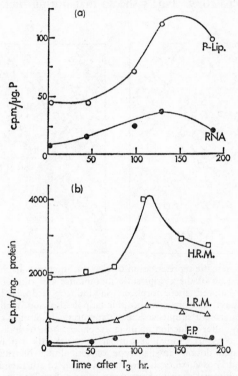

FIG. 4. Coordinated proliferation of endoplasmic reticulum and the increase in hepatic protein synthesis *in vivo* in the heavy rough membranes (densely packed ribosomes attached to membranes) during tri-iodothyronine-induced metamorphosis of young bullfrog tadpoles. (*a*) The appearance of newly synthesized (^{32}P-labelled) heavy rough membrane phospholipids (○) and RNA (●); (*b*) the recovery of labelled nascent proteins formed *in vivo* after a short pulse of radioactive amino acids in heavy rough membranes (□), light rough membranes (△) and free polysomes (●). For details see Tata (1967a).

changes in the types of membrane phospholipids synthesized but it is interesting to note that the simultaneous increase in the formation of the two rough endoplasmic reticulum components is accompanied by an enhanced rate of protein synthesis (Fig. 4b). At the same time, titration of mitochondria-free supernatants with sodium deoxycholate showed that the newly formed ribosomes in induced animals were more tightly bound to membranes of the rough endoplasmic reticulum; that is, they required a higher amount of detergent for their release. It seems that the biosynthetic response to the hormone may be contained in the newly formed ribosomes which may not equilibrate with the existing ribosomal population.

These structural changes are so marked that it is easy to corroborate them by electron microscopy. Fig. 5 shows that during metamorphosis there

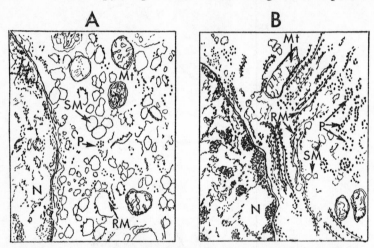

FIG. 5. Schematic representation of electron micrographs showing the reorganization of the cytoplasmic membranes and ribosomes during hormone-induced metamorphosis in bullfrog tadpole liver cells. (a) Hepatocyte in a pre-metamorphic tadpole showing small groups of ribosomes around simple vesicular structures. (b) Similar cell in tadpole 6 days after induction of metamorphosis with tri-iodothyronine, showing dense ribosomal accumulation often attached to a more differentiated double-lamellar type of reticulum. N, nucleus; Mt, mitochondria, SM, smooth membranes; RM, rough membranes; P, typical polysomal chains, rosettes and spirals. See Tata (1967a, b) for details.

is a shift in the distribution of ribosomes from around the simple vesicular membrane structures of the immature larvae to the more complex double lamellar structures more commonly seen in mature tissues. It is interesting that this shift in structural organization of the endoplasmic reticulum coincided with the biochemical observations, mentioned earlier, on the formation of RNA and phospholipid and the synthesis of proteins by polyribosomes. Rather similar coordination of the proliferation of components of the endoplasmic reticulum and protein synthesis has also been observed in the ecdysone-induced development of wing epidermal cells in *Cecropia*

(Wyatt, personal communication) as well as in other systems of rapid growth or functional maturation (see Tata, 1967b, 1968a). The significance of such a coupling of structure and biosynthetic activity may lie in the possibilities that: (a) a response to growth and developmental stimulus is contained in newly formed ribosomes, and (b) that there is a topographical segregation on membranes of the endoplasmic reticulum of differently pre-coded polyribosomes carrying out the synthesis of different groups of proteins. There is of course no direct experimental evidence to verify either of these suggestions but some recent work from our laboratory on the additive effects of growth and developmental hormones in mammalian organs provides an indirect support (Tata, 1968a, b).

In these experiments, when growth hormone, tri-iodothyronine and testosterone were administered in different combinations the increased rates of formation of ribosomes and the membranes of the rough endoplasmic reticulum were coordinated with those of synthesis of different classes of proteins in the liver and seminal vesicles. Because these hormones have quite different latent periods of action the simultaneous administration of any two caused stepwise increases in the rates of proliferation of ribosomes and rough endoplasmic membranes, each burst corresponding in its time-course and magnitude to that induced by each individual hormone. It is assumed in the suggestions made above that an exchange between polysomes and membranes proceeds at a low rate relative to the life-time of these structures.

The role of DNA synthesis is quite obvious in those tissues that are rapidly formed during metamorphosis, such as the limbs and lungs in amphibia and wings in insects. However the possible role of a less impressive increase in the rate of DNA synthesis is not clear in those tissues that do not grow rapidly but undergo a functional maturation, such as the tadpole liver and skin and insect epidermal cells producing new pigments and cuticle. That a relatively small burst of DNA synthesis may be important during metamorphic maturation and needs careful investigation is emphasized by the work of Topper's group on a limited DNA synthesis or cell division as a prerequisite for milk protein synthesis in prolactin-induced development of mammary gland in culture (Lockwood, Stockdale and Topper, 1967).

REQUIREMENT OF BIOSYNTHETIC ACTIVITY FOR TISSUE RESORPTION DURING METAMORPHOSIS

Tissue regression or cell death is an important and integral part of many embryonic developmental processes (Saunders, 1966). During

142   J. R. TATA

amphibian metamorphosis, regression of organs like the tail, gut and gills accompanies the maturation or formation of organs like the liver, limbs and eyes. In insects the ecdysone-induced regression of the salivary gland accompanying development of the cuticle and wings is now a classical system for studying chromosomal puffs (Beermann, 1963; Clever, 1962). The process of amphibian tail resorption, in particular, has been

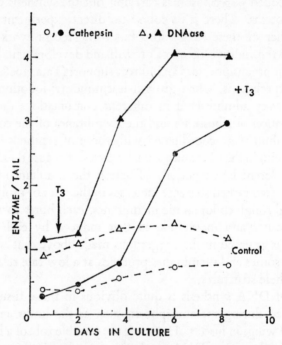

FIG. 6. Accumulation of cathepsin (○, ●) and deoxyribonuclease (△, ▲) in amputated *Rana temporaria* tails maintained in organ culture. Dashed lines are for controls and solid lines are for tails in which regression was induced by the addition of 5×10^{-9} M tri-iodothyronine to the culture medium. For other details see Tata (1966).

studied in detail in Frieden's (1967) and Weber's (1965, 1967) laboratories and these workers have established that the increase in activities of many hydrolases is the basis for the regression. A question that had not been answered until recently was whether or not regression or cell death was also based on a hormonal regulation of biosynthetic processes like that we have seen above for maturation of the liver.

We studied this question in our laboratory by using the technique of thyroid hormone-induced regression of the isolated tadpole tail maintained in organ culture (Tata, 1966). This technique had been already successfully used by Weber (1963) and the induction of regression *in vitro* corresponds in magnitude and speed to that seen in the intact tadpole. Decrease in the size of the amputated tail *in vitro* was accompanied by an increase in the activity of enzymes involved in regression such as cathepsin, phosphatases and deoxyribonuclease (Fig. 6). Earlier work on the comparison of the properties of deoxyribonuclease and cathepsin in regressing

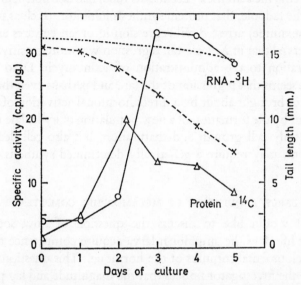

FIG. 7. Burst of additional RNA and protein synthesis during regression of tadpole tails induced in organ culture with tri-iodothyronine added to the medium. O——O, incorporation of [³H]uridine into RNA (12·5 hr after the addition of 10 μc); Δ——Δ, incorporation of ¹⁴C-labelled amino acids into protein (8·8 hr after the addition of 1·6 μc of a mixture of labelled algal amino acids). Incorporation on day 0 is the average value for controls; all other points refer to samples to which tri-iodothyronine (T₃) was added. ..., tail length of controls over the duration of the experiment, ×----×, length of T₃-treated tails showing a marked onset of regression between the second and third day after culture. Curves compiled from data of Tata (1966).

and non-regressing tadpole tails had suggested that a part of the additional enzyme activity following hormone treatment may differ from the basal enzyme present in non-regressing tails (Coleman, 1962; Weber, 1963). It is therefore interesting to note in Fig. 7 that just when regression sets in *in vitro* there is a burst of both RNA and protein synthesis.

To determine whether any of the additional RNA and protein synthesized at the onset of regression was essential for the process itself, we turned to the use of inhibitors of RNA and protein synthesis. From these experiments it was quite clear that inhibition of RNA synthesis with actinomycin D or of protein synthesis with puromycin or cycloheximide completely abolished the regression in tail organ cultures that was induced by tri-iodothyronine. Fig. 8 shows the effect of actinomycin D on tail explants after a week in organ culture. It was also found that inhibition of protein synthesis abolished the hormone-induced increase in hydrolytic enzyme activities. Eeckhout (1966) had also noted, in a different species of the tadpole, that protein synthesis inhibitors (such as puromycin and cycloheximide) arrested tail regression in organ cultures and Weber (1965) observed that in *Xenopus* tail regression was more sensitive than was limb generation to the administration of actinomycin D to the intact tadpole. It seems that regression of the tail, and perhaps even that of other tissues, is not brought about by a direct hormonal activation of lysosomal enzymes but by the formation of a new population of hydrolase molecules. Thus, not only cell growth and maturation, but also cell death during development, may require a genetically determined synthesis of specific proteins.

EARLY ACQUISITION OF METAMORPHIC COMPETENCE

Finally I would like to discuss the question of how soon during embryonic life does the amphibian larva acquire competence to respond to the developmental stimulus of the hormone. This question comes up every time the investigator notices that the magnitude and lag period of a given response varies according to the developmental stage the larva has already reached. It has of course been known for some time that a pre-metamorphic tadpole will respond to thyroid hormone long before it would undergo spontaneous metamorphosis (Moser, 1950; Etkin, 1964; Weber, 1967; Kollros, 1968). This response was largely studied in morphological terms until recently and the acquisition of competence to metamorphose had not been established in biochemical terms. About three years ago, I decided to establish the developmental stage when the amphibian larva acquires competence to respond to thyroid hormones, and focused our attention on the major biosynthetic processes. For this purpose we started with fertilized *Xenopus* eggs and measured the larval capacity to carry out various biochemical processes such as synthesis of RNA, DNA, proteins and phospholipids, uptake of ions and activity of hydrolytic enzymes after exposing the embryos to thyroid hormones at different stages of their development (Tata, 1968c).

Fig. 9 summarizes some of our main findings. It can be seen that a reduction in weight of the tadpole and its capacity to take up radioactive

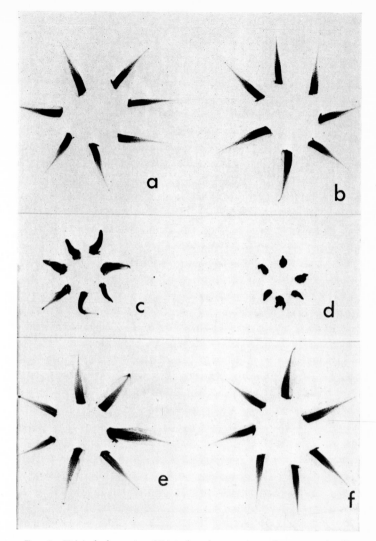

FIG. 8. Tri-iodothyronine (T$_3$)-induced regression of amputated tails of *Rana temporaria* in organ culture. (*a*) control samples on the first day of culture; (*b*) control, after 8 days of culture; (*c*) on the fourth day of culture in medium containing 1 μg T$_3$/ml; (*d*) on the eighth day of culture with T$_3$; (*e*) the same as (*c*) but with 2·5 μg actinomycin D/ml; (*f*) as (*d*) but with actinomycin. See Tata (1966) for details.

To face page 144

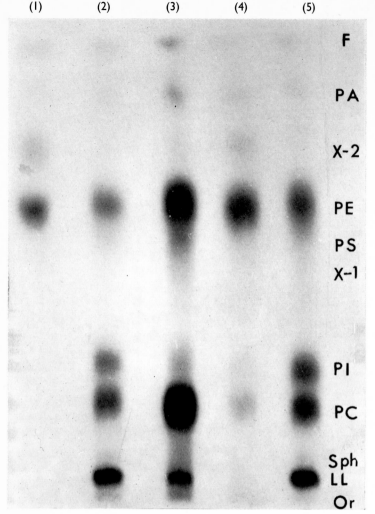

FIG. 11. Autoradiograph of a thin-layer chromatogram of total ^{32}P-labelled phospholipids showing the effects of tri-iodothyronine on larvae at two stages of development. Groups of 40 larvae were killed 16·5 hr after immersion in 75 μc ^{32}P per 250 ml in the presence of antibiotics; the homogenates were treated with 0·4 N HClO$_4$ and phospholipids extracted with methanol-chloroform. Where indicated, metamorphosis was induced by immersion in 500 ml of 2·5 × 10^{-9} M T$_3$ 4 days before killing.

(1) 7-day-old larvae; (2) 7-day-old larvae induced to metamorphose (i.e. T$_3$ treatment begun on day 3); (3) 28-day-old larvae; (4) 13-day-old larvae; (5) 13-day-old induced larvae. Abbreviations: Or, origin of chromatogram; F, front; LL, lysolecithin; Sph, sphingomyelin; PC, phosphatidylcholine; PI, phosphatidyl inositide; PS, phosphatidylserine; PE, phosphatidyl ethanolamine; PA, phosphatidic acid; X-1 and X-2, spots for unidentified ^{32}P-labelled substances. (From Tata, 1968c.)

phosphate accompanied an enhanced rate of synthesis of nucleic acids and protein if the larvae were first exposed to tri-iodothyronine at stages after the first-stage tadpole (Nieuwkoop-Faber stages 36–41). The larvae failed to respond to hormones if these were administered before the second or third day after fertilization. The most novel feature of these studies is the

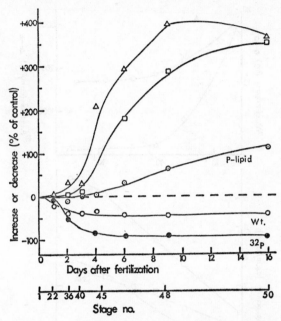

FIG. 9. Summary of some of the multiple phenomena showing the development of early metamorphic competence of *Xenopus* larvae when exposed to $2-5 \times 10^{-9}$ M tri-iodothyronine. The earliest responses, weight loss and reduction in ^{32}P uptake, were detected in larvae that were exposed to the hormone at 2–3 days after fertilization (stages 36–41). The time on the abscissa refers to the stage of development when the larvae were first exposed to the hormone but the measurements were made $4 \cdot 0 \pm 0 \cdot 2$ days after that. The data are derived from different experiments under conditions which varied slightly; the variations for each value are however within ± 12 per cent. O, weight loss, ●, decrease in ^{32}P uptake; △, RNA synthesis (^{32}P incorporation); □, DNA synthesis; ⊙, phospholipids. (From Tata, 1968c.)

very rapid drop in the uptake of orthophosphate ions by the larvae after exposure to thyroid hormones. *Xenopus* larvae normally pick up phosphate ions from water at a slow rate but this can be depressed a hundredfold by 2–3 days after the hormone is administered. This created an obvious complication in studies designed to label phospholipids and RNA with ^{32}P and one which therefore necessitated microinjection of tadpoles with radioactive precursors to follow rates of synthesis of these constituents. Other indices of metamorphosis that were observed in early tadpole stages included overall morphological changes typical of naturally metamorphosing *Xenopus* (80–90 days after fertilization), tail regression and a slight increase in hydrolase activity. Dose–response curves and

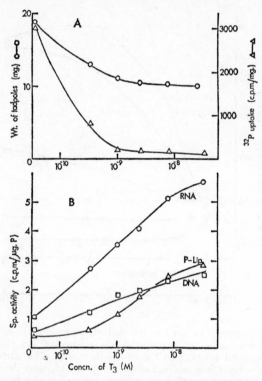

FIG. 10. Response of 8-day-old larvae to different concentrations of tri-iodothyronine. Exposure to the hormone was for 3·3 days and 17·5 μc of ^{32}P were administered 15 hours before death. Corrections were made for changes in ^{32}P uptake in calculating specific activities of nucleic acids and phospholipids. A: effect of T_3 on body weight (○) and absorption of ^{32}P (△). B: effect on the incorporation of ^{32}P into RNA (○), DNA (□) and phospholipids (△). (From Tata, 1968c.)

relative potencies of analogues of thyroxine and tri-iodothyronine were identical in 3-day-old larvae as in more mature tadpoles. In this context, there is an interesting distinction between the dose–response curves for hormonal effects on synthetic processes (nucleic acid, protein and phospholipid) and on catabolic phenomena or phosphate ion uptake (Fig. 10). The latter were more sensitive to low doses of thyroid hormone but exhibited a rapid "saturation" of the sites of action leading to these changes.

On the other hand, the stimulatory effect on synthetic functions had a lower threshold level of hormone but an increasing response was maintained over a wider range of doses. An important point to note is that a period of 2 days after exposure to hormone is necessary for most of the above responses (except for the drop in phosphate ion uptake) to be manifested and for this reason it is not possible to time accurately the onset of a competent response to the hormone.

Studies on hybridization of RNA formed by very early forms of *Xenopus* larvae demonstrated, as for hepatic RNA in late bullfrog tadpoles (Table II), that precocious induction of metamorphosis changed the pattern of distribution of hybridizable and non-hybridizable RNA. Table III shows that RNA extracted from whole embryos had lower hybridization efficiency after exposure to the hormone although the overall rate of RNA synthesis was much enhanced. Once again this is interpreted to mean that possibly small increases in messenger RNA synthesis are swamped by a massive stimulation of ribosomal RNA or RNA of low hybridization efficiency (Wyatt and Tata, 1968). This result therefore differs from findings of hormone-induced increases in hybridization efficiency of RNA from mammalian or avian tissues (Drews and Brawerman, 1967; O'Malley and McGuire, 1968).

TABLE III

HYBRIDIZATION OF WHOLE LARVAL RNA OBTAINED AFTER INDUCTION
OF METAMORPHOSIS IN EARLY STAGES OF *XENOPUS* LARVAE

Metamorphosis was induced in 3-week-old tadpoles of *Xenopus laevis* by immersion in tri-iodothyronine (5 μg/l). Control and treated tadpoles (50–80 per batch) were used for preparation of RNA 20 hr after the injection of 5 μc of [³H]uridine (from Wyatt and Tata, 1968).

Time after T_3 (days)	Specific radioactivity of RNA (counts/min/μg)	Hybridization plateau*	
		Counts/min	Percentage of input
0	87	22	4·8
3	345	68	3·8
5	629	82	2·5

* Hybridization carried out with 5·2 μg of RNA on cellulose nitrate filters coated with 32 μg of *Xenopus* DNA.

There was a qualitative difference in the nature of the phospholipids labelled with ³²P before and after the administration of tri-iodothyronine to 6 to 12-day-old larvae (Fig. 11). *Xenopus* larvae at this age still have a considerable amount of yolk which is likely to be a rich source of some phospholipids. For this reason there is a slow rate of phospholipid synthesis in very young, untreated larvae. Administration of the hormone causes a rapid breakdown of the yolk and enhances the overall rate of phospholipid synthesis. At the same time there was a shift in the pattern of ^{32}P-labelling of phospholipids to one which resembles that seen for membrane phospholipids. Although membranes were not isolated, it is presumed that this shift in pattern reflects a mobilization of phospholipids for the assembly of membranous structures. This is not too surprising because it is known that maturation and differentiation of cells is accompanied by greater structural complexity (Wischnitzer, 1966).

It is quite clear from the above that the *Xenopus* larva acquires a very early competence to exhibit biochemical and morphological responses to the metamorphic stimulus of thyroid hormones. It would be interesting to extend these studies to other amphibian species as well as to insects in order to verify if an early acquisition of competence for late developmental changes is of universal occurrence. A potentially useful application of establishing when competence is acquired is that it facilitates a developmental approach to the problem of hormone "receptors". It is not unreasonable to assume that presence or absence of the right "receptor" substances constitutes the difference between a responsive stage and a non-responsive one. The identification of the initial sites of the interaction between hormones and cell constituents that leads to a chain of events culminating in the physiological effects of hormones is of utmost importance for a fuller understanding of hormone action. In attacking this problem it is useful to recall that most growth and developmental hormones, especially ecdysone and thyroid hormones, are evolutionarily more primitive than some of the sophisticated biochemical mechanisms they are supposed to regulate (Barrington, 1964). For this reason alone, the type of biological regulation seen in hormonal control of metamorphosis is unlikely to arise from a direct interaction of the hormone with each of the individual regulators for the different biochemical mechanisms but rather represents a triggering off of a pre-programmed developmental change, in partially differentiated cells, irrespective of the nature of the programme. Whatever the outcome of future work on these lines, it would not be an exaggeration to say that past work on experimental induction of amphibian and insect metamorphosis has already formed an important basis for the biochemistry of late embryonic development.

SUMMARY

In insects and amphibia the dramatic biochemical and morphological changes that occur during metamorphosis are obligatorily dependent on hormones (ecdysone and thyroid hormones, respectively). Precocious induction with hormones of developmental changes in insect and amphibian larvae reveals a sequence of multiple changes in target cells. Among the earliest responses of the competent cells is a readjustment of permeability barriers to a variety of nutrients and precursors of macromolecules. However, the specificity of developmental changes is most likely to reside in the nature of new or additional species of RNA formed according to a pre-determined pattern. Work on thyroid hormone-induced metamorphosis in bullfrog tadpoles reveals that a relatively long time-interval elapses between an accelerated synthesis of RNA and the appearance of new proteins that characterize the metamorphic change in the liver. During this period there occurs a substantial turnover of ribosomes and their redistribution on membranes of the endoplasmic reticulum. There is also a tight coordination in the formation of new ribosomes (or polyribosomes) and the membranes to which they are attached. Most of these phenomena are common to many other developmental systems. The significance of this reorganization of the developing cell's protein synthesizing machinery may reside in a topographical segregation of precoded polyribosomes engaged in the synthesis of different classes of proteins.

Tissue regression is an integral part of metamorphosis and is as important as growth or differentiation. An analysis of RNA and protein synthesis in thyroid hormone-induced regression of tadpole tails in organ culture has shown that cell death during development requires the formation of new proteins just as it is necessary for those cells that are programmed for further growth and development.

Recent work from our laboratory on the time of acquisition of metamorphic competence has shown that this is acquired by the *Xenopus* larvae at a very early stage of embryonic development. Competence was assessed by biochemical responses to hormone administered at different developmental stages—that is, changes in rates of synthesis of nucleic acids, protein and phospholipid, as well as catabolic changes (water loss, regression) which are characteristic of normal metamorphosis in late embryonic forms.

REFERENCES

BARRINGTON, E. J. W. (1964). *Hormones and Evolution*. London: English Universities Press.

BEERMANN, W. (1963). *Am. Zoologist*, **3**, 23–32.

BERN, H. A., NICOLL, C. S., and STROHMAN, R. C. (1967). *Proc. Soc. exp. Biol. Med.*, **126**, 518–521.

CAMPBELL, P. N. (1965). *Prog. Biophys. molec. Biol.*, **15**, 3–38.
CLEVER, U. (1962). *Chromosoma*, **13**, 385–436.
COHEN, P. P. (1966). *Harvey Lect.*, **60**, 119–154.
COLEMAN, J. R. (1962). *Devl Biol.*, **5**, 232–251.
DREWS, J., and BRAWERMAN, G. (1967). *Science*, **156**, 1385–1386.
EATON, J. E., and FRIEDEN, E. (1968). *Gunma Symposia on Endocrinology*, **5**, 43–53.
EECKHOUT, Y. (1966). *Revue Quest. scient.*, **3**, 377–393.
ETKIN, W. (1964). In *Physiology of the Amphibia*, pp. 427–468, ed. Moore, J. A. New York: Academic Press.
ETKIN, W., and GONA, A. G. (1967). *J. exp. Zool.*, **165**, 249–258.
FINAMORE, F. J., and FRIEDEN, E. (1960). *J. biol. Chem.*, **235**, 1751–1755.
FRIEDEN, E. (1967). *Recent Prog. Horm. Res.*, **23**, 139–186.
GEORGIEV, G. P. (1966). *Prog. Nucl. Acid Res.*, **6**, 259–351.
HENSHAW, E. C. (1968). *J. molec. Biol.*, **36**, 401–411.
KARLSON, P. (1963). *Angew. Chem.*, **2**, 175–182.
KIM, K. H., and COHEN, P. P. (1966). *Proc. natn. Acad. Sci. U.S.A.*, **55**, 1251–1255.
KOLLROS, J. J. (1968). *Ciba Fdn Symp. Growth of the Nervous System*, pp. 179–192. London: Churchill.
LOCKWOOD, D. H., STOCKDALE, F. E., and TOPPER, Y. J. (1967). *Science*, **156**, 945–946.
McCONKEY, E. H., and HOPKINS, J. W. (1965). *J. molec. Biol.*, **14**, 257–270.
MOSER, H. (1950). *Rev. suisse Zool.*, **57**, suppl. 2, 1–144.
NAKAGAWA, H., and COHEN, P. P. (1967). *J. biol. Chem.*, **242**, 642–649.
NAKAGAWA, H., KIM, K. H., and COHEN, P. P. (1967). *J. biol. Chem.*, **242**, 635–641.
O'MALLEY, B. W., and McGUIRE, W. L. (1968). *Proc. natn. Acad. Sci. U.S.A.*, **60**, 1527–1534.
PERRY, R. P., and KELLY, D. E. (1968). *J. molec. Biol.*, **35**, 37–59.
SAMARINA, O. P., LUKANIDIN, E. M., MOLNAR, J., and GEORGIEV, G. P. (1968). *J. molec. Biol.*, **33**, 251–263.
SAUNDERS, J. W., JR. (1966). *Science*, **154**, 604–612.
SCHNEIDERMAN, H. A., and GILBERT, L. I. (1964). *Science*, **143**, 325–333.
SEKERIS, C. E. (1967). *Colloquium Ges. physiol. Chem.*, **18**, 126–151.
SHEARER, R. W., and McCARTHY, B. J. (1967). *Biochemistry, N.Y.*, **6**, 283–289.
TATA, J. R. (1965). *Nature, Lond.*, **207**, 378–381.
TATA, J. R. (1966). *Devl Biol.*, **13**, 77–94.
TATA, J. R. (1967a). *Biochem. J.*, **105**, 783–801.
TATA, J. R. (1967b). *Biochem. J.*, **104**, 1–16.
TATA, J. R. (1968a). *Biochim. biophys. Acta Library*, **11**, 222–235.
TATA, J. R. (1968b). *Nature, Lond.*, **219**, 331–337.
TATA, J. R. (1968c). *Devl Biol.*, **18**, 415–440.
WEBER, R. (1963). *Ciba Fdn Symp. Lysosomes*, pp. 282–300. London: Churchill.
WEBER, R. (1965). *Experientia*, **21**, 665–666.
WEBER, R. (1967). In *The Biochemistry of Animal Development*, vol. II, pp. 227–301, ed. Weber, R. New York: Academic Press.
WILLIAMS, C. M. (1961). *Science*, **133**, 1370.
WISCHNITZER, S. (1966). *Adv. Morphogenesis*, **5**, 131–179.
WYATT, G. R. (1962). In *Insect Physiology*, pp. 23–41, ed. Brookes, V. J. Corvallis: Oregon State University Press.
WYATT, G. R., and LINZEN, B. (1965). *Biochim. biophys. Acta*, **103**, 588–600.
WYATT, G. R., and TATA, J. R. (1968). *Biochem. J.*, **109**, 253–258.

DISCUSSION

Harris: I like your experiments and your interpretation of them very much, Dr Tata, because I like experiments which agree with my own ideas. This proves I am very broad minded! What you are saying about synthesis of the ribosomes fits in very well with what I was saying in my paper. If all the high molecular weight RNA made in the nucleus comes out into the cytoplasm together, then, in the cytoplasm, the average half-life of the RNA which carries the information and the average half-life of any structural RNA components, whatever they may be, must be the same. Any attempt to identify the RNA which carries the information by differences in the rate of its turnover is therefore likely to be a will-of-the-wisp. Since my experiments also suggest that, in the intact cell, you cannot re-programme pre-existing ribosomes, it follows that the only way you can change the programme is by eliminating the existing ribosomes and replacing them with new ones. And that is exactly what your experiments show.

Prasad: Apropos of RNA induced by hormone action, Dr S. J. Segal and his colleagues have suggested that oestrogen-induced uterine RNA mimics the action of oestrogen in stimulating uterine activity. Segal, Davidson and Wada (1965) showed that a uterine RNA fraction from the oestrogen-stimulated rat initiated morphological changes in the uterus characteristic of oestrogen action. Segal, Wada and Schuchner (1965) further showed that the same uterine fraction, synthesized under the regulatory influence of oestrogen, induced implantation of delayed blastocysts in a manner similar to implantation evoked by oestrogen. In their opinion the regulatory action of oestradiol on the uterus is primarily on the biosynthesis of RNA; after this primary effect the hormone is no longer required for the manifestation of oestrogenic action. How does this RNA act in stimulating the uterus: is this stimulation transient, for as long as the RNA is present, or can this RNA mimic oestrogen action in causing hypertrophy and hyperplasia of cells?

Secondly, Mausour (1968) fractionated the RNA synthesized in oestrogen-treated uteri and showed that nuclear RNA and soluble RNA, but not ribosomal RNA, stimulated the uterus of an ovariectomized mouse. He further showed that nuclear and soluble uterine RNA significantly increased the alkaline phosphatase content of the uterus in the ovariectomized mouse; he concluded that exogenous RNA enters the cell and in some way stimulates protein synthesis.

What is the explanation for these findings in terms of the specificity of stimulation of the uterus by oestrogen mediated by these types of RNA?

Tata: I have not myself done work of the type reported by Segal, or by Villee, so that I cannot comment on this.

Prasad: Villee and Fujii (1968) reported that RNA extracted from the seminal vesicles of testosterone-treated immature rats stimulated growth and protein synthesis when instilled into the lumen of the seminal vesicles of three-week-old rats. These results were interpreted as evidence that RNA was involved in mediating the stimulatory action of testosterone. Sherry and Nicoll (1967) showed that the action of prolactin on the crop sac of the pigeon is similarly mediated by RNA.

I am intrigued by the similarity in the action of different hormones reported by these different groups.

Tata: I accept their findings, but I am not so sure about the interpretation.

Bhargava: We recently had occasion to analyse much of the available data on the uptake of exogenous RNA and its biochemical and biological effects, and we were amazed to find that although there are numerous reports dealing with this, there are hardly any in which the RNA used has been characterized. Secondly, we find that the RNA taken up by the cells, particularly heterologous RNA, is degraded more rapidly than the RNA of the recipient cells (Shanmugam and Bhargava, 1966, 1969), which makes me suspicious of the specific biochemical and biological effects which have been reported for exogenous RNA, particularly heterologous RNA.

Salganik: I mentioned earlier (p. 61) that the induction of transcription in animal cells is manifested in the enhanced synthesis of messenger RNA, ribosomal and transfer RNA which provide for the synthesis of appropriate new proteins. The synthesis of ribosomal proteins is enhanced too. Moreover, we found that induction is accompanied by an increase in the number of ribosomes and endoplasmic reticulum membranes (Salganik *et al.*, 1967; Christolubova, 1967). The enhancement of transcription is therefore accompanied by the formation *de novo* of new sets of protein-synthesizing units to provide a new level of translation.

Changes of this kind are seen in rat liver cells 5 hours after the injection of hydrocortisone. One may suppose that the majority of ribosomes and other protein-synthesizing structures in animal cells are engaged by long-lived templates. To provide for the intensive synthesis of new proteins, new protein-synthesizing units must be formed. But when induced transcription (and translation) decrease as a consequence of the elimination of the inducer, the cells acquire their usual appearance as a result of a rather sudden disappearance of part of the ribosomes and endoplasmic reticulum membranes, which have probably become "unemployed" and superfluous. Thus 9–11 hours after hydrocortisone injection, when induced syntheses diminish and then cease as a result of elimination of the hormone, we could see the restitution of their normal ultrastructure in rat liver cells. Apparently, it is more profitable to assemble additional ribosomes and endoplasmic reticulum and disassemble them than to retain such unused structures, and saves a lot of material, energy and space in the cell.

From this point of view, stability of ribosomes and other protein-synthesizing structures is unprofitable and the breakdown of such structures must be provided for in the cell. Indeed, we recently saw that in the course of hydrocortisone induction the activity of RNase and other hydrolases in liver cells increases. It is tempting to suppose that disintegrating enzymes are a rather important part of regulation in differentiated cells.

Tata: I agree.

Burma: Is it not a fact that before tissue regression the hydrolytic enzymes are mostly kept in an inhibited state and this inhibition is withdrawn during regression? Ribonucleases are mostly in the inhibited state. Withdrawal of inhibition might account for the increase in amount of the hydrolytic enzymes.

Tata: This was the idea proposed earlier, that in regression lysosomes or enzymes are being activated that are already there in an inactive state (Weber, 1963). But our tadpole tail culture experiments show that this is unlikely. An uninterrupted synthesis of new protein is essential to induce regression although there is plenty of hydrolytic enzyme in the resting stage.

Fortier: The so-called "positive feedback" effect of thyroxine and tri-iodothyronine in the early developmental stage of the tadpole may be related to experimental evidence suggesting a stimulating effect of low circulating levels of the thyroid hormones on TSH secretion in mammals. This stimulating effect has been inferred from the reported enhancement of the goitrogenic effect of propylthiouracil by small doses of thyroxine, in the rat, whereas larger doses have the opposite effect and depress goitrogenesis (Sellers and Schönbaum, 1962, 1965). I wonder, incidentally, whether the term "positive feedback" is appropriate in this connexion, since the alleged stimulation of TSH secretion by small doses of thyroxine might be ascribed to a general protein anabolic effect of the hormone.

Tata: The term "positive feedback" is used to distinguish the situation from that of the classical negative feedback, but "rebound" may also be a suitable term.

Sarma: Do T3 and T4 have any effect before metamorphosis, during larval development? Earlier reports indicate that T4 exerts a growth-promoting effect on insect larvae (Srinivasan, Moudgal and Sarma, 1955).

Tata: The enhanced rates of synthesis of nucleic acid and protein occur at the first-stage tadpole; that is, between stages 36 and 41 the embryo becomes very responsive to thyroid hormones.

Butcher: Dr Tata, you spoke of early events not necessarily determining later events in hormone action. While it is obvious that a causal relationship is not implicit in an early-late effect, if there is not at least some priming or permissive relationship occurring early then the whole thing is nonsense.

Tata: I have tried to consider various ideas here. One is that of the unitarian school of thought on hormone action which holds that there is only one primary site of action for all hormones. There is no theoretical reason or experimental evidence to suggest that there is only one species of hormone recognition site in the cell. The point I made was that an early event may be totally irrelevant to later events. For example, the relatively immediate action of thyroxine on mitochondria, causing swelling, ion transport changes or even changes in amino acid incorporation, is an interaction unrelated to the chain of later events which result in the preferential synthesis of proteins (see Tata, 1967). It is quite likely that during evolution a variety of adaptive mechanisms have been acquired and that these may act in concert and mutually facilitate one another.

Butcher: We have never implied that fast changes in cyclic AMP levels prove or even indicate that it is involved in later effects of hormones, because we certainly don't believe this: one can only *eliminate* a role for cyclic AMP in this type of study. However, in those systems most thoroughly studied cyclic AMP mimics all the effects of a hormone, for example the effects of ACTH on the adrenal cortex.

However, unless one can prove the continued presence of the hormone, without interaction occurring until a given time, one must assume that the hormone is rapidly or continuously interacting with some component of the cell even though the expression of this interaction may not occur until much later.

Tata: The importance of phenomena associated with cyclic AMP is that we are looking at a primitive reflex to a change that may or may not result in the synthesis of new proteins later on. With some hormones, where we have both rapid metabolic and slow growth-promoting effects, changes in cyclic AMP levels alone may not explain all the actions. On the other hand for adrenaline, which as far as I know has virtually no growth-promoting or enzyme induction effects and has rapid physiological actions, the explanation of adenyl cyclase activation is quite adequate.

Butcher: On the contrary, catecholamines do have growth hormone-like effects, on salivary glands of the rat (Wells, 1967).

Paintal: We have been discussing the positive feedback role of thyrotropic hormone. I am very interested in this, because of the general role of positive feedback in biological systems. Professor Monod believes that positive feedback exists in biological systems, but is it proved that in such cases there is no negative feedback at some stage?

Monod: It has been proved in many organisms that phosphokinase is activated by AMP.

Prasad: Dr Tata has discussed the effect of hormones on diapausing insect larvae and on tadpoles. I would like to mention data obtained with S. Mohla and C. M. S. Dass on the effect of oestrogen on the "diapausing" mammalian embryo, the blastocyst. The blastocyst of the rat implants between days 5 and 6 post-coitum but implantation can be delayed by bilateral ovariectomy on day 3 post-coitum followed by daily treatment with 2–4 mg of progesterone. During this period of delayed implantation, development of the embryo is arrested in the blastocyst stage; such blastocysts can be activated and induced to implant by giving a small quantity of oestrogen. During the period of delayed implantation the blastocysts are in a state of "diapause" when metabolic activities are at a low ebb. We have studied the metabolic changes in the delayed blastocysts by instilling [³H]cytidine, [³H]phenylalanine and [³H]thymidine into the uterine lumen and autopsying the animals at specific time intervals after giving 1 μg of oestradiol.

Minimal RNA synthesis occurs in the blastocyst and uterus during delayed implantation. Protein synthesis appears to be minimal in the delayed uterus, whereas the delayed blastocyst shows no evidence of protein synthesis. Eighteen hours after oestrogen had been given the blastocyst showed enhanced RNA synthesis, markedly so in the cells of the inner cell mass and to a lesser degree in the trophoblast. Enhancement of protein synthesis in the blastocyst follows a similar pattern.

Our next interest was to determine the time-sequence of action of oestrogen on the blastocyst and uterus. If we expose delayed blastocysts *in utero* to 15 minutes of pulse-labelling with [³H]cytidine and calculate the percentage of labelled subepithelial stromal cells, we get a clear picture of the time-sequence of action of oestrogen. In the delayed rat, 6·1 per cent of the stromal cells are labelled (Table I); this increases gradually to 10·5 per cent at 3 hours, to 61·3 per cent (at 6 hours), and declines (27·5 per cent) at 18 hours. On the other hand, oestrogen optimally enhances RNA synthesis in the blastocyst within one hour and this is maintained up to 18 hours. Oestrogen treatment results in enhanced synthesis of DNA in the blastocyst by 42 hours, at a time when the uterus is as yet unresponsive; DNA synthesis is activated in the uterus at about 50 hours after oestrogen treatment.

TABLE I (PRASAD)

TIME-SEQUENCE OF ACTION OF OESTRADIOL ON THE INCORPORATION OF [³H]CYTIDINE IN THE UTERUS AFTER 15 MINUTES OF PULSE LABELLING

Treatment	Epithelial cells: degree of labelling	Subepithelial stromal cells		Blastocyst: degree of labelling
		Degree of labelling	Percentage of cells labelled	
Control (delay)	+N*	+N	6·1	+
Oestrogen† 1 hour	+N	+N	3·1	++
Oestrogen 3 hours	+N	+N	10·5	++
Oestrogen 6 hours	+N	++N	61·3	++
Oestrogen 12 hours	+N	++N	65·5	++
Oestrogen 18 hours	+N	+N	27·5	++

+ A few grains per cell.
++ Large number of grains per cell.
* N, label predominantly in the nucleus.
† 1 μg of oestradiol-17β administered subcutaneously.
Delayed implantation was produced by bilateral ovariectomy on day 3 post-coitum and administration of 4 mg progesterone/day. Five μC of [³H]cytidine were instilled into the uterine lumen 15 minutes before autopsy on day 9 of delayed implantation. The blastocysts and uterus were processed for autoradiography, using NTB³ nuclear track emulsion and 15 days' exposure.

These studies show that oestradiol has a differential time-sequence of action on the uterus and blastocyst. The blastocyst which is lying free in the uterine lumen is stimulated much earlier than the activation of similar processes in the uterus. It is likely that oestrogen may be acting directly on the blastocyst and stimulating its synthetic activities.

Harris: I wonder whether I can draw attention to an experiment on a plant system which is relevant to a good deal of this discussion? This is an experiment with the unicellular alga *Acetabularia* (Spencer, 1968). The nucleus of this alga can be readily removed, and, under certain conditions, the enucleate cell will regenerate a perfectly normal cap, after this has been amputated. The cap is a distinctive feature of *Acetabularia*. It has been shown that if you give the plant hormone, kinetin, to *Acetabularia* cells they will form their caps prematurely (Zetsche, 1963). Spencer showed that if you take out the nucleus and then give kinetin, the enucleate cell still forms its cap prematurely. We know from the classical experiments of Hämmerling (for a review see Hämmerling, 1963) that the morphology of the cap is determined by nuclear genes, so that we are not dealing with chloroplast control. So here, it seems to me, is one case in which there is decisive evidence about where the hormone is acting. It is quite clear that, whatever the hormone may be doing to nuclear RNA in the nucleated cell, whether it is increasing or decreasing RNA synthesis, the specific phenotypic effect of the hormone, that is, the acceleration of the growth and morphological development of the plant, takes place even when the nucleus is not there. This is the only case that I know of in which there has been a decisive demonstration of this kind.

REFERENCES

CHRISTOLUBOVA, N. B. (1967). *Genetika*, No. 11, 47.
HÄMMERLING, J. (1963). *A. Rev. Pl. Physiol.*, **14**, 65.
MAUSOUR, A. M. (1968). *Acta endocr., Copenh.*, **57**, 465–472.
SALGANIK, R. I., CHRISTOLUBOVA, N. B., KIKNADZE, I. I., MOROSOVA, T. M., GRYASNOVA, I. M., and VALEEVA, F. S. (1967). In *Structure and Functions of Cell Nuclei*, p. 20. Moscow: Nauka.
SEGAL, S. J., DAVIDSON, O. W., and WADA, K. (1965). *Proc. natn. Acad. Sci. U.S.A.*, **54**, 782–787.
SEGAL, S. J., WADA, K., and SCHUCHNER, E. (1965). *47th Annual Meeting Endocrine Soc.*, p. 26, Abstract No. 12.
SELLERS, E., and SCHÖNBAUM, E. (1962). *Acta endocr., Copenh.*, **40**, 39–50.
SELLERS, E., and SCHÖNBAUM, E. (1965). *Acta endocr., Copenh.*, **49**, 319–330.
SHANMUGAM, G., and BHARGAVA, P. M. (1966). *Biochem. J.*, **99**, 297.
SHANMUGAM, G., and BHARGAVA, P. M. (1969). *Indian J. Biochem.*, **6**, 64–70.
SHERRY, W. E., and NICOLL, C. S. (1967). *Proc. Soc. exp. Biol. Med.*, **126**, 824–829.
SPENCER, T. (1968). *Nature, Lond.*, **217**, 62.
SRINIVASAN, V., MOUDGAL, N. R., and SARMA, P. S. (1955). *Science*, **122**, 644.
TATA, J. R. (1967). *Biochem. J.*, **104**, 1–16.
VILLEE, C. A., and FUJII, T. (1968). In *Endocrinology*, pp. 125–130, ed. Gual, C. Amsterdam: Excerpta Medica Foundation. International Congress Series No. 157.
WEBER, G. (1963). *Ciba Fdn Symp. Lysosomes*, pp. 282–300. London: Churchill.
WELLS, H. (1967) *Am. J. Physiol.*, **212**, 1293–1296.
ZETSCHE, K. (1963). *Planta*, **59**, 624.

INTERCELLULAR CONTROL OF INTRACELLULAR METABOLIC ACTIVITY

P. M. Bhargava

Regional Research Laboratory, Hyderabad, India

INTERCELLULAR material and cell contact are two of the major parameters characterizing intercellular organization in higher organisms. It seems reasonable to assume that they influence the biochemical properties and, indirectly, the functions expressed in the tissue, of the component cells. I should like to present an experimental approach towards understanding what these properties might be and to describe some results obtained by following this approach. I shall then briefly discuss possible implications of intercellular controls in two phenomena, ageing and carcinogenesis.

EXPERIMENTAL APPROACH

Perfusion of a tissue largely removes blood and circulating humoral factors. Dispersion of the perfused tissue to a single cell suspension results in removal of the intercellular material and lessening of the cell contact. Provided that the cell suspensions are prepared by a technique that causes minimal non-specific damage to the cells, any differences in the biochemical behaviour of the cell suspensions and the tissue slices would be likely to be due to the removal of one or the other, or both, of the above two influences present in the organized tissue. That this is so could be confirmed by reaggregating primary cell suspensions *in vitro* and showing that these changes are reversed. There is now considerable evidence to suggest that during such reaggregation of tissue cells, intercellular material is synthesized (*inter alia*, Overton, 1968; Glasier, Richmond and Todd, 1968) and, of course, cell contact reestablished. Whether or not changes observed on dispersion of the tissue to a single cell suspension are due to the lessening of cell contact can be found by determining, under appropriate conditions, whether or not these changes are dependent on cell concentration in a shaken cell suspension, in which the extent and magnitude of cell contact will depend on the concentration of the cells.

Thus the experimental approach involves: (*a*) development of a suitable method for preparing single cell suspensions; (b) a comparative study of the biochemical properties of the cell suspension, using perfused slices as a control; (c) reaggregation of the cells to determine whether the biochemical changes observed on dispersion of the tissue are reversed; and (d) a study of the effect of cell concentration on the biochemical changes observed on dispersion of the tissue.

I shall describe experiments done in our laboratory following this approach which indicate that intercellular material and cell contact may be involved in control of some of the permeability properties of the organized cells.

POSSIBLE REGULATION OF PERMEABILITY BY INTERCELLULAR MATERIAL AND CELL CONTACT

Preparation of liver cell suspensions

Our work has been largely confined to rat liver cell suspensions. These were prepared according to the method of Jacob and Bhargava (1962) in which the tissue is perfused with calcium-free Locke's solution containing citrate and then dispersed in a tissue disperser consisting of a conical rubber pestle and a hard glass tube. The final suspension consists predominantly (more than 95 per cent) of parenchymal cells with very little cell debris. The cells look intact under a phase contrast microscope, respire significantly, synthesize protein and RNA, and transport chylomicrons, fatty acids and free and esterified cholesterol (Iype and Bhargava, 1965; Hayek and Tipton, 1966; Bhargava and Bhargava, 1962; Friedman and Epstein, 1967; Jacob and Bhargava, 1965; Higgins, 1967; Higgins and Green, 1966, 1967; R. Reddy, G. Shanmugam, M. A. Siddiqi, L. F. Hussain and P. M. Bhargava, unpublished results). On average, 90 per cent of the parenchymal cells of the tissue are recovered in the cell suspension (Iype and Bhargava, 1963; Iype, Bhargava and Tasker, 1965). The method uses no enzyme, disadvantages of the use of which have been pointed out elsewhere (Bhargava, 1968). The parenchymal cells obtained in suspension retain the ability to recognize their own kind *in vivo* and thus go specifically to the liver when injected intravenously into another rat of the same inbred strain (Bhargava and Bhargava, 1968).

Significant biochemical differences between liver cell suspensions and perfused slices and the role of intercellular material

We have observed the following differences between the liver cell suspensions and the perfused slices; the latter were obtained from the same animal from which the cell suspensions were prepared.

TABLE I

TRANSPORT OF ^{14}C-LABELLED NUCLEIC ACID BASES INTO RAT LIVER SLICES AND PARENCHYMAL
CELL SUSPENSIONS AT $37°$ C

Base and its derivatives in the acid-soluble fraction
(counts/min/mg slices or cell suspension protein)

Time of incubation (hours)	[14]C]Orotic acid		[14]C]Adenine		[14]C]Uracil	
	Slices	Cell suspension	Slices	Cell suspension	Slices	Cell suspension
0	3	3	57	2	16	2
1	255	4	880	9	171	2
2	700*	4*	700	4	—	3

* 3 hours.

The concentration of the radioactive precursors in the incubation medium (Ca^{++}-free Krebs-Ringer phosphate buffer) was 8×10^4 counts/min ($3\cdot64$ μmole)/ml for orotic acid; 1×10^6 counts/min ($15\cdot8$ μmole)/ml for adenine; and $1\cdot4 \times 10^6$ counts/min ($2\cdot75$ μmole)/ml for uracil. (Renuka Naidu and P. M. Bhargava, unpublished results.)

(1) The cells in suspension, unlike the slices, cannot transport orotic acid, adenine and uracil (Table I). Thus orotic acid is not incorporated at all into the RNA of the cell suspensions, unlike that of the tissue slices (Jacob and Bhargava, 1965), and the cell suspensions, unlike the slices, cannot catabolize uracil (Jacob and Bhargava, 1964).

(2) The cells, unlike the slices, can take up ribonuclease. The ribonuclease taken up initially degrades the ribosomal RNA in such a way that the overall structural integrity of the ribosomes is not affected (Kumar and Bhargava, 1969). Thus the 40 and 60s ribosomal subunits isolated from the cells in suspension incubated with ribonuclease give the normal sedimentation pattern in a sucrose density gradient, although their RNA is degraded (Fig. 1).

(3) The cells, unlike the slices, can also take up macromolecular, homologous RNA or *E. coli* RNA (Fig. 2; Shanmugam and Bhargava, 1966, 1969) and highly polymerized *E. coli* DNA (Fig. 3).

(4) Actinomycin D is transported into the cells in suspension at a much faster rate than into the slices (Shanmugam and Bhargava, 1968).

(5) The cells in suspension, like the slices, can synthesize albumin, but unlike the tissue, appear unable to secrete it into the medium (Table II).

TABLE II

SECRETION OF SERUM ALBUMIN BY RAT LIVER PARENCHYMAL CELL SUSPENSIONS

Time of incubation (hours)	Specific activity of albumin (counts/min/mg)	
	Cells	Medium
0	66	77
3	311	85

The cells were incubated in Ca^{++}-free Krebs-Ringer phosphate buffer with a mixture of ^{14}C-labelled amino acids ($0\cdot04$ μmoles (6×10^4 counts/min)/ml) and the albumin fraction was isolated by alcohol extraction of the trichloracetic acid precipitate of the cells. Albumin was estimated by precipitation with antisera in the equivalence zone. (Layeek F. Hussain and P. M. Bhargava, unpublished results.)

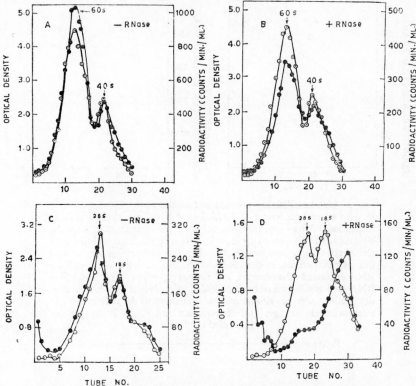

FIG. 1. Sucrose density gradient patterns of ribosomes (A and B) and RNA (C and D) from ^{32}P-labelled rat liver parenchymal cell suspensions incubated for 20 minutes at 37° C in Ca^{++}-free Krebs-Ringer phosphate buffer in the absence (A and C) and in the presence (B and D) of ribonuclease (1 µg/ml). ●, radioactivity (due to ribosomes or RNA from the incubated cells); ○, optical density (due to unlabelled carrier ribosomes or RNA from rat liver). (Kumar and Bhargava, 1969.)

(6) The cells in suspension do not leak out any free amino acid during preparation (M. A. Siddiqi and P. M. Bhargava, unpublished results), but they leak out more than half of their free nucleotides (Jacob, 1964) (Table III).

TABLE III

AVERAGE FREE AMINO ACID AND FREE NUCLEOTIDE CONTENT OF RAT
LIVER SLICES AND PARENCHYMAL CELL SUSPENSIONS

Tissue preparation	Free amino acids* (µg)	Free nucleotides† (mg)
Slices	3·54	1·00
Cell suspension	5·05	0·41

* Per mg dry weight.

† Contained in 1 g wet weight of slices, or in parenchymal cell suspension derived from 1 g wet weight of the tissue; in the latter case, a correction has been made for the cells broken during the preparation of the cell suspension. (Jacob, 1964; M. A. Siddiqi and P. M. Bhargava, unpublished results.)

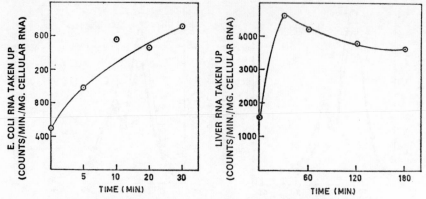

FIG. 2. Uptake of highly purified and undegraded, ^{32}P-labelled, total
E. coli (●) and rat liver (○) RNA by rat liver parenchymal cells in
suspension. After incubation for the specified time at 28°c with the
[32]P-labelled RNA [49–56 μg (24 000–48 000 counts/min)/ml] in Ca[++]-
free Krebs-Ringer phosphate buffer, the tissue preparation was washed
free of radioactivity in the cold and the radioactivity in the cellular
RNA estimated (Shanmugam and Bhargava, 1966). The ^{32}P-labelled
RNA used was shown to be undegraded by analytical and sucrose density gradient ultracentrifugation, and contained all the radioactivity in
internucleotide linkage. (G. Shanmugam and P. M. Bhargava, unpublished results.)

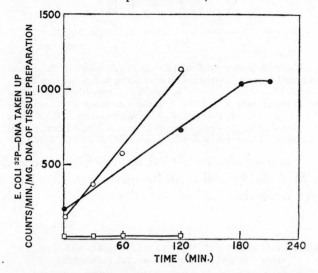

FIG. 3. Uptake of highly purified and polymerized, ^{32}P-labelled E. coli
DNA by rat liver parenchymal cells in suspension (○, Expt 1; ●, Expt
2) and rat liver slices (□, Expt 1). After incubation for the specified
time at 37°c with the ^{32}P-labelled DNA [213 μg (3320 counts/min)/ml]
in Ca[++]-free Krebs-Ringer phosphate buffer, the tissue preparation was
washed free of radioactivity in the cold and the radioactivity in the
cellular DNA estimated. (G. Shanmugam and P. M. Bhargava, unpublished results.)

These observations suggest that dispersion of a tissue leads to specific permeability changes of all the four possible types: substances which normally can get into the organized tissue do not get into the cells in suspension; substances which normally do not get into the organized tissue can get into the cells in suspension; substances which are normally secreted from the cells in the organized tissue cannot be secreted by the

TABLE IV

SUMMARY OF MAJOR PERMEABILITY DIFFERENCES OBSERVED BETWEEN RAT LIVER SLICES AND PARENCHYMAL CELL SUSPENSIONS

Material	Slices	Cell suspension
Orotic acid, adenine, uracil	Go in	Do not go in
Ribonuclease	Does not go in	Goes in
Macromolecular RNA (liver, E. coli)	Does not go in	Goes in
Macromolecular DNA	Does not go in	Goes in
Actinomycin D	Goes in	Goes in at a much faster rate
Serum albumin	Comes out	Does not come out
Free nucleotides	Do not come out	Come out partially

(Modified from Bhargava, 1968.)

cells in suspension; and substances which are normally retained within the cells in the organized tissue can no longer be retained by the cells in suspension. These permeability changes, summarized in Table IV, do not appear to have any relationship to the molecular weight of the substances involved. None of these changes is dependent on the concentration of the cells, and they are mostly qualitative. This suggests that one of the functions of intercellular material may be to regulate some of the permeability properties of the organizing cells, presumably by interaction with the membrane.

Role of cell contact

We have found that as the concentration of liver parenchymal cells increases in a cell suspension, the rate of transport of amino acids, for example of ^{14}C-L-histidine (Fig. 4), into the cell decreases; their incorporation into protein also decreases correspondingly. Within the range of cell concentrations used, respiration was not affected. There was no evidence of accumulation of toxic products in the medium, and the specific activity of histidine outside the cell seemed to stay constant for all the cell concentrations used; further, the amount of [^{14}C]histidine taken up by the cells, even at the highest cell concentration used, represented a very small fraction of the [^{14}C]histidine present in the medium. A similar dependence of transport of amino acids (or their incorporation into protein), on cell concentration, has been observed by other workers in the case of spermatozoa (Bhargava, Bishop and Work, 1959), liver cells (Friedman and Epstein, 1967), ascites tumour cells (M. A. Siddiqi and P. M. Bhargava, unpublished work), and *E. coli* (Rao and Bhargava, 1965). The transport of phosphate into normal and tumour cells has also been shown to depend on cell concentration (Blade and Harel, 1965; Blade, Harel and Hanania, 1966).

These observations suggest that when mammalian cells establish contact with each other there may be a reduction in their ability to transport certain essential nutrients. Loewenstein and Kanno (1964) have also shown that the permeability of epithelial cells is altered at the cell junctions.

Reaggregation of liver cells

In spite of many attempts, it has not been possible for us to reaggregate adult rat liver parenchymal cells in suspension *in vitro* under conditions which do not damage the cells. Reaggregation of these cells has however been reported to occur in 24 hours on injecting the cell suspension in the intraperitoneal cavity of another rat (Laws and Stickland, 1961). We have confirmed this observation, and using labelled cells, have shown that the aggregation in the intraperitoneal cavity is quantitative. It should

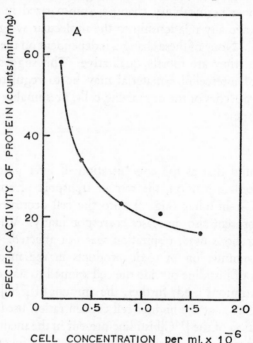

FIG. 4a

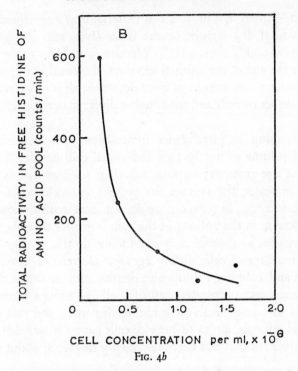

FIG. 4. Incorporation of [^{14}C]histidine into the protein (A) and free amino acid pool (B) of rat liver parenchymal cells suspended at varying cell concentration in 6 ml of Ca^{++}-free Krebs-Ringer phosphate buffer and incubated at 37°C for 3 hours with 0·0056 μmole (22 000 counts/min)/ml of the precursor. Histidine was isolated from the free amino acid pool by paper chromatography. (M. A. Siddiqi and P. M. Bhargava, unpublished results.)

therefore be interesting to investigate the permeability of these aggregates to nucleic acid bases, ribonuclease, RNA, DNA and actinomycin D, and the secretion of serum albumin and leakage of nucleotides from the cells in the aggregates. Until this is done, the conclusion that one of the functions of intercellular material is to exercise control over some of the permeability properties of the cell membrane, and that the permeability changes observed on dispersion of the tissue to a single cell suspension are not non-specific changes due to damage to the cells but may be related to the removal of intercellular material, must be considered at most tentative.

IMPLICATIONS FOR AGEING

These results suggest that intercellular material may have some role in regulating intracellular properties through control of permeability. We have recently shown that in liver this material may be as much as 40 per cent of the total dry weight of the tissue (Iype and Bhargava, 1963; Iype, Bhargava and Tasker, 1965). We also found that as the weight of the liver (or the age of the animal) increases, the number of parenchymal cells per gramme wet weight of liver decreases; thus older animals have a smaller number of cells and more non-cellular material per unit weight of liver.

We have, using an interference microscope, also measured the dry weight and volume of nearly 1500 individual cells derived from rats of five different age groups (3–24 months). Fig. 5 shows that as the age of the animal increases, the average dry weight and volume of the paren-chymal cells increase; however, they do not increase in the *same* propor-tion; the increase in the volume of the cells is greater than the increase in their dry weight, so that there is a reduction in the average density of the liver parenchymal cells during ageing. Correlation coefficients be-tween mass and volume, volume and density, and mass and density, for each age group taken separately, and for all age groups taken together, also showed that larger cells have higher dry mass, and that except for 3-month-old animals, larger or heavier cells have a lower density (Table V). It is significant that rats become sexually mature at about $2\frac{1}{2}$ months

TABLE V

CORRELATIONS BETWEEN DRY MASS, VOLUME AND DENSITY OF RAT LIVER PARENCHYMAL CELLS
FROM RATS OF DIFFERENT AGE GROUPS

Age (months)	Number of cells	Parameters correlated*	Correlation coefficient†	t value‡	Whether significant or not	Significance level (per cent)
3	401	M and V	0·53	12·60	Yes	0·1
		V and D	0·04	0·72	No	—
		M and D	0·08	0·14	No	—
6	359	M and V	0·75	21·60	Yes	0·1
		V and D	0·59	13·60	Yes	0·1
		M and D§	0·25	3·91	Yes	0·1
9 and 10	343	M and V	0·81	25·20	Yes	0·1
		V and D	0·59	13·35	Yes	0·1
		M and D	0·17	3·08	Yes	1·0
12	298	M and V	0·90	35·60	Yes	0·1
		V and D	0·68	16·10	Yes	0·1
		M and D	0·45	8·67	Yes	0·1
24	56	M and V	0·77	8·76	Yes	0·1
		V and D	0·80	9·68	Yes	0·1
		M and D	0·57	5·10	Yes	0·1

* M, dry mass; V, volume; D, density.
† Calculated on an IBM 1620 computer.
‡ $t > 120 = 3\cdot29$ (at 0·1 per cent); 2·58 (at 1 per cent). $t_{40} = 3\cdot55$ (at 0·1 per cent).
§ For 227 cells.

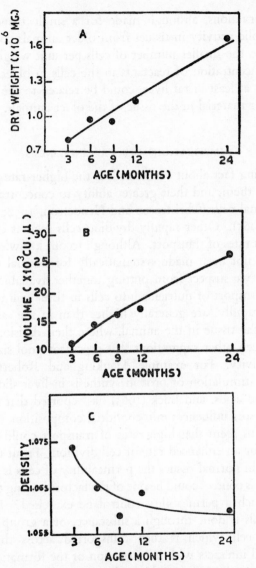

FIG. 5. Variation with age in the average dry weight (A), volume (B), and density (C), of rat liver parenchymal cells. Dry weight was determined by the extinction transfer method using an American Optical Co. Baker interference microscope. (Layeek F. Hussain, B. Gopinath and P. M. Bhargava, unpublished results.)

and that ageing in this animal probably begins some time between 3 and 6 months.

These observations, although made for a single tissue, suggest that lower metabolic activity in tissues from older animals could be, at least in part, due to the smaller number of cells per unit weight of tissue and the lower concentration of reactants in the cells in older animals. This phenomenon, at least in rat liver, could be related to the larger amount of intercellular material in the tissue of the older animals.

IMPLICATIONS FOR CARCINOGENESIS

An interesting fact about cancer cells is the higher rate of transport of nutrients into them, and their greater ability to concentrate them, compared to normal cells (Christensen and Henderson, 1952; Johnstone and Scholefield, 1965). Other rapidly dividing cells such as embryonic cells also have high rates of transport. Although to our knowledge such comparisons have not been made systematically for normal cells grown in tissue culture, we suspect from putting together available evidence that the rates of transport of nutrients into cells in tissue culture, where they are dividing rapidly, are generally higher than in the same cells when organized in the tissue in the animal, where their division is controlled. There have been other suggestions that precursor pool size can regulate metabolic activity. For example, Hanking and Roberts (1965) have demonstrated stimulation of protein synthesis in liver slices by elevated levels of amino acids, and Luck (1965) has reported that the size of the pool of precursors influences mitochondrial composition.

It would thus seem that high rates of transport could be one of the prerequisites for an enhanced rate of cell division. From this, one could postulate that in normal tissues the permeability of cells is regulated, and that a loss of this control could be one of the factors leading to malignancy.

How can such a permeability control be exercised? It is logical to assume that this is done through a substance, or a group of substances, made in the cell which is then transported across the cytoplasmic membrane and interacts with it. Cessation of the formation of or alterations in these substances could lead to a malignant transformation. Suggestions that the cell surface (which has been considered as an important regulator of enzymic processes within the cell; Best, 1960) is involved in carcinogenesis have been made repeatedly, for example recently by Haughton and Amos (1968).

Since the ability to make such substances must be inherited from cell to cell, it is most likely that, if they exist, the primary information for their synthesis resides in nuclei or mitochondria.

I want to propose here the following model (Fig. 6) of chemical carcinogenesis in which a central role of mitochondrial DNA is envisaged, based on these considerations. The cell may contain two types of mitochondria with overlapping but not identical information; only type 1 is considered capable of producing the postulated substance(s) (RNA or protein?) which is eventually transported across the cytoplasmic membrane and, as a component of the intercellular material, has a role to play

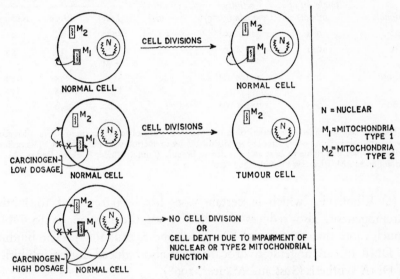

FIG. 6. A new scheme of chemical carcinogenesis.

in regulating permeability. The carcinogenic agent could result in the eventual deletion of type 1 mitochondria from the cell. The chemical carcinogens, for example, might preferably go to type 1 mitochondria, bind to their DNA, and prevent their replication, which, over several cell divisions, would lead to their disappearance from the cell.

This model is supported by the following observations.

(1) All carcinogens which have been studied for their binding to nucleic acids have been shown to bind to DNA of the susceptible tissue(s) of susceptible animals, and, in most cases, much less to non-susceptible tissues (Brookes, 1966; Miller and Miller, 1966; Farber, 1968). For example, urethane (0·5–1·0 mg/g body weight) gives tumours of lung, and at a lower incidence, of kidney and liver, with a 100 per cent frequency

in certain strains of mice, and with lower frequency in rats. Table VI shows its greater binding to DNA of tissues from mouse, and among the tissues studied, preferentially to lung. The binding of the carcinogen to DNA was found to increase dramatically from sub-carcinogenic to carcinogenic dosages.

TABLE VI

BINDING OF [^{14}C]URETHANE TO DNA OF RAT AND MOUSE TISSUES *in vivo*

Animal	Urethane Dose (mg/g wt of animal)	Urethane Specific activity (counts/min/mg)	Time (hours)	Specific activity of DNA (counts/min/mg) Lung	Liver	Kidney
Rat	0·1	5·7 × 10^5	24	222	22	41
Rat	0·1	5·7 × 10^5	24	530	15	55
			48	94	85	39
Mouse	0·1	1·1 × 10^6	24	9520	75	629
			48	1360	125	340
Mouse	0·5	7·2 × 10^4	12	2160	140	270

[^{14}C]urethane was injected intraperitoneally and DNA was isolated from the tissue (pooled from several animals in some cases) after the specified time by extraction with cold phenol. The radioactivity in DNA was stable to heat and covalently bound. (Sumati Bhide, E. P. K. Reddy, M. A. Siddiqi and P. M. Bhargava; unpublished results.)

(2) Riboflavin, which in certain cases has been reported to inhibit carcinogenesis, also reduces the binding of the carcinogen to DNA (much more than to protein) (Dingman and Sporn, 1967). The binding to DNA of carcinogenic hydrocarbons is also reduced during inhibition of DNA synthesis (Suss and Maurer, 1968).

(3) Carcinogens can bind to DNA *in vitro*, enzymically or non-enzymically, with little specificity (Malkin and Zahalsky, 1966; Gower and Sims, 1968; King and Phillips, 1968; Enomoto *et al.*, 1968). The observed specificity with respect to species, tissue and intracellular site in the tissue could be due to differences in permeability to the carcinogen.

(4) Co-carcinogenesis, for example by croton oil (Frei and Harsono, 1967), has been suggested to be due to an increase in the permeability of the cells to the carcinogen in the presence of the co-carcinogen.

(5) Urethane has been shown to go preferentially to mitochondria when given in carcinogenic doses (Berenblum *et al.*, 1958; S. Bhide, E. P. K. Reddy, M. A. Siddiqi and P. M. Bhargava, unpublished results). The model predicts that the primary, essential lesion in chemical carcinogenesis is in mitochondrial DNA, where it would lead to a deletion of type 1 mitochondria—an irreversible change. It does not preclude

reversible effects of the carcinogen on DNA, like the single-strand scissions reported to occur in DNA with 4-hydroxylaminoquinoline oxide (Sugimura, Otake and Matsushima, 1968).

(6) The model predicts that malignant tumours would be deficient in type 1 mitochondria and therefore the carcinogen causing them would not bind to their DNA when administered in minimal carcinogenic dosage. We have observed that the binding at 24 hours of [^{14}C]urethane (0·1 mg/g weight of the animal) to the DNA of urethane-induced lung tumours is less than 5 per cent of the binding to the DNA of normal lung tissue in Swiss mice. Cells transformed *in vitro* with carcinogenic hydrocarbons have been shown to be resistant to higher concentrations of the carcinogen (Borenfreund *et al.*, 1966).

(7) Some carcinogens exhibit acute toxicity at high dosage, the site of action being the same tissue as that in which tumours are caused. Thus a single dose of aflatoxin causes fatal injury to liver in susceptible animals; when the same dose is spread over a longer period, liver tumours are obtained. It can be postulated that at low doses aflatoxin binds the DNA of only type 1 mitochondria; at higher concentrations, it may bind also to the DNA of type 2 mitochondria and nuclei. This is amenable to experimental verification.

(8) The number of mitochondria in tumour cells is well known to be less than that in the normal cells.

(9) Attardi and Attardi (1968) have recently shown that mitochondria synthesize a RNA which is transported on to the cell membrane.

(10) Mitochondrial populations and their DNA have been shown to be heterogeneous (Matile and Bahr, 1968; Vinograd, 1968).

(11) The cell surface of tumour cells is known to differ from that of the normal cell type from which the tumour is derived, in physical (e.g. charge), chemical and immunological properties. For example, Emmelot, Visser and Benedetti (1968) have shown that globular knobs (50–60 Å) present on liver parenchymal cell membranes are absent from cells of certain hepatomas, and Hakomori, Teather and Andrews (1968) have reported that haematosides, which in normal cells are masked by a protein cover removable by trypsin, are unmasked in cells transformed to malignancy.

(12) There has been a host of reports, although most of them are subject to some criticism, in which the uptake of normal cell DNA by tumour cells has been shown to result in a loss or reduction of malignancy (for review, see Bhargava and Shanmugam, 1970). Reports that DNA of tumour cells transforms normal cells into malignant cells are however not

172 P. M. BHARGAVA

in conformity with these. Material from adult connective tissue has been reported to inhibit the growth of several human tumours *in vitro* (Parshley, 1965).

SUMMARY

Two of the parameters which characterize intercellular organization are intercellular material and cell contact. In liver, intercellular material may be as much as 40 per cent of the total dry weight of the tissue. Dispersion of a tissue to a single cell suspension largely reduces these two influences. A comparison of the properties of the organized tissue with those of single cell suspensions, prepared from the tissue under appropriate conditions, could therefore give a clue to the controls, if any, exercised by the presence of intercellular material and cell contact. Evidence has been obtained which suggests that the organization of parenchymal cells in liver results in regulation of transport of certain materials, both from the outside to the inside of the cell and from the inside of the cell to the outside. It is postulated that such a permeability control may be involved in regulation of certain specific—even tissue-specific—functions in higher organisms. Some aspects of the possible role of these controls in two processes, ageing and carcinogenesis, are discussed. Evidence is presented which suggests that the lower metabolic activity in tissues from older animals could be partly due to the smaller number of cells per unit weight of tissue and lower concentration of reactants in the cells of older animals; these could be related to the larger amount of intercellular material in the older animals. A model of carcinogenesis is presented in which a central role is envisaged for mitochondrial DNA.

Acknowledgement

The work from this laboratory mentioned in this paper was carried out by Dr P. T. Iype, Dr S. T. Jacob, Dr (Mrs) Kamalini Bhargava, Dr G. Shanmugam, Mr R. Reddy, Mr V. Kumar, Mr M. A. Siddiqi, Mr E. P. K. Reddy, Miss Layeek Hussain, Mr B. Gopinath, Mrs Shyamala Rao, Mrs Renuka Naidu and Mr K. K. Gudibanda, all of this laboratory, and Dr (Mrs) Sumati Bhide of the Indian Cancer Research Centre, Bombay.

REFERENCES

ATTARDI, G., and ATTARDI, B. (1968). *Proc. natn. Acad. Sci. U.S.A.*, **61**, 261.
BERENBLUM, I., HARAN-GHERA, N., WINNICK, R., and WINNICK, T. (1958). *Cancer Res.*, **18**, 181.
BEST, J. B. (1960). *Int. Rev. Cytol.*, **9**, 129.
BHARGAVA, K., and BHARGAVA, P. M. (1962). *Life Sciences*, **1**, 477.
BHARGAVA, K., and BHARGAVA, P. M. (1968). *Expl Cell Res.*, **50**, 515.
BHARGAVA, P. M. (1968). *Sci. Cult., Calcutta*, **34** (Supplement), 105.
BHARGAVA, P. M., BISHOP, M. W. H., and WORK, T. S. (1959). *Biochem. J.*, **73**, 247.
BHARGAVA, P. M., and SHANMUGAM, G. (1970). *Prog. Nucl. Acid Res.*, in preparation.

BLADE, E., and HAREL, L. (1965). *C. r. hebd. Séanc. Acad. Sci.*, Paris, **260**, 6464.
BLADE, E., HAREL, L., and HANANIA, N. (1966). *Expl Cell Res.*, **41**, 473.
BORENFREUND, E., KRIM, M., SANDERS, F. K., STERNBERG, S. S., and BENDICH, A. (1966). *Proc. natn. Acad. Sci. U.S.A.*, **56**, 672.
BROOKES, P. (1966). *Cancer Res.*, **26**, 1994.
CHRISTENSEN, H. N., and HENDERSON, M. E. (1952). *Cancer Res.*, **12**, 229.
DINGMAN, C. W., and SPORN, M. B. (1967). *Cancer Res.*, **27**, 933.
EMMELOT, P., VISSER, A., and BENEDETTI, E. L. (1968). *Biochim. biophys. Acta*, **150**, 364.
ENOMOTO, M., SATO, K., MILLER, E. C., and MILLER, J. A. (1968). *Life Sciences*, **7**, 1025.
FARBER, E. (1968). *Cancer Res.*, **28**, 1859.
FREI, J. V., and HARSONO, T. (1967). *Cancer Res.*, **27**, 1482.
FRIEDMAN, T., and EPSTEIN, C. J. (1967). *Biochim. biophys. Acta*, **138**, 622.
GLASIER, R. M., RICHMOND, J. E., and TODD, P. (1968). *Expl Cell Res.*, **52**, 43, 71.
GOWER, P. L., and SIMS, P. (1968). *Biochem. J.*, **110**, 159.
HAKOMORI, S. I., TEATHER, C., and ANDREWS, H. (1968). *Biochem. biophys. Res. Commun.*, **33**, 563.
HANKING, B. M., and ROBERTS, S. (1965). *Biochim. biophys. Acta*, **104**, 427.
HAUGHTON, G., and AMOS, D. B. (1968). *Cancer Res.*, **28**, 1839.
HAYEK, D. H., and TIPTON, S. R. (1966). *J. Cell Biol.*, **29**, 405.
HIGGINS, J. A. (1967). *J. Lipid Res.*, **8**, 636.
HIGGINS, J. A., and GREEN, C. (1966). *Biochem. J.*, **99**, 631.
HIGGINS, J. A., and GREEN, C. (1967). *Biochem. J.*, **104**, 26P.
IYPE, P. T., and BHARGAVA, P. M. (1963). *Life Sciences*, **2**, 311.
IYPE, P. T., and BHARGAVA, P. M. (1965). *Biochem. J.*, **94**, 284.
IYPE, P. T., BHARGAVA, P. M., and TASKER, A. D. (1965). *Expl Cell Res.*, **40**, 233.
JACOB, S. T. (1964). Ph.D. thesis, Agra University, Agra, India.
JACOB, S. T., and BHARGAVA, P. M. (1962). *Expl Cell Res.*, **27**, 453.
JACOB, S. T., and BHARGAVA, P. M. (1964). *Biochim. biophys. Acta*, **91**, 650.
JACOB, S. T., and BHARGAVA, P. M. (1965). *Biochem. J.*, **95**, 568.
JOHNSTONE, R. M., and SCHOLEFIELD, P. G. (1965). *Adv. Cancer Res.*, **9**, 144.
KING, C. M., and PHILLIPS, B. (1968). *Science*, **159**, 1351.
KUMAR, B. V., and BHARGAVA, P. M. (1969). *Life Sciences*, in press.
LAWS, J. O., and STICKLAND, L. H. (1961). *Expl Cell Res.*, **24**, 240.
LOEWENSTEIN, W. R., and KANNO, Y. (1964). *J. Cell Biol.*, **22**, 565.
LUCK, D. J. L. (1965). *J. Cell Biol.*, **24**, 445.
MALKIN, M. F., and ZAHALSKY, A. C. (1966). *Science*, **154**, 1665.
MATILE, PH., and BAHR, G. F. (1968). *Expl Cell Res.*, **52**, 301.
MILLER, J. A., and MILLER, E. C. (1966). *Pharmac. Rev.*, **18**, 805.
OVERTON, J. (1968). *J. Cell Biol.*, **39**, 101a.
PARSHLEY, M. S. (1965). *Cancer Res.*, **25**, 387.
RAO, S., and BHARGAVA, P. M. (1965). *J. gen. Microbiol.*, **40**, 219.
SHANMUGAM, G., and BHARGAVA, P. M. (1966). *Biochem. J.*, **99**, 297.
SHANMUGAM, G., and BHARGAVA, P. M. (1968). *Biochem. J.*, **108**, 741.
SHANMUGAM, G., and BHARGAVA, P. M. (1969). *Indian J. Biochem.*, **6**, 64–70.
SUGIMURA, T., OTAKE, H., and MATSUSHIMA, T. (1968). *Nature, Lond.*, **218**, 392.
SUSS, R., and MAURER, H. R. (1968). *Nature, Lond.*, **217**, 752.
VINOGRAD, J. (1968). *Fedn Proc. Fedn Am. Socs exp. Biol.*, **27**, 462.

DISCUSSION

Fortier: Some observations of Professor Selye on the effect of purely physical factors on the morphological and functional differentiation of connective tissue cells are possibly relevant to your paper, Dr Bhargava. Selye's experiments involved the insertion of glass tubes of different lengths and diameters in the subcutaneous tissue of rats. A connective tissue bridge was formed within the tube, establishing a connexion between the two ends, and the nature of the tissue thus formed ranged from Wharton's jelly to actual bone, including bone marrow components, according to the length and diameter of the inserted tube (Selye, 1959; Selye, Lemire and Bajusz, 1960; Selye, Prioreschi and Barath, 1961). Would you rule out the possibility that similar physical or mechanical forces are involved in some of the intercellular control mechanisms described in your experiments on isolated or aggregated cells?

Bhargava: Our experiments certainly do not rule this out, but unless we are able to show that reaggregation *in vitro* reverses these changes we cannot even be certain that they are a function of removal of intercellular influences. They could be merely due to a non-specific damage to the cells during dispersion of the tissue.

Paintal: You mentioned that your liver cell suspension homed to the liver when injected intravenously. What is the size of these cells and how do they get past the pulmonary capillaries?

Bhargava: This problem has been studied for tumour cells which disperse in the blood stream when metastases form. Such cells have been shown to pass through narrow capillaries by elongating (Zeidman, 1961), and this could happen to the liver cells too. There is also the special feature of liver that the sinusoidal lining is discontinuous, thus allowing the injected cells access to the hepatic cords of the recipient animal (for references, see Bhargava and Bhargava, 1968).

Salganik: Dr Bhargava, are you sure that your methods establishing the permeability of your cells to ribonuclease are sensitive enough? It is known that RNA in the cell is protected; we established that viral RNA inside cells is sensitive to ribonuclease but the RNA of the cells themselves is not damaged. The permeability of cell cultures and whole animal cells to ribonuclease has been established in several laboratories.

Bhargava: However, we do find a difference between slices and cell suspensions.

Siddiqi: You described a curious effect whereby the ribosomes remained intact but their RNA was degraded after treatment of cells in suspension with ribonuclease. Have you treated ribosomes alone with ribonuclease?

Bhargava: This has actually been done; the phenomenon is not new. If one takes ribosomes from say *E. coli*, reticulocytes or HeLa cells and treats them with ribonuclease (0·1–150 μg/ml), one gets particles which seem to contain degraded RNA but are morphologically intact (Tissieres *et al.*, 1959; Shakulov, Aitkhozhin and Spirin, 1963; Santer and Smith, 1966; Gierer, 1963; Fenwick, 1968). Santer and Smith have also reported that although ribonuclease-treated ribosomes from *E. coli* may have about 10 per cent less RNA and also less protein than untreated ribosomes they give the same pattern, on density-gradient centrifugation, as is obtained with untreated 80s, 60s or 40s particles. We therefore cannot rule out a small loss of RNA and protein from the ribosomes on treatment with ribonuclease in our experiments. If we increase the concentration of ribonuclease all we get is fewer ribosomes, but whatever ribosomes we obtain show subunit sedimentation coefficient values of 60s and 40s in sucrose density gradients. This has led us to believe that the effect of ribonuclease on the ribosomes in liver cells proceeds in two stages. First it cleaves a limited number of internucleotide linkages in such a way that the polyribonucleotide fragments continue to be held together, so that the morphological appearance of the particles is not significantly altered. Then a critical stage is reached when the ribosome falls apart.

Tata: I wonder whether there is not a trivial explanation for this result in that ribonuclease bound to ribosomes would be removed by deproteinizing agents.

Bhargava: I do not believe this explanation is true, as we have done controls in which ribonuclease was added after the incubation; in these controls the ribosomal RNA was not degraded. Further the unlabelled carrier RNA, which was added after the incubation in the experiments I have reported, was not degraded (see Fig. 1, p. 161). RNA was isolated in the presence of polyvinyl sulphate and bentonite, which should take care of ribonuclease.

Siddiqi: Dr Bhargava, I am sceptical about your interpretation of the effect of cell concentration on the uptake of an amino acid (histidine). How can you infer anything from that about the role of cell contact?

Bhargava: In these experiments the cells were shaken at 100 oscillations/min at concentrations ranging from 0·1 to 10×10⁶ cells per ml in 6 ml of the medium in a 25 ml conical flask, so presumably more of the transitory contacts were made by the cells at the higher concentrations. I agree that this is not exactly the kind of contact one finds between cells in tissues.

Harris: In your experiments on permeability changes after dissociation of cells you found that adenine, uracil and orotic acid all failed to enter isolated cells, yet these cells were making RNA, although not taking in the bases. Are they taking in nucleosides? Do you extract intracellular adenine derivatives?

Bhargava: Yes, nucleosides are taken in. In our experiments on the uptake of bases we isolated the total intracellular free nucleotide pool and looked for the labelling of its various components.

Harris: If you get no adenine derivatives in the cell labelled from adenine and no uracil derivatives from uracil that is most extraordinary, if the cell is making RNA.

Bhargava: Adenine, uracil and orotic acid are, of course, not the normal precursors of RNA nucleotides!

White: I would assume that these phenomena of entry and exit of various substances are energy-requiring processes. If you subjected these cells to anaerobic conditions or deleted their sources of energy, would you lose this specificity of ingress and egress of substrates?

Bhargava: We haven't done this but it is quite possible. What we have studied are actively metabolizing cells which respire and oxidize the usual substrates.

White: It may be noted that in the case of the rat thymocyte, one can demonstrate differences in the selective permeability of the plasma membrane and the nuclear membrane. Thus, intact thymocytes permit rapid entry of uridine, and the base is rapidly phosphorylated and utilized for RNA synthesis, whereas thymocyte nuclei do not utilize the free base. In contrast, the nuclei rapidly utilize the nucleotides (which are not utilized by the whole cells), and will not take up the free bases. Apparently phosphorylation in cytoplasm is essential for the entry of the bases into the nucleus. These differences in substrate utilization by whole cells and nuclei afford an index of the success which one has had in preparing nuclei free of cytoplasm.

The size of the cell population would be a very critical factor in this selectivity of uptake. In the studies of Eagle (1958-1959) with various mammalian cells in culture, it was observed that whether or not a particular amino acid was essential in the medium for continuing cell growth was influenced by the density of the cell population. With lower cell population numbers, the rate of loss of non-essential amino acids to the medium appeared to accelerate, leading to evidence of an apparent need to supplement the medium with an amino acid that had not been considered as "essential". Do variations in substrate transfer appear in your system as you alter cell population densities?

Bhargava: The specific activity of the free amino acids outside does not change significantly with a change in cell concentration, and the qualitative permeability changes that I have mentioned are independent of cell concentration.

Sarma: Did you compare these findings on the effect of intercellular factors with those from cells prepared in another way, say by trypsin treatment?

Bhargava: We don't like trypsin treatment for studies on *primary* cell suspensions, for many reasons (Bhargava, 1968).

Harris: In culturing mammalian cells one often observes a phenomenon that is analogous to the old observation in bacteria known as the nidus effect. Sometimes, in order to get cells to grow in culture, you have to have a certain concentration of cells in one place, and if you disperse them they don't get going. This is also a well-known effect in animal cells. It is currently being studied by Harry Rubin and his colleagues (Rein and Rubin, 1968).

REFERENCES

BHARGAVA, K., and BHARGAVA, P. M. (1968). *Expl Cell Res.*, 50, 515.
BHARGAVA, P. M. (1968). *Sci. Cult.*, *Calcutta*, 34 (Supplement), 105.
EAGLE, H. (1958–1959). *Harvey Lect.*, 54, 156–175.
FENWICK, M. L. (1968). *Biochem. J.*, 107, 481.
GIERER, A. (1963). *J. molec. Biol.*, 6, 148.
REIN, A., and RUBIN, H. (1968). *Expl Cell Res.*, 49, 666.
SANTER, M., and SMITH, J. R. (1966). *J. Bact.*, 92, 1099.
SELYE, H. (1959). *Nature, Lond.*, 184, 701–703.
SELYE, H., LEMIRE, Y., and BAJUSZ, E. (1960). *Wilhelm Roux Arch. EntwMech.*, 151, 572–585.
SELYE, H., PRIORESCHI, P., and BARATH, M. (1961). *Oncologia, Basel*, 14, 65–72.
SHAKULOV, R. S., AITKHOZHIN, M. A., and SPIRIN, A. S. (1963). *Biokhimiya*, 27, 631.
TISSIERES, A., WATSON, J. D., SCHLESINGER, D., and HOLLINGSWORTH, B. (1959). *J. molec. Biol.*, 1, 221.
ZEIDMAN, E. (1961). *Cancer Res.*, 38, 21.

RECENT STUDIES ON THE FEEDBACK CONTROL OF ACTH SECRETION, WITH PARTICULAR REFERENCE TO THE ROLE OF TRANSCORTIN IN PITUITARY–THYROID–ADRENOCORTICAL INTERACTIONS

CLAUDE FORTIER, FERNAND LABRIE, GEORGES PELLETIER, JEAN-PIERRE RAYNAUD, PIERRE DUCOMMUN, ALFONSO DELGADO, ROBERT LABRIE AND MY-ANH HO-KIM

Laboratoires d'Endocrinologie, Département de Physiologie, Faculté de Médecine, Université Laval, Québec, Canada

THE feedback control of adenohypophysial activity, indicated by the reciprocal relationship between the secretion of pituitary tropic hormones and the level of their corresponding target-gland hormones in the blood, has been chiefly considered in situations involving the components of a single functional axis (Brown-Grant, 1967; Fortier, 1966) and scant attention has been paid to the possible involvement of this homeostatic mechanism in the interactions between the adenohypophysis and two or more of its target glands. Studies by our group over the last few years on the interactions between pituitary, thyroid, adrenal cortex and gonads have brought to light the key role of a corticosteroid-binding protein of plasma, variously known as transcortin (Sandberg, Slaunwhite and Antoniades, 1957) or CBG (Daughaday, 1958), in the feedback-mediated adjustment of pituitary-adrenocortical function to changes in the secretory activity of the thyroid and gonads (Labrie, 1967; Labrie et al., 1968; Pelletier, 1969).

ROLE OF TRANSCORTIN IN THE ADJUSTMENT OF PITUITARY–ADRENOCORTICAL SECRETION TO THYROID ACTIVITY IN THE RAT

Influence of thyroid hormone on pituitary-adrenocortical activity

At an early stage of the analysis of the pituitary-thyroid-adrenal cortex triad, the influence of the thyroid hormone on parameters of pituitary-adrenocortical activity was assessed in the rat. Four weeks after thyroidectomy, depression of this activity was shown by lower pituitary ACTH

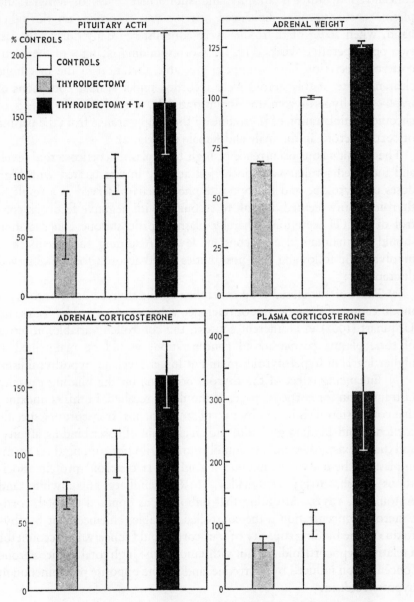

FIG. 1. Effects of thyroidectomy (one month before killing) and of daily administration of thyroxine (T_4, 20 µg/100 g body weight) for the last two post-operative weeks on parameters of pituitary-adrenocortical activity in the male rat.

concentration, adrenal atrophy and subnormal levels of adrenal and plasma corticosterone. Enhanced activity was recorded, on the other hand, when 20 μg of thyroxine were administered daily during the last two post-operative weeks (Fig. 1). Corresponding changes of the corticosterone secretion rate without appreciable alteration of the metabolic clearance rate of this steroid were recorded under similar conditions of hypo- and hyperthyroidism, as illustrated by a study of the effects of chronic administration of thyroxine on the disappearance from the plasma of corticosterone in the male and female rat (Fig. 2).

The simultaneous occurrence of high total plasma corticosteroid levels and increased pituitary-adrenocortical activity in rats treated with high doses of thyroxine and the opposite phenomena observed as a result of thyroidectomy seemed difficult to reconcile with negative feedback control of ACTH secretion, whereby plasma corticosterone concentration should be maintained at a constant level. Assuming that feedback is involved, the following two possibilities could account for the observed discrepancy:

(*a*) Reset of the closed-loop controller by the thyroid hormone. By a modification of the set-point or reference value, as suggested by Yates and Urquhart (1962) in a different context, the controlled variable, in terms of total plasma corticosteroid concentration, would be maintained at higher levels in hyperthyroidism and at lower levels in hypothyroidism.

(*b*) Enhancing effect of the thyroid hormone on the binding capacity of transcortin for corticosterone. It should be recalled, in this connexion, that corticosterone is bound by two plasma proteins: transcortin, a specific corticosteroid-binding globulin, and an albumin of lesser binding affinity, and that it exists, therefore, in equilibrium distribution among three forms in plasma: the native or unbound steroid and two steroid–protein associations (Bush, 1957; Daughaday, 1958; Sandberg, Slaunwhite and Antoniades, 1957). Assuming that unbound, as opposed to total, corticosterone concentration is the controlled variable, enhancement by thyroxine of the binding capacity of transcortin could conceivably account for a relative hypocorticoidism, notwithstanding the high total corticosterone concentration induced by thyroxine, and for the opposite phenomenon in hypothyroidism.

Effect of thyroid hormone on transcortin

To test this latter hypothesis the distribution of corticosterone between its free, albumin-bound and transcortin-bound fractions was studied by the combined use of gel filtration and equilibrium dialysis, at 37°c,

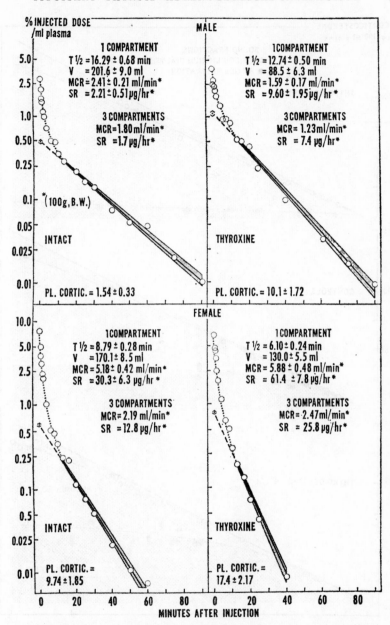

FIG. 2. Effects of thyroxine (20 μg/100 g body weight once a day for 10 days) on the plasma disappearance characteristics (T½, half-life; V, virtual volume of distribution; MCR, metabolic clearance rate) of a tracer dose of tritium-labelled corticosterone and on the secretion rate (SR) of the steroid, in male and female rats, at steady state. Calculations according to one and three-compartment models.

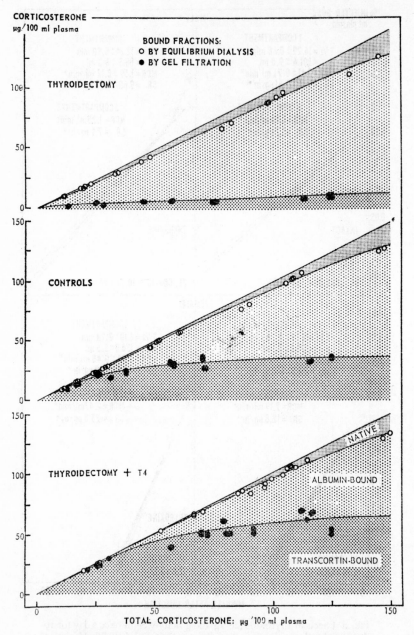

FIG. 3. Effects of thyroidectomy and of thyroxine (T4) administration on the binding capacity of transcortin for corticosterone. Plasma samples from each group were incubated at 37°C with increasing concentrations of ^{14}C-labelled corticosterone.

following incubation with ^{14}C-labelled corticosterone of pooled plasma samples from intact, thyroidectomized and thyroxine-treated animals.

As shown by the distribution of corticosterone between its three forms for increasing concentrations of the steroid, the binding capacity of transcortin was depressed by thyroidectomy and markedly enhanced by the chronic administration of thyroxine (Fig. 3).

Since the binding capacity of transcortin is a function of its total number of binding sites and of its association constant for corticosterone, it was of interest to analyse in these terms the observed effect of thyroxine. This was done by applying the general equations of the law of mass action describing the equilibrium between a corticosteroid and two binding proteins to the data provided by two related methods, one involving dialysis alone and the other using combined dialysis and gel filtration (Labrie, 1967). As shown in Fig. 4, which illustrates the latter approach, the effect of thyroxine was found to be exerted on the number of transcortin binding sites rather than on its association constant for corticosterone.

According to Seal and Doe (1962), transcortin provides one corticosteroid-binding site per molecule. The number of binding sites, which directly reflects the concentration of this globulin, was determined in subsequent experiments by gel filtration at $4°$c, following incubation of individual plasma samples (1 ml) with a tracer amount of [4-^{14}C]corticosterone and a saturating amount of the unlabelled steroid ensuring a concentration of 150 μg/100 ml of plasma (Labrie et al., 1968). This method takes advantage of the absence of dissociation of the transcortin-corticosterone complex during its passage through the column at $4°$c, because of the higher transcortin-corticosterone equilibrium constant recorded at this temperature (Labrie, 1967).

The effect of thyroxine on the number of transcortin binding sites, already evident after two days of thyroxine administration (Fig. 5), appears to be specific to transcortin, since opposite changes of total plasma proteins were observed under similar experimental conditions (Labrie, 1967).

Opposite effects of thyroid and adrenal cortical hormones on transcortin

In view of the higher plasma corticosterone levels induced by giving thyroxine, it was of interest to dissociate the effects of these hormones on transcortin binding. As shown by the increased number of transcortin binding sites induced by adrenalectomy and by the enhanced stimulating effect of thyroxine on this parameter observed as a result of hypophysec-

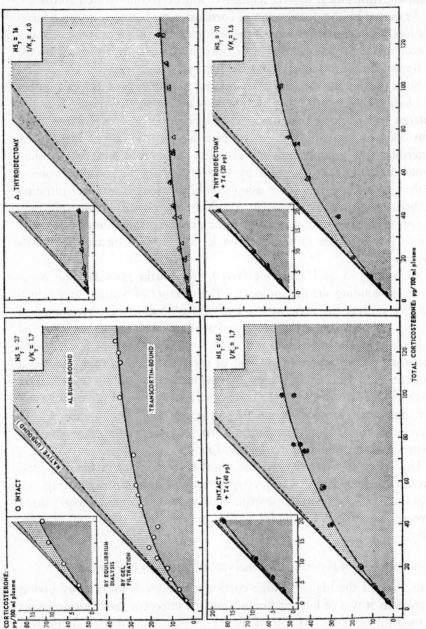

FIG. 4 (*see opposite page*)

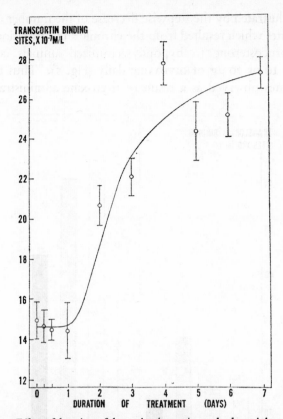

FIG. 5. Effect of duration of thyroxine (20 μg/100 g body weight, once a day) treatment on the number of transcortin binding sites determined by dextran gel filtration at 4° c.

tomy and of adrenalectomy (Fig. 6), corticosteroids were found, in agreement with other reports (Westphal, Williams and Ashley, 1962; Seal and Doe, 1965; and Gala and Westphal, 1966), to depress the binding capacity of transcortin. The opposite effects of corticosterone and of thyroxine

FIG. 4. Effects of thyroidectomy (one month before killing) and of thyroxine (T4) administration (14 days) in intact and thyroidectomized rats on the distribution of plasma corticosterone between its unbound, albumin-bound and transcortin-bound fractions for increasing concentrations of the total steroid, as determined by the combined use of equilibrium dialysis and gel filtration at 37° c, following incubation of pooled plasma samples with a tracer amount of [14C]corticosterone and graded doses of the unlabelled steroid. Recorded values for the number of transcortin binding sites (NS$_T$ expressed in μg of the transcortin-bound fraction per 100 ml of plasma, equivalent in molar concentration to μg × 0·289 × 10^{-7} moles/l) and the transcortin–corticosterone association constant (K$_T$) were derived from general equations of the law of mass action.

are further illustrated by the stepwise decrease of the number of transcortin binding sites which resulted from the chronic administration of graded doses of corticosterone to hypophysectomized animals concurrently treated with 10 or 20 μg of thyroxine daily (Fig. 7). Thus the increase of binding sites observed as a result of thyroxine administration in the

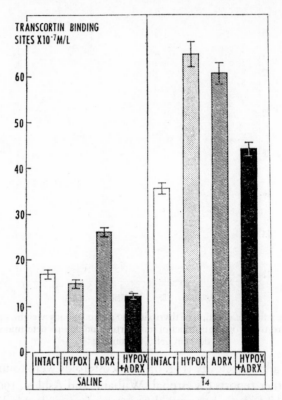

FIG. 6. Effects of hypophysectomy, adrenalectomy and thyroxine (T4; 20 μg/100 g body weight, once a day for 7 days) administration, singly or in combination, on the number of transcortin binding sites.

intact animal represents the algebraic summation of two opposite effects: enhancement by thyroxine and depression by corticosterone, the effect of thyroxine being predominant in the rat.

Correlation of the effects of thyroid hormone on pituitary-adrenocortical secretion, transcortin binding and the equilibrium distribution of corticosterone in plasma

What is the physiological significance of the thyroxine-induced enhancement of the binding capacity of transcortin, in terms of negative feedback control of ACTH secretion? To answer this question, the effects of thyroidectomy and of graded doses of thyroxine were ascertained on relevant parameters.

A progressive increase of adrenal weight, and hence a corresponding enhancement of ACTH secretion, was observed with increasing doses of thyroxine (Fig. 8). Chronic administration of graded doses of thyroxine also resulted in stepwise increases of total and of transcortin-bound plasma

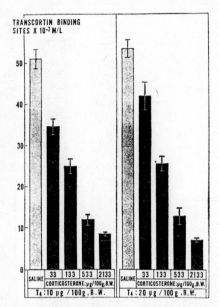

FIG. 7. Effect of increasing doses of corticosterone, administered for 7 days, on the number of transcortin binding sites in hypophysectomized rats respectively receiving 10 and 20 μg of thyroxine/100 g body weight, once a day, over the same period.

corticosterone, *without detectable alteration of the unbound fraction of the steroid* (Fig. 9).

This strongly suggests that the unbound, as opposed to the total, plasma corticosterone concentration is the variable under feedback control, and that the binding capacity of transcortin is responsible in the rat for the adjustment of ACTH and corticosterone secretion to changes in thyroid activity.

It would appear that by increasing the number of transcortin binding sites thyroxine tends to decrease free corticosterone, the controlled variable. This, in turn, calls into play a feedback-mediated increase of pituitary-adrenocortical secretion which, in view of the unaltered metabolic

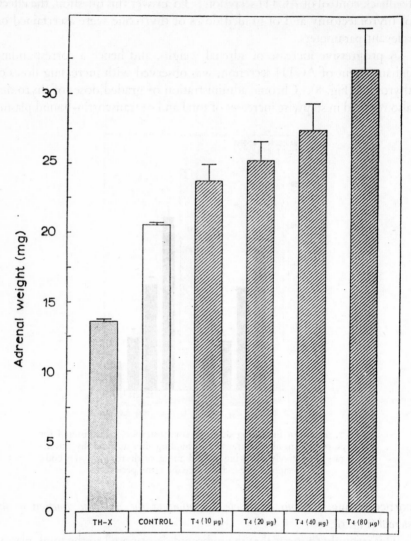

FIG. 8. Effects of thyroidectomy (TH-X) and of chronic administration
of graded doses of thyroxine (T4) on adrenal weight in the rat.

clearance rate of corticosterone, results in an elevation of the total concentration of this steroid in plasma which is proportional to the increase in the number of transcortin binding sites.

A new dynamic equilibrium determined by increased transcortin binding and characterized by a *virtual* depression of the free corticosterone

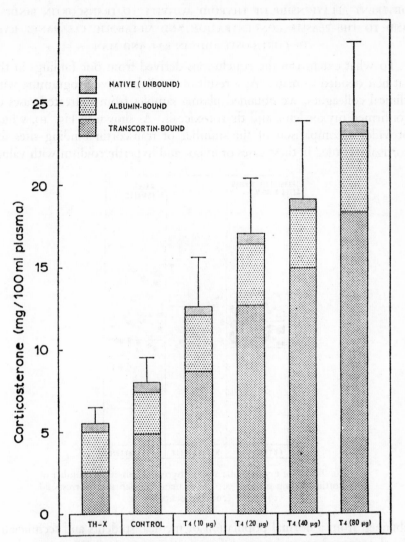

FIG. 9. Effects of thyroidectomy (TH-X) and of chronic administration of graded doses of thyroxine (T4) on the total plasma corticosterone concentration and on the distribution of corticosterone between its three fractions under basal conditions in the rat.

level compensated by accelerated rates of ACTH and corticosterone secretion thus provides a satisfactory explanation for the simultaneous occurrence of high total plasma corticosteroid levels and increased adrenal cortical activity in the hyperthyroid rat, and for the opposite changes observed as a result of thyroidectomy.

7*

DIFFERENT RELATIONSHIP OF THYROID ACTIVITY TO TRANSCORTIN BINDING
AND TO THE PLASMA CONCENTRATION AND METABOLIC CLEARANCE RATE
OF CORTICOSTEROIDS IN RAT AND MAN

To what extent can the conclusions derived from our findings in the rat be extended to man? As a result of a collaborative programme with clinical colleagues, we obtained plasma samples from untreated cases of confirmed myxoedema and thyrotoxicosis. As shown in Fig. 10, which provides a comparison of the number of transcortin binding sites for cortisol recorded in these cases of hypo- and hyperthyroidism with values

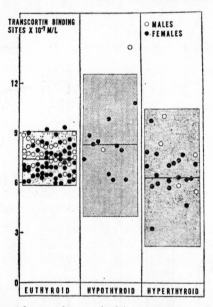

Fig. 10. Effects of myxoedema and of thyrotoxicosis on the number of transcortin binding sites for cortisol in man. Shaded areas correspond to 95 per cent confidence limits.

obtained from a normal population of medical students and technicians, the three populations do not differ significantly. Furthermore, in direct contrast with our findings in the rat, there is a suggestive trend to respectively depressed and enhanced binding capacities of transcortin in hyperthyroid and hypothyroid patients. In agreement with other reports (Levin, Daughaday and Bremer, 1955; Mikulaj and Nemeth, 1958; Peterson, 1958; Felber *et al.*, 1959), no difference was recorded in the three groups with regard to the resting plasma corticosteroid levels, the mean cortisol concentrations being 19 μg/100 ml in the euthyroid group, 20 in the hypothyroid and 18 in the hyperthyroid. It has been clearly established, on the other hand, by Samuels and co-workers (1957) and by Peterson (1958) that, although the plasma cortisol level is unaltered in man by the functional state of the thyroid, both the metabolic clearance and the secretion rates of this corticosteroid are increased in the hyperthyroid and decreased in the hypothyroid patients.

The reverse situation holds in the rat where, as previously noted, the resting plasma corticosterone concentration is markedly affected by thyroid activity, whereas the metabolic clearance rate of this steroid is not. Furthermore, in the rat, the binding capacity of transcortin indicates a relationship to the functional state of the thyroid which is essentially different from that observed in man. Relating these observations, the following questions may be answered:

(a) Why is the functional state of the thyroid differently related to the binding capacity of transcortin in rat and man?

(b) Why is the metabolic clearance rate of corticosteroids differently affected by the thyroid hormone in these two species?

Comparison of the depressing effects of corticosterone and of cortisol on transcortin

It should be recalled, in connexion with the first question, that since the number of transcortin binding sites depends on the opposite effects of thyroxine and adrenal cortical hormones, it will be affected by a shift in their equilibrium. It was possibly relevant, therefore, that different corticosteroids, corticosterone and cortisol respectively, are predominantly secreted in rat (Bush, 1953; Holzbauer, 1957) and man (Peterson, 1957). Are the two steroids equally potent in depressing the binding capacity of transcortin?

In an experiment in adrenalectomized rats with or without prior treatment with thyroxine, the depressing effect of cortisol on transcortin binding was found to be appreciably greater than that of corticosterone in the saline-treated animals, and even more so in the thyroxine-treated rats (Fig. 11). This provides a promising clue and it is tempting to suggest that the greater depressing effect of cortisol may account for two of the observed discrepancies between rat and man; the enhancing effect of thyroxine on transcortin binding being neutralized in man by the antagonistic effect of the corticosteroid and the consequently unaltered transcortin being consistent with the unchanged resting cortisol level in this species. Preliminary results of a comparative study on the effects of thyroxine on relevant parameters in other species predominantly secreting corticosterone (rabbit) and cortisol (hamster) are in agreement with this view (Labrie and Pelletier, unpublished observations).

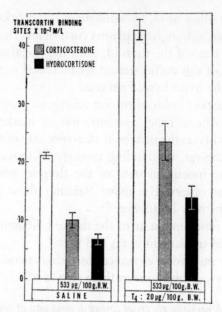

FIG. 11. Comparative effects of hydrocortisone (cortisol) and corticosterone on the number of transcortin binding sites in the adrenalectomized rat treated or not with thyroxine (T4).

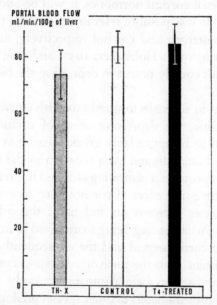

FIG. 12. Effects of thyroidectomy (one month before killing) and of thyroxine (T4) administration (10 μg/100 g body weight, twice daily for 10 days) on the portal blood flow in the male rat.

It appears likely that in man, where the plasma cortisol level and trans-cortin binding capacity are independent of the thyroid, the changes of the cortisol secretion rate associated with hypo- and hyperthyroidism are determined by thyroxine-induced alterations of the metabolic clearance rate of this steroid.

ROLE OF TRANSCORTIN IN THE ADJUSTMENT OF THE METABOLIC CLEARANCE RATE TO THYROID ACTIVITY

With regard to the different effects of the thyroid hormone on the metabolic clearance rate of corticosteroids in rat and man, it should be recalled that, since the liver is the chief, if not the exclusive, inactivating site for corticosteroids, three factors are liable to affect the metabolic clearance of these hormones: (a) liver blood flow, (b) reducing capacity of the enzymic systems of the liver and (c) rate of transfer across cell membranes affecting the access of corticosteroids to the hepatic inactiva-tion sites. To what extent are these factors influenced by the thyroid hormone?

Effect of thyroid hormone on portal blood flow

In agreement with the reported independence of hepatic blood flow from thyroid activity in man (Myers, Brannon and Holland, 1950), neither thyroidectomy nor thyroxine administration were found to affect portal blood flow measured in the rat by means of an electromagnetic flow-meter (Fig. 12), which, according to Lacroix and Leusen (1965), accounts for most of the blood supply to the liver.

Effect of thyroid hormone on the in vitro reducing activity of the liver

It has been reported by Yates, Urquhart and Herbst (1958) and by Tomkins and McGuire (1960) that thyroxine pretreatment markedly increases the reducing activity of incubated liver slices, whereas thyroid-ectomy has the opposite effect. The enhancing effect of thyroxine pre-treatment on the ring A reduction of corticosterone by rat liver slices, confirmed in our laboratory (Fig. 13), made it difficult to explain why, contrary to what is observed for cortisol in man, the metabolic clearance rate of corticosterone is not similarly affected by thyroid hormone in the rat.

Relationship between transcortin binding and corticosteroid inactivation by the liver

Since, in the light of corticosterone disappearance studies and of direct estimates of corticosterone extraction during a single passage through the liver, the metabolic clearance rate of this steroid in the adult male rat was found to represent less than 50 per cent of the total hepatic blood flow (Labrie, 1967), the lack of manifestation *in vivo* of the thyroxine-induced enhancement of the liver reducing activity observed *in vitro* could only be ascribed to a third factor: reduced access of corticosterone to the inactivation sites of the liver. This led us to consider the possibility that the access of corticosteroids to the inactivation sites is conditioned by the relative binding capacities for these steroids of competing plasma and

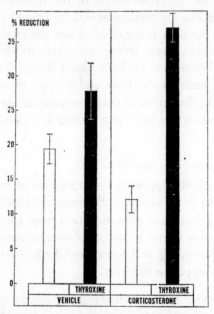

FIG. 13. Effects of thyroxine (20 μg/100 g body weight, daily) and corticosterone (1566 μg/100 g body weight, twice daily), administered for 7 days to rats hypophysectomized one month previously, on ring A reduction of corticosterone by incubated liver slices.

liver proteins. In man, where transcortin binding is not affected by the thyroid hormone, the increase in liver reducing activity associated with hyperthyroidism is reflected by a corresponding increase in the metabolic clearance rate of cortisol. In the rat, on the other hand, the thyroxine-induced increase of the binding capacity of transcortin would interfere with the access of corticosterone to the liver inactivation sites and would thereby mask or neutralize the enhancement by thyroxine of the liver reducing capacity. Tait and Burstein (1964) used a similar line of reasoning to account for the decreased metabolic clearance rate of cortisol observed by Nelson and co-workers (1963) as a result of oestrogen administration in man, and so did Sandberg and Slaunwhite (1963) to explain the inhibition of cortisol metabolism resulting from the addition of purified transcortin to a liver homogenate incubated in a protein-free medium.

Experiments in which the isolated livers of intact, thyroidectomized and thyroxine-treated rats were perfused with blood from similarly

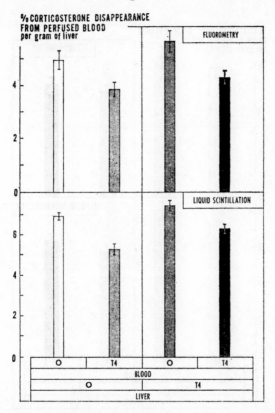

Fig. 14. Dissociation, through perfusion (at a rate of 8 ml/min for 10 min) of the isolated livers of intact and thyroxine-treated rats with blood from similarly treated animals in appropriate "donor-recipient" permutations, of the blood and liver components in the effect of thyroxine (T4, 10 μg/100 g body weight, twice daily for 10 days) on corticosterone inactivation, expressed as the percentage disappearance of the labelled or unlabelled steroid (respectively determined by liquid scintillation and fluorometry) added to the perfused blood.

treated animals, in appropriate "donor-recipient" permutations, were aimed at testing the above hypothesis. This experimental arrangement made it possible to increase or to depress selectively the binding capacity of transcortin or the corticosteroid-reducing activity of the liver and to assess the effects of these alterations on the inactivation of a uniform amount of corticosterone added to the blood of donors (adrenalectomized 24 hours before killing) with a tracer amount of the tritium-labelled steroid, by fluorometric determination and liquid scintillation counting of corticosterone in chloroform extracts of the incoming blood and of the liver effluents. The percentage of corticosterone extracted from the

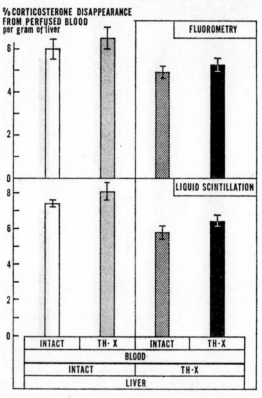

FIG. 15. Dissociation, according to procedure outlined for Fig. 14, of the blood and liver components in the effect of thyroidectomy (TH-X, performed 7 weeks before the experiment) on corticosterone inactivation.

perfused blood was related to the number of transcortin binding sites determined by gel filtration at 4° c.

In agreement with previous findings, the number of transcortin binding sites was increased by thyroxine treatment from a control value of $9 \cdot 21 \pm 0 \cdot 26$ to $17 \cdot 0 \pm 0 \cdot 61 \times 10^{-7}$ moles/l (corresponding to $59 \cdot 0 \pm 2 \cdot 1 \mu g/$ 100 ml) and depressed by thyroidectomy from $11 \cdot 2 \pm 0 \cdot 46$ for controls to $7 \cdot 76 \pm 0 \cdot 25 \times 10^{-7}$ moles/l.

Corticosterone inactivation, significantly depressed when the blood from thyroxine-treated donors was perfused through the livers of untreated rats, was increased when the blood of untreated donors was perfused through the livers of thyroxine-treated recipients and an algebraic summation of these opposite effects was recorded when the blood of thyroxine-treated donors was perfused in similarly treated recipients (Fig. 14). Converse alterations of corticosterone inactivation were observed when thyroidectomy was substituted for pretreatment with thyroxine (Fig. 15).

It may be inferred from these results that, in agreement with our hypothesis, the reduced access of corticosterone to the liver inactivation sites resulting from the increased transcortin binding induced in the rat by thyroxine cancels the enhancement by this hormone of the liver reducing capacity. The different effects of thyroxine on the metabolic clearance rate of corticosteroids in rat and man can thus be related to its influence on transcortin binding in the two species which, in its turn, is apparently determined by the nature of the predominantly secreted corticosteroid and the presence (cortisol) or absence (corticosterone) of a 17α-hydroxyl grouping on its D ring.

DISCUSSION

It appears paradoxical, from a teleological standpoint, that the different modalities of the feedback-mediated adjustment of pituitary-adrenocortical secretion to thyroid activity in rat and man should depend on a minor structural characteristic of the secreted corticosteroid, which sets the delicate equilibrium between the antagonistic effects of thyroid and adrenal cortical hormones on transcortin binding. This explanation suggested by our findings should not, however, be considered as definitive until it is thoroughly assessed in the light of clinical data.

Though an increased binding capacity of transcortin was observed in one case of adrenal cortical insufficiency (Doe, Fernandez and Seal, 1964) and a depressed capacity in a few cases of Cushing's syndrome (De Moor et al., 1966), the scant relevant literature does not indicate that Addison's disease and hyperadrenocorticism are generally accompanied, in man, by significant alterations of transcortin (Doe, Fernandez and Seal, 1964; De Moor et al., 1966; Murray, 1967). In view of the possible risks involved in comparing the effects of exogenous thyroxine in normal subjects and in nontreated cases of hypopituitarism and of Addison's disease, clarification of this issue will probably await the concurrent assessment of thyroid and of

198 CLAUDE FORTIER *et al.*

adrenal cortical activity, as related to the binding capacity of transcortin, in cases of hypo- and of hyperadrenocorticism.

Whether an increase of the number of transcortin binding sites, as in the rat, or of the metabolic clearance rate, as in man, is responsible for creating a virtual depression of the unbound fraction of the corticosteroid, the controlled variable, the same end-result is achieved in both species and pituitary-adrenocortical activity is adjusted, through negative feedback, at a higher level.

Since transcortin, as well as Δ-4-hydrogenase, which is presumably involved in the thyroxine-induced enhancement of the liver reducing capacity, is produced in the liver, it may be speculated, beyond direct experimental evidence, that the overall influence of the thyroid on pituitary-adrenocortical activity is exerted by stimulating the synthesis by the liver of only one of these proteins (Δ-4-hydrogenase) in man, or of both in the rat.

SUMMARY

Studies on the role of corticosteroid binding by transcortin in the adjustment of pituitary-adrenocortical secretion to thyroid activity justify the following conclusions:

(1) The increase of the number of transcortin binding sites observed as a result of the administration of thyroxine to the intact rat represents the algebraic summation of two opposite effects: enhancement by thyroxine and depression by corticosterone, with predominance of the stimulating effect of thyroxine.

(2) The unbound, as opposed to total, corticosterone concentration is the variable under feedback control and, in view of the unaltered metabolic clearance rate of corticosterone in the rat, the binding capacity of transcortin is entirely responsible, in this species, for the adjustment of ACTH and corticosterone secretion to altered thyroid activity.

(3) The independence of the metabolic clearance rate of corticosterone from thyroid activity, in the rat, is ascribed to the reduced access of corticosterone to the liver inactivation sites, which results from the increased transcortin binding induced by thyroxine and neutralizes the enhancement by this hormone of the liver reducing capacity.

(4) In man, on the other hand, where, presumably as a result of the observed cancellation by cortisol of the enhancing effect of thyroxine on transcortin binding, the binding capacity of this protein is independent of thyroid activity, the increase in liver reducing activity associated with hyperthyroidism is reflected by a corresponding increase in the metabolic

clearance rate of cortisol which, by inducing a virtual depression of the controlled variable, results in a feedback-mediated enhancement of pituitary-adrenocortical activity.

Acknowledgements

The investigations reported in this paper were supported by grants from the Medical Research Council of Canada (MT-1205 and 1555) and the U.S. Air Force Office of Scientific Research (AF-AFOSR-511-65, 511-67 and 69-1627), and were carried out during the tenure of MRC research fellowships by three of the authors (F.L., G.P. and A.D.). We are indebted to Professor Asher Korner for information on the liver perfusion technique developed in his laboratory.

REFERENCES

Brown-Grant, K. (1967). In *Regulation and Control in Living Systems*, pp. 176–255, ed. Kalmus, H. New York: Wiley.

Bush, I. E. (1953). *J. Endocr.*, **9**, 95–101.

Bush, I. E. (1957). *Ciba Fdn Colloq. Endocr. Hormones in Blood*, pp. 263–285. London: Churchill.

Daughaday, W. H. (1958). *J. clin. Invest.*, **37**, 519–522.

De Moor, P., Steeno, O., Brosens, I., and Hendrikx, A. (1966). *J. clin. Endocr. Metab.*, **26**, 72–78.

Doe, R. P., Fernandez, R., and Seal, U. S. (1964). *J. clin. Endocr. Metab.*, **24**, 1029–1039.

Felber, J. P., Reddy, W. J., Selenkow, H. A., and Thorn, G. W. (1959). *J. clin. Endocr. Metab.*, **19**, 895–906.

Fortier, C. (1966). In *The Pituitary Gland*, vol. 2, pp. 195–234, ed. Harris, G. W., and Donovan, B. T. London: Butterworths.

Gala, R. R., and Westphal, U. (1966). *Endocrinology*, **79**, 67–76.

Holzbauer, M. (1957). *J. Physiol., Lond.*, **139**, 306–315.

Labrie, F. (1967). *Interactions hormonales et rôle de la transcortine dans l'ajustement de l'activité hypophyso-surrénalienne.* Thèse de doctorat es sciences, Université Laval, Québec.

Labrie, F., Pelletier, G., Labrie, R., Ho-Kim, M.-A., Delgado, A., MacIntosh, B., and Fortier, C. (1968). *Annls Endocr.*, **29**, 29–43.

Lacroix, E., and Leusen, J. (1965). *J. Physiol., Paris*, **57**, 115–216.

Levin, M. E., Daughaday, W. H., and Bremer, R. (1955). *J. Lab. clin. Med.*, **45**, 833–840.

Mikulaj, L., and Nemeth, S. (1958). *J. clin. Endocr. Metab.*, **18**, 539–542.

Murray, D. (1967). *J. Endocr.*, **39**, 571–591.

Myers, J. P., Brannon, E. S., and Holland, B. C. (1950). *J. clin. Invest.*, **29**, 1069–1077.

Nelson, D. H., Tauney, H., Mestman, G., Gieschen, V. W., and Wilson, L. D. (1963). *J. clin. Endocr. Metab.*, **23**, 261–265.

Pelletier, G. (1969). *Transcortine et interactions hormonale.* Thèse de doctorat es sciences, Université Laval, Québec.

Peterson, R. E. (1957). *J. biol. Chem.*, **225**, 25–37.

Peterson, R. E. (1958). *J. clin. Invest.*, **37**, 736–743.

Samuels, L. T., Brown, H., Eik-Nes, K., Tyler, F. H., and Dominguez, O. V. (1957). *Ciba Fdn Colloq. Endocr. Hormones in Blood*, pp. 208–232. London: Churchill.

Sandberg, A. A., and Slaunwhite, W. R. (1963). *J. clin. Invest.*, **42**, 51–54.

Sandberg, A. A., Slaunwhite, W. R., and Antoniades, H. N. (1957). *Recent Prog. Horm. Res.*, **13**, 209–267.

Seal, U. S., and Doe, R. P. (1962). *J. biol. Chem.*, **237**, 3136–3140.

SEAL, U. S., and DOE, R. P. (1965). *Steroids*, 5, 827–841.
TAIT, J. F., and BURSTEIN, S. (1964). In *The Hormones*, vol. 5, pp. 441–557, ed. Pincus,
  G., Thimann, K. V., and Astwood, E. B. New York: Academic Press.
TOMKINS, G. M., and McGUIRE, J. S. (1960). *Ann. N.Y. Acad. Sci.*, 86, 600–604.
WESTPHAL, U., WILLIAMS, W. C., and ASHLEY, B. D. (1962). *Proc. Soc. exp. Biol. Med.*,
  109, 926–929.
YATES, F. E., and URQUHART, J. (1962). *Physiol. Rev.*, 42, 359–443.
YATES, F. E., URQUHART, J., and HERBST, A. L. (1958). *Am. J. Physiol.*, 195, 373–380.

DISCUSSION

Tata: Have you thought of doing these experiments in birds? There is a possibility that oestrogen may induce transcortin in birds, by analogy with the effects of oestradiol on avian thyroxine-binding globulin (both transcortin and thyroxine-binding globulin are α-globulins).

Fortier: We have not studied transcortin in birds. However, we could confirm, in the rat, the stimulating effect of oestrogens on the corticosteroid-binding capacity of transcortin demonstrated in man by Daughaday, Hozak and Biederman (1959) and by Sandberg and Slaunwhite (1959). We found, furthermore, that this effect is entirely mediated by the pituitary–thyroid axis, since it is prevented by either hypophysectomy or thyroidectomy and that, conversely, a single dose of oestradiol (4 μg/100 g body weight), which induces a marked increase of the number of transcortin binding sites in the intact male rat, concurrently results in a pronounced and long-lasting rise in the plasma concentration of TSH and thyroxine. It was inferred from a concomitant study of pituitary-adrenocortical activity, under these conditions, that pituitary–thyroid activation and the consequent enhancement of the transcortin binding capacity are essential links in pituitary–adrenocortical adjustment to the level of circulating oestrogens in the rat (Labrie *et al.*, 1968).

Tata: Thyroxine causes proliferation of microsomal membranes and since the reductase is a microsomal enzyme you might increase the rate of steroid reduction. Did you compare the time-course of the increased rate of corticosteroid reduction and increased transcortin binding activity after the administration of thyroxine? Is it possible to observe a dissociation of the two functions?

Fortier: As mentioned in the paper, we have confirmed the enhancement by thyroxine pretreatment of the ring A reduction of corticosterone by rat liver slices (see Fig. 13, p. 194). We have not, however, studied the time-course of this effect of thyroxine.

White: These findings of Professor Fortier are another example of a kind of homeostatic relationship between thyroxine and adrenocortical steroid activity. This is also true with regard to lymphoid tissue. For example, whereas the adrenal steroids produce involution of lymphoid tissue, the thyroid hormones stimulate proliferation of this tissue (Dougherty, 1952). Apparently, the levels of secretory activity of these two systems seem to counterbalance one another. Since in your studies the concentration of unbound corticoids does not change, these hormones are probably not the variable parameter of your studies, despite the surprising suggestion that thyroxine somehow prevents access of corticoids to liver cells. Is it possible that the hypothalamic regulator has been set at a different level of sensitivity? Perhaps the corticoids have less access to hypothalamic cells, as they do to liver cells, in the hyperthyroid state and therefore you are not getting the feedback mechanism working in circumstances when you might expect it to work?

Fortier: We interpret our findings as entirely consistent with the operation of feedback without any resetting of the set-point or reference value of the closed-loop controller. The observed constancy of free corticosterone, under our experimental conditions, which does not exclude minor undetectable fluctuations of the controlled variable, may be considered to be analogous to the relative constancy of temperature achieved by an efficient thermostatically controlled heating system, the efficiency of the control being inversely related to the magnitude of the observed deviation (error) from the set-point. It is inferred from our findings that by increasing the binding capacity of transcortin, thyroxine tends to decrease free corticosterone, the controlled variable, and thus calls into play a feedback-mediated increase of pituitary–adrenocortical activity which, in view of the unaltered metabolic clearance rate of corticosterone, results in an elevation of the total concentration of this steroid in plasma which is proportional to the increase in the number of transcortin binding sites and allows the concentration of free corticosterone to approximate to the reference value of the controller.

White: Although the level of the free steroid does not change, you might perhaps have an alteration in sensitivity in terms of the critical concentration required to trigger the secretory mechanism.

Harris: How do you assay the binding sites, Professor Fortier?

Fortier: The number of binding sites was originally determined by applying the general equations of the law of mass action to the data provided by two related methods, respectively involving equilibrium dialysis alone and combined dialysis and gel-filtration, at 37° c, following incubation at 37° c of pooled plasma samples with a tracer amount of ^{14}C-labelled corticosterone and graded amounts of the unlabelled steroid.

In connexion with gel-filtration, appropriate correction factors were used for (1) dilution of the plasma and (2) dissociation of the transcortin–corticosterone complex during its passage through the column (Labrie, 1967).

Monod: So you don't equilibrate the columns with corticoids before you pass the complex through?

Fortier: In agreement with accepted procedure (De Moor *et al.*, 1962; Seal and Doe, 1963; Quincey and Gray, 1963) the dextran gel columns were not pre-equilibrated with corticosterone—hence the necessity for the correction factors at 37° c. With this technique, in view of the relatively loose binding of corticosterone with albumin, the albumin-bound fraction is initially retained with the free fraction by the dextran gel, which acts as a molecular sieve, and the transcortin-bound fraction is therefore recovered in the first peak of the effluent, whereas the free, as well as the original albumin-bound, fractions constitute the second peak. A third method developed by Labrie and used in subsequent studies for determining the number of transcortin binding sites involved prior saturation of the binding sites through incubation of individual plasma samples with an appropriate amount of unlabelled corticosterone and a tracer amount of the ^{14}C-steroid, and takes advantage of the absence of dissociation of the transcortin–corticosterone complex during its passage through the column at 4° c, in view of the higher equilibrium constant recorded at this temperature (Labrie *et al.*, 1968).

Monod: What is the association constant of transcortin and corticosterone?

Fortier: As recorded from a number of determinations, the mean equilibrium constant at 37° c (310°AB) is $1 \cdot 75 \times 10^7$ moles/l.

Edelman: The free energy of binding (ΔF) would be $-10 \cdot 26$ Kcal/mole.

Monod: Then I understand, because if you pass the protein over a column which is not equilibrated you would expect everything to go out, but with a very high association constant like this presumably the rate of dissociation is so small that it doesn't have time to get away.

Fortier: As I said, the association constant is still markedly higher at 4° c, and the dissociation of the transcortin–corticosteroid complex during its passage through the column is completely abolished at this temperature. After saturation of the transcortin binding sites the number of occupied sites (transcortin-bound fraction) determined under these conditions corresponds to the total number of binding sites, and correction factors for dilution of the plasma or dissociation of the steroid–protein complex are no longer required (Labrie, 1967).

Edelman: The high binding constant of 10^7 moles/l at $37°$c raises the whole question of the function of proteins like transcortin. If they bind with such a high affinity one wonders whether they are transporting the hormone at all, because there is then the problem of how they release the steroid to the site at which it acts. You could argue that the hormones work while they are on the protein, which would be interesting in itself, but you could also argue that they are *not* transporting proteins and may be doing something else. Has there been speculation of this kind, or any direct experiment to indicate what would happen if you put the pure protein into an animal?

Fortier: A case in point is the inhibition of cortisol reduction reported by Sandberg and Slaunwhite (1963) as a result of the addition of purified transcortin to a liver homogenate incubated in a protein-free medium. A similar interpretation was inferred from our liver perfusion experiments (Figs. 14 and 15, pp. 195 and 196) in which we showed that, in the rat, the thyroxine-induced increase in the number of transcortin binding sites interfered with the access of the steroid to the inactivating sites and neutralized the enhancement of the reducing capacity of the liver associated with experimental hyperthyroidism.

Monod: The question could be put this way: is transcortin a buffer, a chelator, or a transporter? Of course it might be a bit of all three, but from the association constant of 10^7 I would think that it is more a chelator than anything else, unless there is some mechanism by which the transcortin can deliver the corticosterone to the target tissue and at the same time preserve it from destruction by the liver; that is not inconceivable. The experimental evidence shows that the transcortin does act as a sort of chelator with respect to the liver enzymes. The question is, does it act as chelator also with respect to the receptors in the target tissue?

Fortier: One would assume from our liver perfusion experiments that the free fraction only would enter the target tissue and that, according to mass action, the initial equilibrium between the free and transcortin-bound steroid would be promptly restored. The binding protein would thus ensure a readily available supply of the steroid to the target tissues and could be considered with equal justification as a chelator and as a transporter, in addition to its role in the feedback-mediated adjustment of ACTH secretion to the level of thyroid and gonadal hormones.

Monod: What is the specificity of transcortin?

Fortier: Transcortin, an α_1-glycoprotein with a molecular weight of 53 000 (Doe and Seal, 1965), has a high degree of specificity for corticosterone and cortisol, since slight molecular alterations of these corticosteroids markedly decrease their ability to displace ^{14}C-labelled corticosterone and cortisol from the transcortin binding sites (Daughaday, 1957). However, a slight binding affinity of transcortin for progesterone has been reported (Westphal, 1965).

Talwar: In the serum it appears that there are two types of steroid-binding proteins. Albumin can bind large quantities of almost all steroids. The binding is not selective and has low association constants. Then there are proteins such as transcortin which are present in very small amounts ($0 \cdot 6$–$0 \cdot 9$ μmole/litre; Murphy and Pattee, 1963), have high specificity for the ligand and bind the steroid tightly. The latter property is akin to receptors for oestradiol in the uterus. It is possible that the tightly binding proteins present in the serum leak out of the target tissues. Serum does represent to some extent the content of the cells also. They may have a function other than the mere transport of the steroid.

Tata: The question of the definition of circulating proteins that bind hormones is an important one. From all the available evidence transcortin, as other hormone-binding proteins in the blood, acts as a "buffer" rather than carrying the hormone into the target cell. It has been shown that the rate at which hormones are available to tissues is inversely proportional to the amount of serum binding protein present outside.

Monod: I would hope that you were wrong, Dr Tata! The reason is that everybody is thinking of acceptors, actual targets of hormones, and when one sees such a high binding constant one is tempted to ask whether this protein may not be *both* a chelator–buffer with respect to a vast number of systems which have fairly low affinity for the hormone, and *also* a transporter able specifically to go to the target tissue.

Tata: This depends on what the association constant for the intracellular binding sites is, which we don't know yet; but the little evidence one has in tissue homogenates or isolated mitochondria is that it is extremely high (see Tata, 1962).

Monod: Suppose you inject the transcortin already saturated with cortisol; does one have a physiological effect of the cortisol on the target tissue or not? If it depends on the association constant you would have very little effect; while if it acts as a transporter and finds the right target tissue and therefore increases the local concentration of the hormone you will have a positive effect.

Tata: Some years ago Shellabarger and I (1961) did an experiment in birds (chicken, duck). They lack the α_1-thyroxine-binding globulin, with the result that the bird will not discriminate between thyroxine and its analogues (with respect to both biological activity and the rate at which they can be metabolized or will enter into tissues). When we acutely infused birds with human α_1-thyroxine-binding globulin and thyroxine or its analogues, the animal handled these hormones as do humans or other mammals. Experiments in which different amounts of hormones were added to the same amount of binding protein suggested that we were dealing with only the free component of the hormone diffusing into the tissue.

Jagannathan: Will any of the enzymes which act on corticosteroids, such as the dehydrogenases, act in the presence of transcortin *in vitro*, and if they do, will the binding constant change appreciably, because that would offer a mechanism for release of the hormone from transcortin?

Fortier: I can only refer again to the reported inhibition of cortisol reduction by a liver homogenate as a result of the addition of transcortin to the protein-free medium (Sandberg and Slaunwhite, 1963); which, in the light of our liver perfusion experiments, suggests that transcortin interferes with the access of the corticosteroid to the inactivating sites. I don't understand your implication of an effect of enzymes on the binding constant.

Harris: If there is free corticosteroid in the plasma, the flow of the free corticosteroid out of the blood vessels into the extracellular spaces must be faster than the flow of the protein-bound material. It seems highly improbable that the bound form could be more available to the tissues than the free form. Dr Tata's argument and the argument that Professor Monod has made from the binding kinetics seems to me to be the right one. What we understand about the vascular membrane makes it very improbable that the form bound to protein could get across to the tissues before the free form.

Fortier: If it did, we would be at a loss to explain the results of our liver perfusion experiments.

Bhargava: Professor Monod, wouldn't you expect anyway that the proteins which bind circulating constituents present in very low amounts would have high binding constants?

Monod: I am speculating on this point because of this extremely high binding constant. I would think that with these fantastically insoluble compounds circulating in the blood with this high concentration of all sorts of proteins, and it is well known that any steroid will stick to any protein with fairly high binding constants of the order of 10^5 or more, these figures for free steroid must be wrong; there cannot be any free steroids. There may be some steroid loosely bound.

Bhargava: Is there any correlation, Dr Tata, between the affinity of the material binding to the protein and the amount in which it occurs in the blood?

Tata: Robbins and Rall (1957, 1960) have reviewed evidence in man and experimental animals that the transporting proteins are acting as "buffers". This is why it is so interesting that in those animals that lack the high-affinity binding protein it is extremely difficult to measure circulating hormones in the blood.

Fortier: The hamster which, incidentally, predominantly secretes cortisol, and in which the number of transcortin binding sites was found in my laboratory to be markedly lower $(2 \cdot 9 \times 10^{-7}$ moles/l; Pelletier, 1969) than in the rat $(13 \times 10^{-7}$ moles/l; Labrie, 1967), also has, under resting conditions, a cortisol secretion rate of only $0 \cdot 5$ μg/100g body weight/hr (Frenkel et al., 1965), as compared to a corticosterone secretion rate of $1 \cdot 7$ μg/100 g body weight/hr in the rat (Pelletier, Labrie and Fortier, 1967).

With regard to the existence of the free or unbound form of the corticosteroid in plasma which is queried by Professor Monod, I believe that our findings based on equilibrium dialysis at 37° c (Figs. 3 and 4, pp. 182 and 184) are in general agreement with those of other workers in this field (Westphal, Ashley and Selden, 1961; Quincey and Gray, 1963; Gala and Westphal, 1964).

White: Professor Monod's suggested experiment should be possible to do. One gets a good dose–response relationship between the concentration of circulating corticoids and the lymphoid tissue response. Therefore, it should be possible with this end-point to establish the relative quantity of corticoids bound to transcortin and that which is free, by making use of the lymphoid tissue response in the adrenalectomized animal.

Monod: The interesting possibility is whether the transcortin besides being a buffer, which it obviously is, is also a means of transporting specifically the hormone to the right target tissue. It is a very intriguing possibility, which is immediately suggested by the high association constants.

Harris: Is there any transcortin in the extracellular fluid? What are the relative concentrations of transcortin in the vascular and extracellular fluids?

Fortier: Sandberg and co-workers infer from their data that 50 per cent of the transcortin pool is extra-plasmatic and that the ratio of transcortin to total proteins in the various extracellular fluids is of the same order as in plasma (Sandberg et al., 1964).

Harris: If it is an α-globulin, its passage to the tissues would be slower than that of albumin; it would be a very slow process compared to the flow of the free form or even the albumin-bound form. If the free corticoid finds its way to the cell surface, and there seems little doubt that it does, I can't see the point of having a protein to help it find its way to the cell.

Monod: What are the most specific targets?

Fortier: The effect of corticosteroids on gluconeogenesis in the liver may be considered fairly specific.

Harris: Liver is not a very good example for the discussion of facilitated transport, because the sinusoids lining the liver spaces are fenestrated, so that the cells are in direct contact with the blood. There is not even a vascular barrier of the kind you would have in another organ.

Monod: These high binding constants remind one of the oestrogen receptors that Professor Talwar helped to discover which have an extremely high binding constant, of the order of 10^{11} or 10^{12}. The binding constant of 10^7 found for transcortin is already pretty high, and nature doesn't do these things except for some good reason.

White: Presently available data indicate that the uterine "receptor protein" which binds oestrogen is probably not identical with the plasma protein which binds oestrogen (Gorski *et al.*, 1968).

Monod: This is precisely why I feel that it may be something akin to these high-affinity receptors.

Talwar: If it is true that for cortisol to exert its effect it needs to be bound to proteins with such high affinity, one of the mechanisms could be that only those tissues which can interact with or pick up this *protein* will respond to this hormone.

Monod: This is exactly what I mean. The only initial evidence for the oestrogen receptor was the fact that there is a specific pumping or concentration effect of oestradiol in target tissues, because the receptor is there. Another mechanism for the same result would be to have the receptor circulating in the blood but having specific affinity for certain tissues.

Tata: One has also to consider intracellular "buffers" in hormone-sensitive tissues.

Harris: We don't really know that this hormone has to get *into* the cell; it might only have to reach the cell surface. In the case of ACTH it appears that the hormone merely has to reach the cell surface to exert its effect (Schimmer, Ueda and Sato, 1968).

Talwar: One locus of action of some other protein hormones also appears to be at the membrane, as for example insulin (Levine and Goldstein, 1955), and growth hormone (see our paper, p. 115).

Tata: Localization in itself is not of much use if one does not know the site of action of the hormone.

208   DISCUSSION

Jagannathan: These binding constants are for aqueous systems; if a small amount of phospholipid is added, is the binding constant markedly altered?

Fortier: The binding constant alluded to by Dr Jagannathan refers to the result of determinations made in plasma, a phospholipid-containing medium (83 mg/100 ml, with a 95 per cent range of 36–130 mg/100 ml in the rat; Spector, 1956). Remarkable agreement was demonstrated, moreover, by my group, through computer simulation, between the corticosteroid-binding protein equilibria predicted by the equations of the law of mass action for one ligand (corticosterone) and two binding proteins (transcortin and albumin) and observed equilibria under steady-state conditions (Tremblay, Normand and Fortier, 1969; Tremblay, 1969).

REFERENCES

DAUGHADAY, W. H. (1957). *J. clin. Invest.*, **36**, 881.
DAUGHADAY, W. H., HOZAK, I., and BIEDERMAN, O. (1959). *J. clin. Invest.*, **38**, 998–999.
DE MOOR, P., HEIRWEGH, K., HEREMANS, J. F., and DECLERCK-RASKIN, M. (1962). *J. clin. Invest.*, **41**, 816–827.
DOE, R. P., and SEAL, U. S. (1965). In *Endocrinology*, p. E153, ed. Mason, A. S., and Redecilla, A. M. Amsterdam: Excerpta Medica Foundation. International Congress Series No. 99.
DOUGHERTY, T. F. (1952). *Physiol. Rev.*, **32**, 379.
FRENKEL, J. K., COOK, K., GEADY, H. J., and PENDLETON, S. K. (1965). *Lab. Invest.*, **14**, 142–156.
GALA, R. R., and WESTPHAL, U. (1964). *Fedn Proc. Fedn Am. Socs exp. Biol.*, **23**, 357.
GORSKI, J., TOFT, D., SHYAMALA, G., SMITH, D., and NOTIDE, A. (1968). *Recent Prog. Horm. Res.*, **24**, 45.
LABRIE, F. (1967). *Interactions hormonales et rôle de la transcortine dans l'ajustement de l'activité hypophyso-surrénalienne.* Thèse de doctorat es sciences, Université Laval, Québec.
LABRIE, F., PELLETIER, G., LABRIE, R., HO-KIM, M.-A., DELGADO, A., MACINTOSH, B., and FORTIER, C. (1968). *Annls Endocr.*, **29**, 29–43.
LEVINE, R., and GOLDSTEIN, M. S. (1955). *Recent Prog. Horm. Res.*, **11**, 343–380.
MURPHY, B. P., and PATTEE, C. J. (1963). *J. clin. Endocr. Metab.*, **23**, 459.
PELLETIER, G. (1969). *Transcortine et interactions hormonale.* Thèse de doctorat es sciences, Université Laval, Québec.
PELLETIER, G., LABRIE, F., and FORTIER, C. (1967). *Proc. Can. Fedn Biol. Socs*, **10**, 66.
QUINCEY, R. V., and GRAY, C. H. (1963). *J. Endocr.*, **26**, 509–516.
ROBBINS, J., and RALL, J. E. (1957). *Recent Prog. Horm. Res.*, **13**, 161.
ROBBINS, J., and RALL, J. E. (1960). *Physiol. Rev.*, **40**, 415.
SANDBERG, A. A., and SLAUNWHITE, W. R. (1959). *J. clin. Invest.*, **38**, 1290–1297.
SANDBERG, A. A., and SLAUNWHITE, W. R. (1963). *J. clin. Invest.*, **42**, 51–54.
SANDBERG, A. A., WOODRUFF, M., ROSENTHAL, H., NIENHOUSE, S., and SLAUNWHITE, W. R. (1964). *J. clin. Invest.*, **43**, 461–466.
SCHIMMER, B. P., UEDA, K., and SATO, G. H. (1968). *Biochem. biophys. Res. Commun.*, **32**, 806.
SEAL, U. S., and DOE, R. P. (1963). *Endocrinology*, **73**, 371–376.
SHELLABARGER, C. J., and TATA, J. R. (1961). *Endocrinology*, **68**, 1056–1058.

SPECTOR, W. S. (ed.) (1956). *Handbook of Biological Data*, p. 53. Philadelphia and London: Saunders.

TATA, J. R. (1962). *Recent Prog. Horm. Res.*, **18**, 221–259.

TREMBLAY, R. (1969). *Facteurs régulateurs du táux de clearance métabolique de la corticostérone chez le rat.* Thesè de doctorat es sciences, Université Laval, Québec.

TREMBLAY, R., NORMAND, M., and FORTIER, C. (1969). *Proc. Can. Fedn Biol. Socs*, **12**, 11.

WESTPHAL, U. (1965). In *Endocrinology*, p. E54, ed. Mason, A. S., and Redecilla, A. M. Amsterdam: Excerpta Medica Foundation. International Congress Series No. 99.

WESTPHAL, U., ASHLEY, B. D., and SELDEN, G. L. (1961). *Archs Biochem.*, **92**, 441–448.

THE ROLE OF THE THYMUS GLAND IN THE HORMONAL REGULATION OF HOST RESISTANCE*

ABRAHAM WHITE AND ALLAN L. GOLDSTEIN

Department of Biochemistry, Albert Einstein College of Medicine,
Yeshiva University, Bronx, New York

AT the outset, I wish to acknowledge gratefully, on behalf of my colleagues and myself, the invitation from the Ciba Foundation and Dr Wolstenholme to participate in this symposium. Also, I would like to express particularly our indebtedness to a member of this conference, Dr J. F. A. P. Miller, who, together with workers in other laboratories (Archer and Pierce, 1961; Fichtelius, Laurell and Phillipsson, 1961; Miller, 1961; Arnason et al., 1962; Good et al., 1962; Jankovic, Waksman and Arnason, 1962), provided the initial data establishing the essential role of the thymus for the development of host immunological competence in neonatal animals. These initial observations of the effects of neonatal thymectomy were described as the beginning of the golden age of "thymology" (Miller, 1967). Since these initial reports, a voluminous number of new publications has appeared implicating the thymus in the development of host resistance. However, before these latter very significant observations of the past decade, there had been a period of over 100 years of recorded investigations of the chemistry, physiology and clinical significance of the thymus. It will not be possible to make adequate reference to this vast literature here, but we have recently summarized the pertinent data supporting the concept of an endocrine role of the thymus (White and Goldstein, 1968). The significant literature relevant to the functions of the thymus in immunological phenomena has been compiled recently in a scholarly review by Miller and Osoba (1967). These reviews, together with the publication of three recent symposia on the role of the thymus in health and disease (Defendi and Metcalf, 1964; Good and Gabrielson, 1964; Wolstenholme and Porter, 1966) and monographs by Metcalf (1966) and by Hess (1968), provide a substantial background for

* Text of paper given by Professor White as an open lecture at the All-India Institute of Medical Sciences, New Delhi, during the symposium.

students and investigators seeking a basis for orientation in the subject of thymology.

In this paper we shall have as one primary objective the presentation of some of the data from our laboratory in support of the general thesis that the thymus should be added to the accepted list of endocrine glands. A second objective will be to provide data which indicate that by virtue of the production of one or more humoral factors, the thymus plays a fundamental role in the regulation of the structure and functions of lymphoid tissue. Thirdly, as a consequence of its regulatory influence on lymphoid tissue, experimental data can be assembled which establish a humoral role for the thymus in regulating certain of the functions that have been attributed to the thymus and the lymphatic system, including host immunological competence.

THE THYMUS AS AN ENDOCRINE ORGAN

The first experimental removal of the thymus in laboratory animals was achieved in 1845 by Restelli (see White and Goldstein, 1968) who operated on 98 animals, including sheep, dogs and calves. Only six of these animals (four sheep, one calf and one dog) survived the operation. Of these, all died from 9 to 23 days post-operatively as a consequence of infection. Over the subsequent years, a great number of diverse functions was claimed for the thymus (see Park and McClure, 1919) and a variety of polar and nonpolar cell-free thymic extracts and fractions were described with biological activities (White and Goldstein, 1968). The latter included an influence on growth, regulation of blood glucose and calcium and phosphorus metabolism, and an interplay with various endocrine glands, notably the adenohypophysis. It may be noted that certain of these earlier reports of diverse biological actions of thymic preparations could have some validity in the light of knowledge of the embryological and developmental history of the thymus, supplemented by more recent investigations. Thus, the reported effects of thymic extracts on calcium and phosphate metabolism (Bracci, 1905; Schwartz, Price and Odell, 1953; Potop, Boeru and Mreană, 1966) might be a reflection of the similar embryological origins of certain cell types, for example, C cells of the thymus, the parathyroids, and in certain species, the ultimobranchial body. Indeed, calcitonin, the calcium and phosphate regulatory hormone whose presence has been established in the parathyroids, the thyroid, and the ultimobranchial body, has been isolated recently from the human thymus (Galante et al., 1968). The presence of this hormone, therefore, in crude thymic extracts, could provide a basis for previous reports of the role of the thymus in calcium and phosphate metabolism. Again, the established roles of thyroidal and growth hormones in accelerating the proliferation of lymphoid tissue (Dougherty, 1952), coupled with the involution of this tissue under the influence of the secretions of the adrenocorticotropic–adrenal cortical axis (Dougherty and White, 1945; Dougherty, 1952; White, 1947–48), lend credence to the hypothesis that adenohypophyseal mechanisms may interrelate with the thymus and/or thymic factors in the regulation of lymphoid tissue structure and function.

Our earlier studies designed to elucidate the role of the lymphocyte in antibody production and in normal immune phenomena led us to the hypothesis that lymphoid structures themselves might participate in humoral mechanisms concerned with immune globulin synthesis. Our initial studies of the role of lymphoid tissue in antibody production (White and Dougherty, 1946; White, 1947–48) led to the demonstration that normal lymphocytes contain gamma globulin (White and Dougherty, 1945) and that these lymphocytes (Dougherty, Chase and White, 1944) as well as malignant proliferating lymphocytes from immunized animals both contain and have the capacity to synthesize antibody (Dougherty, White and Chase, 1945). These observations, coupled with the role of such lymphocytes in passive immunity (Chase, 1945, 1953), contributed to the understanding of the functional significance of the lymphocyte in antibody production and in immune phenomena.

The hypothesis that lymphoid tissue might have a humoral role developed from a series of studies in collaboration with Dr T. F. Dougherty. These experiments demonstrated that lymphoid cells are a prime target of the action of the 11-oxygenated adrenal cortical steroids (Dougherty and White, 1945; White, 1947–48; Dougherty, 1952). The dramatic dissolution of lymphoid tissue produced by these steroid hormones, considered in the light of the known "feedback" action of target tissue steroid hormones on the secretory activity of the adenohypophysis, led us to postulate that lymphoid tissue might produce one or more factors which could repress or modulate ACTH-adrenal cortical secretory levels. The initial and simplest test of this hypothesis appeared to be the examination of lymphoid tissue extracts for their ability to induce changes characteristic of adrenal cortical suppression—namely, a state of hypoadrenalcorticalism or of "functional adrenalectomy".

In 1949 we described (Roberts and White, 1949) the fractionation of calf, rabbit and rat thymic tissue by the cold ethanol technique devised by Cohn and his colleagues for the separation of the plasma proteins (Cohn *et al.*, 1946). It was demonstrated that a protein fraction from calf thymic tissue, when administered to normal rats over a ten-day period, produced an increase in the weight of lymphoid tissue and an absolute increase in the numbers of blood lymphocytes. In 1956, Gregoire and Duchateau (Gregoire and Duchateau, 1956) obtained from thymic tissue of rats and pigs exposed to a lethal dose of radiation, a cell-free fraction which when administered to normal rats produced a lymphocytosis and an increase in lymphoid tissue weight. In the same year, Metcalf (1956) described the preparation of an aqueous extract of thymic tissue, which when injected into newborn mice produced a lymphocytosis. The factor responsible for this activity was termed by Metcalf "the lymphocyte stimulating factor (LSF)". The lymphocytopoietic activity of similar crude extracts has been summarized elsewhere (White and Goldstein, 1968).

The principal consequences of neonatal thymectomy are: (1) a deficiency in the lymphocyte population of the blood and the lymphoid tissues; (2) a loss of the ability of the operated animal to elicit most cell-mediated immune responses; (3) the development of a syndrome described as "wasting disease" because the thymectomized animals fail to grow normally after weaning, exhibit atrophic symptoms, fail to survive beyond a period of several months post-operatively, and have a diminished ability to make antibodies to some antigens.

Since the initial studies of neonatal thymectomy in mice, rats and rabbits, it has been established in many other experimental animals, including guinea pigs, hamsters, chickens, opossums, dogs and calves, and clinically in the human, that experimental removal or failure of the thymus to develop in the newborn will produce a diminished population of lymphocytes in lymphoid tissue, as well as in the blood. In addition, host immunological competence either fails to develop normally or is markedly depressed. More recently, it has been recognized that thymectomy in older animals may lead to a slow diminution of lymphocyte populations, with attendant alteration of phenomena which are dependent on normal numbers and functions of these cells (Metcalf, 1965; Miller, 1965; Taylor, 1965). The demonstration that thymic transplants, as well as thymic tissue in Millipore diffusion chambers, were capable of preventing wasting disease and restoring immunological responsiveness in neonatally thymectomized animals of several species (reviewed by Miller and Osoba, 1967) rekindled interest in a possible humoral function of the thymus (reviewed by White and Goldstein, 1968). Lymphocyto-poietic activity and partial alleviation of some of the deficiencies of neonatally thymectomized mice with cell-free thymic extracts have been reported by several laboratories (reviewed by White and Goldstein, 1968).

BIOASSAY OF CELL-FREE THYMIC EXTRACTS

In our own studies, we used initially syngeneic mouse thymic tissue. As the need for larger quantities of tissue developed, we turned to rat thymus and, subsequently, to the calf as a source of thymic tissue for fractionation studies. The fractionation procedure described below has been developed with calf thymus but is equally applicable to thymic tissue of other species. All procedures, except where indicated, are carried out at $5°$c. The calf thymic tissue is homogenized with $0·15$ M-NaCl, followed by a low-speed centrifugation which removes insoluble cellular debris. If the supernatant fraction is then centrifuged at $105000 \times g$ for approximately 60 min, the clear supernatant solution exhibits lymphocytopoietic activity when administered daily to adult CBA mice (Klein, Goldstein and White, 1965, 1966). The treated mice show an increase in lymph node weight, expressed as milligrammes per gramme of body weight, and an increase in the numbers of circulating lymphocytes. The active soluble component of this crude thymic fraction was subsequently further purified and designated as thymosin (Goldstein, Slater and White, 1966). Thymosin has also been found to stimulate incorporation of precursors of DNA, RNA and protein into lymph nodes of injected mice, reflecting at a biochemical level the gross increases in lymph node weight seen after administration of thymosin to adult CBA mice.

Data suggesting a stimulatory action of thymic fractions on lymphocytopoiesis in normal immunologically competent animals must be interpreted with some caution. Injection of preparations with a high content of foreign protein into test animals may induce a proliferation of lymphoid tissue characteristic of the response to any antigenic stimulus. We have indeed found this to be the case when adult CBA mice are injected with thymic fractions other than those from syngeneic donors. The degree of antigenicity of such thymic fractions is influenced by the quantity of protein administered and the immunological competence of the recipient animal. Thus, the more highly purified thymosin fractions, when administered in microgramme quantities ($< 10 \mu g$), induce lymphoid tissue proliferation not only in normal mice (see below) but also in mice whose immunological responsivity has been depressed by prior exposure to X-irradiation, as well as in either adrenalectomized, germfree or neonatally thymectomized mice (unpublished results). However, at higher doses, the most active preparations of thymosin presently available, and which have the properties of a protein, are antigenic. Indeed, an antiserum to thymosin has been prepared (Hardy *et al.*, 1968, 1969). Antigenic activity, however, can be dissociated from certain of the other biological activities of thymosin to be considered below.

A detailed histological study has been made by Dr T. F. Dougherty of the University of Utah School of Medicine of the tissues of adult CBA mice injected subcutaneously daily with 3 mg of a thymosin fraction for seven days and sacrificed 24 hr after the last injection (unpublished results). Dr Dougherty has reported (personal communication) extensive changes

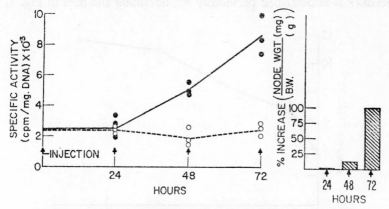

FIG. 1. Lymph node weights and the incorporation *in vivo* of [³H]-thymidine (10 μc per animal) into the total DNA of pooled axillary, brachial, and inguinal lymph nodes of adult Swiss-Webster CD-1 mice after daily injections (indicated by arrows) of (● — ●) a calf thymic fraction (fraction 1A) equivalent to 2·4 mg protein (each injection) dissolved in 0·25 ml 0·15 M-NaCl, and (○ - - - ○) 0·15 M-NaCl, 0·25 ml. Animals sacrificed 24 hr after the last injection; each animal received the dose of tritiated thymidine 1½ hr before sacrifice. Each point in the figure represents the pooled lymph nodes of four mice. (From Goldstein, Slater and White, 1966.)

in the cellular populations of both the lymph nodes and spleens in animals treated with these active extracts. There was a marked proliferation of lymphoid cells in the lymph nodes; many of the cells were in active mitosis. Indeed, rather surprising mitotic activity was observed in single free lymphocytes in the loose connective tissue. In the spleen of thymosin-treated animals there was extensive replacement of the erythroid elements by lymphoid cell populations.

The time-course of the stimulatory action of a thymosin fraction (Fraction 1A) on the incorporation of a labelled precursor ([³H]thymidine) into the total DNA of lymph nodes of young mice is depicted in Fig. 1. The data indicate that there is present in thymic extracts a protein fraction which under these experimental conditions stimulates the incorporation of a labelled precursor into lymph node DNA.

On the basis of the data depicted in Fig. 1, we have developed an *in vivo* assay for use in the purification of thymosin (Goldstein, Slater and White, 1966; unpublished results). Adult CBA mice from our own colony are injected subcutaneously daily for three days with the material to be assayed. Twenty-two and a half hours after the third injection of the test preparation each mouse receives an intraperitoneal injection of 10 μc of [³H]thymidine (specific activity 3·0 c/m-mole). The subsequent procedure is as described previously for obtaining the data in Fig. 1. It

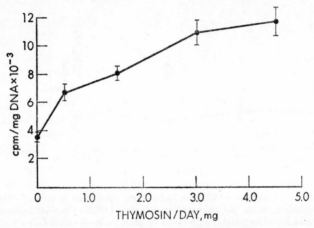

FIG. 2. Effect of varying quantities of a thymosin fraction (fraction 3) in the standard 72 hr *in vivo* assay system depicted in Fig. 1, except that the mice used were 60-day-old CBA/W male animals from our own colony. Each point represents the mean value of pooled lymph nodes of 12 mice in three individual assays. I=range of values.

is apparent from those data that after a single injection there is little stimulation of lymphocytopoiesis. After two injections there appears to be some lymphocytopoietic response, and after three injections there is more than a doubling of the incorporation of tritiated thymidine into the DNA of lymph nodes of thymosin-injected mice as compared with control animals treated with saline. It is also apparent from the right-hand portion of Fig. 1 that the lymph node weight begins to increase after two injections of thymosin, and after three injections is significantly larger than in saline-injected control mice. The time-interval between injection of an active thymic fraction and measurable lymphoid tissue proliferation is shortened by approximately 24 hr in the adrenalectomized mouse (unpublished results).

Fig. 2 depicts data obtained in an examination of a possible dose-response relationship in the *in vivo* assay procedure. This assay has been useful in the purification of the lymphocytopoietic activity of calf thymic extracts.

PREPARATION AND PROPERTIES OF THYMOSIN

The procedure now in use for the extraction and purification of thymosin includes several modifications of that described previously (Goldstein, Slater and White, 1966). The thymosin activity in the high-speed supernatant fraction (105 000 × **g**) described above is remarkably heat stable, in contrast to the heat lability of the lymphocyte stimulating factor described by Camblin and Bridges (1964), Metcalf (1966), and by Trainin, Burger and Kaye (1967). The relative heat stability of our preparation allows us to subject the above described high-speed supernatant fraction to a heat step (>80° c for 15 min) which precipitates approximately 80 per cent of the contaminating proteins. The latter are removed by high-speed centrifugation (> 20 000 × **g**), and the thymosin activity in the supernatant solution is precipitated by slowly pouring the latter with stirring into 10 volumes of cold acetone (−30° c). The precipitated thymosin is collected on a Buchner funnel and purified further by fractionation with ammonium sulphate, followed by successive column chromatography on DEAE-cellulose and Sephadex G-100. The most highly purified preparation available at this time represents an approximately 200-fold purification in comparison with the activity, per milligramme of protein, of the initial high-speed supernatant fraction (see above). The fraction obtained from the Sephadex G-100 chromatography step will produce a significant increase in the incorporation of tritiated thymidine into lymph node DNA when injected in a concentration of 10 μg of protein per day for three days in the *in vivo* assay.

Preliminary evidence suggests that during the described fractionation of thymosin, an inhibitor of lymphocytopoiesis is removed. This is being explored further.

Chemical studies of the purified thymosin fraction have yielded the following data (unpublished observations). Polyacrylamide gel electrophoresis at pH 8·3 in Tris-glycine buffer reveals one major and two minor components, all migrating toward the anode. On the basis of calibrated Sephadex gel chromatography, the active component appears to have a molecular weight of less than 100 000 and exhibits some tendency to aggregate. The preparation has less than one per cent of carbohydrate and a trace of lipid which is not essential for biological activity.

This activity is either identical to, or closely associated with, a protein. The preparation at this stage appears to have nothing unusual in its amino acid composition. The purified material, as described also for the earlier fractions, is relatively heat stable and does not contain nucleic acid. Activity is unchanged after digestion with either ribonuclease or deoxyribonuclease, but is slowly destroyed by digestion with proteolytic enzymes.

Thymosin lymphocytopoietic activity has also been demonstrated in thymic tissue of a variety of mammals, including the rat, the rabbit, the guinea pig, the hog, and the human. It may also be of interest to report that while the general concept has prevailed that the functional roles of the thymus are important primarily for the newborn and for younger animals before puberty, we have found that the quantity of thymosin which can be isolated per gramme of thymus in older animals, such as old cows and steers, is as great as or greater than that isolated from the thymus of young calves (unpublished results). These observations are in harmony with data suggesting that the humoral agent of the thymus is not elaborated by the mature thymocyte but probably by the epithelial cells. In the thymus of the older animal, the relative proportion of epithelial cells to mature thymocytes is higher than in the thymus of the younger animal. Also, these observations are relevant for the developing concept of a functional role for the thymus in the adult animal.

Similar stimulation of incorporation into lymph node RNA occurs with a precursor such as [³H]uridine, and into the total protein of pooled lymph nodes with either leucine or phenylalanine as labelled precursor, following administration of thymosin for three days to mice, as described above.

ROLE OF THYMOSIN IN HOST IMMUNOLOGICAL COMPETENCE IN IMMUNOLOGICALLY DEFICIENT MICE

Studies of the properties of thymosin in biological systems in which the thymus has been implicated indicate that a number of the essential functions ascribed to the thymus or thymocytes can be mimicked by our cell-free preparation of thymic tissue. One of the obvious experiments was concerned with whether or not thymosin could replace thymic function in the neonatally thymectomized mouse, and thus moderate the severity of symptoms developing in such operated animals as a result of diminished host immunological competence. These studies have been conducted in collaboration with Dr Y. Asanuma (Asanuma, Goldstein and White, 1969).

It is worthwhile to emphasize that we have used for most of our studies mice of the CBA strain which have been carried for a number of years in our laboratory. These animals derived originally from Dr L. C. Strong at Yale University School of Medicine and subsequently from Dr T. F. Dougherty of the University of Utah School of Medicine. We have recently established unequivocally that another line of CBA mice, namely that from the Jackson laboratories at Bar Harbor, Maine, and designated as CBA/J, differs significantly from our line of CBA animals in life-span, size of lymphoid tissue, and the relative resistance of this lymphoid tissue to whole-body X-irradiation. We make this point here to alert others to the need to designate clearly the source of CBA mice used experimentally, even though they are of the same strain. For the above reasons, we have recently designated our line of CBA mice as CBA/W.

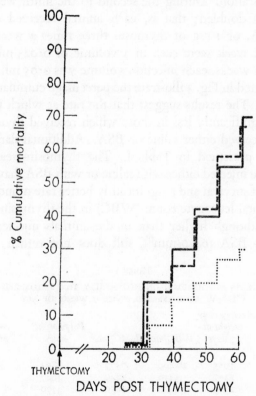

FIG. 3. Cumulative mortality in neonatally thymectomized CBA/W mice. ——, animals injected with saline; – – – –, animals injected with BSA; · · · ·, animals injected with thymosin (fraction 3; Goldstein, Slater and White, 1966). See text for other details. (From Asanuma, Goldstein and White, 1969.)

Our CBA/W mice, when subjected to thymectomy within 24 hours after birth, have a high incidence of mortality, reaching 80 per cent by nine weeks post-operatively. The 20 per cent of the operated animals that survive 63 days or longer look quite normal, although their body weight is less than normal. These survivors have been shown to be immunologically impaired and are highly susceptible to infections. Wasting disease, as described initially by Miller (1961), begins to be discernible shortly after weaning—that is, at approximately 30 days post-thymectomy—and progresses as the mice age.

Neonatally thymectomized CBA/W mice were divided into three groups. Each animal received, beginning three days after thymectomy, an intraperitoneal injection three times a week during the first week of either saline, 0·5 mg of bovine serum albumin (BSA), or 0·5 mg of a thymosin preparation. During the second to the ninth weeks, the dose of protein was doubled; that is, each animal received either saline, 1·0 mg of BSA, or 1 mg of thymosin three times a week. Injections during the first week were each in a volume of 0·025 ml; during the second to ninth weeks, each injection volume was 0·05 ml.

The data plotted in Fig. 3 illustrate the percentage cumulative mortality of the animals. The results suggest that the rate at which the mice succumbed was significantly less in those which received thymosin than in those which received either saline or BSA. Additional data obtained in this study are presented in Table I. The thymosin-treated mice, in contrast to those injected either with saline or with BSA, have a markedly increased rate of survival and a significantly better rate of increase in body weight. The total leucocyte count (WBC) in the thymosin-treated mice (7983/mm³), although higher than in the animals injected with saline (5476/mm³) or BSA (6420/mm³), still does not return to a normal

TABLE I

EFFECTS OF THYMOSIN ON NEONATALLY THYMECTOMIZED
CBA/W MICE TREATED UNTIL 9 WEEKS OF AGE

Treatment*	Average body weight at 63 days	WBC/mm^3	LMC/mm^3	Polymorphs/ mm^3	Survival Number	Percentage
Sham operated (14)†	19·5	10 469	7577	2892	14	100
Saline (24)	12·7	5476	1429	4047	8	33
BSA (20)	13·5	6420	1650	4770	6	30
Thymosin (40)	16·5	7983	3258	4725	28	70

* 1st week: 0·50 mg protein, three time a week; 2nd-9th week: 1·0 mg protein, three times a week.

† Numbers in parentheses are number of animals in each group.

(From Asanuma, Goldstein and White, 1969.)

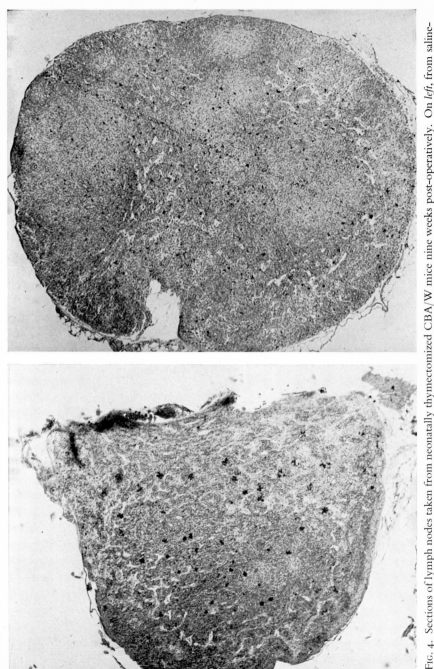

FIG. 4. Sections of lymph nodes taken from neonatally thymectomized CBA/W mice nine weeks post-operatively. On *left*, from saline-treated mouse; note depletion of small lymphocytes, general "washed out" appearance and absence of follicular development. On *right*, from thymosin-treated mouse; note increased population of small lymphocytes and evidence of follicular development. (From Asanuma, Goldstein and White, 1969.)

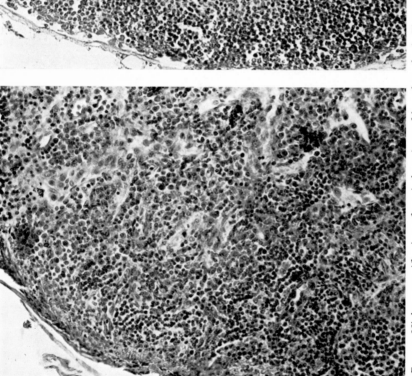

Fig. 5. Higher magnification of cortical sections of lymph nodes in Fig. 4. Note in section from saline-treated animal, on *left*, hyperplasia of reticular epithelial elements. On *right*, section from thymosin-treated mouse shows well-developed follicle with evident germinal centre and numerous small lymphocytes in paracortical region. (From Asanuma, Goldstein and White, 1969.)

($10\,469/\mathrm{mm}^3$) value under these experimental conditions. This is also reflected in the absolute numbers of lymphocytes (LMC) which, although higher in the thymosin-treated animals ($3258/\mathrm{mm}^3$) than in those injected with saline ($1429/\mathrm{mm}^3$) or BSA ($1650/\mathrm{mm}^3$), did not reach normal ($7577/\mathrm{mm}^3$) levels. Histological examination of the lymph nodes and spleens of the thymosin-treated mice also shows significant restoration of lymphoid cell populations, particularly in the cortical areas ("thymic dependent") of lymph nodes and spleen (Figs. 4 and 5).

Fig. 6 is a photograph of two neonatally thymectomized mice from the above experiment, nine weeks after operation. The animal on the left was treated with thymosin three times a week; the one on the right, which shows wasting, received a similar quantity of BSA. Once wasting begins, the animal does not survive more than two to three weeks.

These data obtained with neonatally thymectomized mice support previous evidence from a number of laboratories (reviewed by Miller and Osoba, 1967) that a neonatally thymectomized mouse does not grow at a normal rate, shows depletion of its lymphoid cells, and dies prematurely as a consequence of infection. This has led us to examine two aspects of the role of the thymus in host immunological competence, namely, its influence on cell-mediated immune processes as reflected in the homograft response, and humoral antibody production as revealed by the capacity to synthesize 19s antibody. These studies have been conducted in collaboration with Drs M. A. Hardy and J. Quint of our Department of Surgery and Dr J. R. Battisto of our Department of Microbiology and Immunology (Goldstein et al., 1969a).

CBA/W mice were thymectomized within 24 hours of birth, divided into groups, and treated as indicated in Table II. Animals surviving at 9 weeks of age received at that time an allograft of skin from A/J strain

TABLE II

INFLUENCE OF THYMOSIN ON ALLOGRAFT SURVIVAL IN
NEONATALLY THYMECTOMIZED CBA/W MICE*

Group	Number of mice in group	Allograft survival‡ 32 days after grafting
Thymectomized + saline	10	7
Thymectomized + BSA†	4	2
Thymectomized + spleen fraction†	4	4
Thymectomized + thymosin†	22	1
Sham thymectomized	14	0

* Thymectomy within 24 hr of birth.
† 1st week: 0·50 mg protein, three times a week; 2nd–9th week: 1·0 mg protein, three times a week.
‡ A/J strain skin.

(From Goldstein et al., 1969a.)

8*

mice. The table shows the number of allografts which survived 32 days after the skin graft—that is, until the mice were 95 days of age. At 91 days of age, each animal received an intraperitoneal injection of 1 ml of a 2·5 per cent suspension of sheep erythrocytes (SRBC). Four days later the animals were sacrificed and the number of antibody-forming cells per spleen determined using the Jerne assay (Jerne and Nordin, 1963). The data in Table II show that neonatally thymectomized animals treated with either saline, BSA, or a calf spleen fraction do not reject allografts readily. A normal rejection time by CBA/W mice of an A/J allograft is approximately 14 days. In groups treated with saline, BSA and the spleen fraction, most of the grafts were still intact 32 days after grafting, with evidence of normal hair growth. The several mice that did reject their grafts did so much later than sham-thymectomized controls—that is, between days 22 and 28. In striking contrast, only one of the 22 animals treated with thymosin retained the allograft for 32 days after grafting. Most of the thymosin-treated mice rejected their grafts within 16 days. Fig. 7 shows two animals from the group providing the data in Table II; the photograph was taken at 95 days of age. These results support the suggestion that, in this experimental design, a thymosin preparation did indeed function in lieu of the thymus in the thymectomized animal in endowing the host with specific capacity to reject a skin allograft.

The number of antibody-producing cells within a single experimental group as measured by the Jerne plaque assay shows a wide degree of variability. This variability has also been encountered routinely by other investigators. Despite this, the plaque assay has utility in detecting antibody production at an early time before appearance of measurable circulating titres. The data plotted in Fig. 8 indicate that the number of antibody-producing cells per spleen in the thymectomized control animals treated either with saline or with BSA are in the low range, as reported previously by others. Although the mean number of antibody-producing cells in the plaque assays for the animals treated with thymosin (3399) is slightly elevated above these values for the mice injected with saline (845) or BSA (892), the wide variability within each group prevents the data from having significance with the number of animals presently available. These studies are being repeated with both larger numbers of animals and larger doses of thymosin.

At present, therefore, it may be concluded from these data that at a time when the ability of the neonatally thymectomized, thymosin-treated mice appeared to be relatively normal with regard to their capacity to reject a histoincompatible skin graft, their response to an antigenic

FIG. 6. Photograph of neonatally thymectomized CBA/W mice at nine weeks of age. On *left*, thymosin-treated mouse; on *right*, BSA-treated mouse. (From Asanuma, Goldstein and White, 1969.)

FIG. 7. Photograph of two neonatally thymectomized 95-day-old CBA/W mice grafted with A/J skin 32 days previously. On *left*, thymosin-treated mouse; note total rejection of allograft with scar tissue remaining. On *right*, saline-treated mouse; note allograft survival with obvious hair growth. (From Goldstein *et al.*, 1969a.)

To face page 222

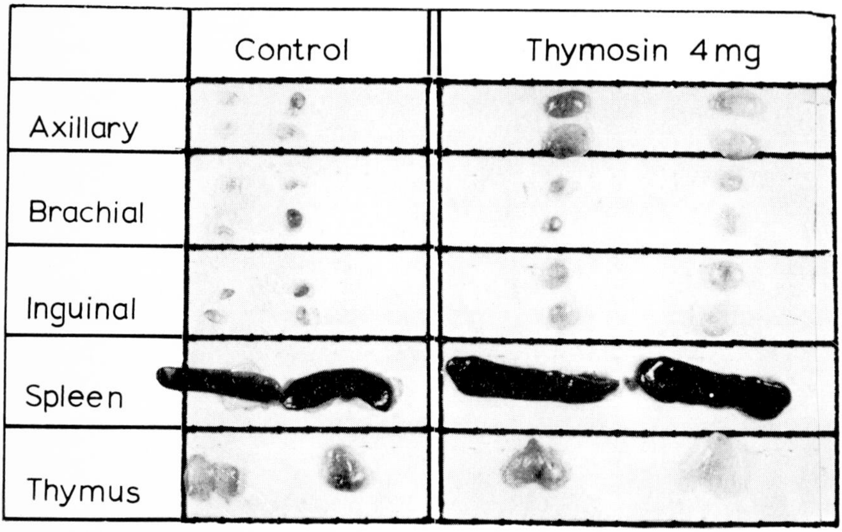

	Control	Thymosin 4mg
Axillary		
Brachial		
Inguinal		
Spleen		
Thymus		

FIG. 9. Photograph of gross lymphoid tissue from each of four CBA/W mice exposed to 700 R whole-body radiation. Tissues on *left* from each of two mice injected daily with 0·2 ml 0·15 M-NaCl subcutaneously for ten days after irradiation. On *right*, tissues from each of two mice similarly irradiated but then injected daily for ten days with 4 mg thymosin in 0·2 ml 0·15 M-NaCl.

challenge of SRBC did not appear to be significantly restored. Whether or not there is a primary influence of thymosin on 19s antibody production remains to be explored.

Another approach to the study of the role of the thymus in immunological competence in adult animals is the use of thymectomy together

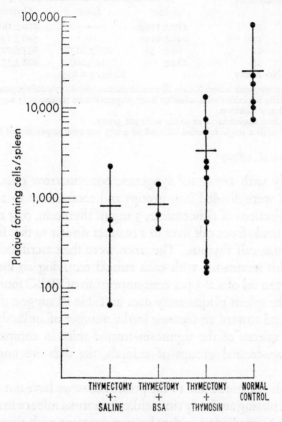

FIG. 8. Individual values of 19s antibody-producing cells (plaque forming) in spleens of neonatally thymectomized CBA/W mice treated with either saline, BSA or thymosin. (From Goldstein et al., 1969a.)

with a lethal dose of X-irradiation, followed by administration of syngeneic bone marrow cells. This procedure permits the animals to survive but leaves them essentially immunologically incompetent. Table III presents the results of a study using this experimental approach in an attempt to elucidate further the influence of thymosin on 19s antibody production (Goldstein et al., 1969a). These A/J adult mice were thymectomized one week before X-irradiation with 800 R and then injected

TABLE III

PLAQUE FORMING CELLS (PFC) TO SHEEP ERYTHROCYTES IN
SPLEENS OF ADULT THYMECTOMIZED LETHALLY IRRADIATED AND
BONE MARROW RESTORED A/J MICE*

Days post irradiation†	Number of PFC per spleen from mice treated with:		
	Saline	Liver	Thymosin
10	119±110‡	—	270±198‡
14	241±114	—	540±148
21	66± 56	276±215‡	633±223
28	215± 85	354±207	828±332
No X-ray		10 897±6 855	

* Thymectomy one week before X-ray; all treated mice received 1·7 × 10⁷ syngeneic bone marrow cells; up to 21 daily injections with saline or liver preparation or thymosin (3 mg protein/mouse).
† 800 R total-body radiation.
‡ Means± standard deviations; five to ten mice per group.
Immunization with a single injection of 1 ml of a 2·5 per cent suspension of SRBC four days before sacrifice.

(From Goldstein et al., 1969a.)

intravenously with $1·7 \times 10^7$ syngeneic bone marrow cells. Following this, animals were divided into groups and received up to 21 daily sub-cutaneous injections of either saline, 3 mg of thymosin, or 3 mg of a liver preparation made from calf liver in a manner similar to that for preparing thymosin from calf thymus. The mice were then sacrificed at intervals following this treatment, with each animal receiving an intraperitoneal injection of 1·0 ml of a 2·5 per cent suspension of SRBC four days before sacrifice. The spleen plaque assay data in Table III suggest that although there is a trend toward an increase in the number of antibody-producing cells in the spleens of the thymosin-treated mice in comparison to the saline or liver-treated groups of animals, the data are not statistically significant.

It might also be noted that to the present time we have not succeeded in affecting circulating antibody titres either in normal mice or in normal mice subjected to X-irradiation, either by pretreatment with thymosin before antigenic challenge or by giving thymosin during the period of immuniza-tion (unpublished results). These observations, taken together with the skin allograft studies presented above and others considered below, indicate that although thymosin does influence cell-mediated immune phenomena, there are as yet no significant data supporting a role for thymosin in humoral antibody phenomena.

EFFECTS OF THYMOSIN ON X-IRRADIATED MICE

For some time we have been interested in the effects of ionizing radiation on lymphoid tissue, which is one of the most radiosensitive tissues of the body. A number of years ago we were able to elucidate the role of the adrenal cortex in radiation-induced lymphoid tissue involution (Dougherty and White, 1946). With this background of information, we have recently completed an initial study of the influence of thymosin on the rate of regeneration of lymphoid tissue in animals subjected to whole body X-irradiation (Goldstein et al., 1969b). CBA/W mice were subjected to either 400 R or 700 R whole-body irradiation. After irradiation the mice were divided into groups, one of which received subcutaneous injections of 0·15 M-saline, a second 3 mg of BSA, and a third 3 mg of thymosin. Injections were given daily, with the first within two hours of exposure to X-ray. Animals were sacrificed at daily intervals after irradiation, each mouse receiving an intraperitoneal injection of 10 μc of either [^{3}H]deoxycytidine or [^{3}H]thymidine 1½ hr before sacrifice. Pooled axillary, inguinal and brachial lymph nodes were weighed on a torsion balance and the DNA was isolated from these pooled lymph nodes and its radioactivity determined as described previously. In addition, thymus, lymph nodes and spleen were taken for histological studies and for radioautography.

The effect of thymosin and BSA on lymph node weight of mice exposed to 400 R total-body X-irradiation is shown in Table IV. At 24 hr after irradiation there is a significant decrease in the ratio of lymph node to body weight of mice in all experimental groups in comparison to unirradiated mice injected with saline. At 48 and 72 hr after irradiation

TABLE IV

EFFECT OF INJECTION OF THYMOSIN AND OF BOVINE SERUM ALBUMIN ON LYMPH NODE WEIGHT OF CBA/W MICE AFTER EXPOSURE TO 400 R TOTAL-BODY X-IRRADIATION

Type of injection*	Number of animals in experiment	Duration of experiment post X-ray (hr)	Total protein injected daily (mg)	Number of injections	Lymph node weight (mg/gm body weight)†	Percentage change
0·15 M-NaCl	8	‡	—	3	1·46	—
0·15 M-NaCl	16	24	0	1	1·10	−26
BSA§	8	24	3	1	0·56	−49
Thymosin	16	24	3	1	1·01	−8
0·15 M-NaCl	16	48	0	2	0·79	—
BSA	8	48	3	2	0·68	−14
Thymosin	16	48	3	2	1·03	+30
0·15 M-NaCl	24	72	0	3	0·90	—
BSA	8	72	3	3	1·13	+26
Thymosin	24	72	3	3	1·50	+67

* Each injection subcutaneously in a volume of 0·2 ml.
† Pooled axillary, inguinal and brachial lymph nodes from groups of 4 mice.
‡ Unirradiated animals.
§ BSA, bovine serum albumin.

(From Goldstein et al., 1969b.)

these values remain lower than those seen in unirradiated mice. In contrast, the irradiated mice treated with thymosin after irradiation show, by 48 hr after exposure, a more rapid rate of restoration of lymph node weight and at 72 hr this has returned to normal values.

The effect on the weight of mouse lymph nodes and spleen of giving thymosin to animals before their exposure to 400 R whole-body X-irradiation is shown in Table V. The nodes and spleens of irradiated mice that had received thymosin before irradiation are smaller than pretreated, unirradiated thymosin controls, but are significantly larger than lymphoid tissues from saline-treated irradiated controls. Indeed, the nodes of animals pretreated with thymosin are almost as large at 72 hr after irradiation as those of control, saline-injected unirradiated mice.

TABLE V

EFFECT OF PRIOR INJECTION OF THYMOSIN ON LYMPH NODE AND SPLEEN WEIGHT OF CBA/W MICE EXPOSED TO 400 R WHOLE-BODY X-IRRADIATION

Type of injection*	Duration of experiment post X-ray (hr)	Number of injections	Lymph node weight (mg/gm body weight)†	Percentage change	Spleen weight (mg/gm body weight)†	Percentage change
0·15 M–NaCl	0	7	1·63	—	4·09	—
Thymosin	0	7	3·79	+133	5·15	+26
0·15 M–NaCl	4	7	1·79	—	3·59	—
Thymosin	4	7	3·11	+74	4·15	+16
0·15 M–NaCl	24	8	1·24	—	2·90	—
Thymosin	24	8	2·72	+119	3·81	+31
0·15 M–NaCl	48	9	1·13	—	2·14	—
Thymosin	48	9	2·77	+145	2·78	+30
0·15 M–NaCl	72	10	1·0	—	2·37	—
Thymosin	72	10	1·93	+93	3·18	+34

* Each injection subcutaneously in a volume of 0·2 ml. Thymosin-treated animals received 3 mg protein daily, beginning seven days before radiation.

† Each value represents the mean of a group of eight animals; lymph node weights represent pooled axillary, inguinal and brachial lymph nodes.

(From Goldstein et al., 1969b.)

Fig. 9 is a photograph of selected lymphoid tissue from CBA/W mice which had been exposed to 700 R whole-body irradiation ten days previously and then treated daily subcutaneously, beginning within two hours of exposure to X-rays, with either thymosin (4 mg per day) or saline.

Fig. 10 illustrates data for the effect of exposure of mice to 400 R whole-body X-irradiation and the influence of thymosin on the incorporation of [³H]deoxycytidine into lymph node DNA over a subsequent 72 hr period. Fig. 11 shows data for a similar study in which the radiation dose was increased to 700 R and in which [³H]thymidine was used as the labelled precursor of DNA during a subsequent 168 hr period. Also, in this second study the daily dose of thymosin was increased to 4 mg per animal. It is apparent in both studies that administration of thymosin to CBA/W mice after irradiation accelerated the rate of lymphoid tissue proliferation as reflected in the incorporation of radioactivity into lymph node DNA. The classical "overshoot" phenomenon of lymphoid tissue regeneration following X-irradiation (Nygaard and Potter, 1960) is observed in both the control and thymosin-treated animals.

Certain of the histological data from these irradiation studies are briefly summarized here. A more detailed description will appear elsewhere (Goldstein et al., 1969b). Within 24 hr of irradiation in control

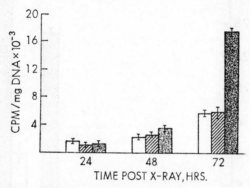

FIG. 10. Effect of injecting thymosin and bovine serum albumin (BSA) on the incorporation in vivo of [³H]deoxycytidine into lymph node DNA of CBA/W mice exposed to 400 R whole-body X-ray irradiation. Conditions of radiation: 400 R delivered as 109 R per min; 280 kV; 20 mA; HVL, 1·94 mm Cu; TSD, 52 cm. Each bar represents the average value for a group of eight mice. Other experimental details in text. □ Saline injection; ▨ bovine serum albumin injection, 3 mg/day; ▩ thymosin injection, 3 mg/day. I=range of mean values. (From Goldstein et al., 1969b.)

animals there is a marked cellular destruction, with nuclear debris and pyknosis in spleens, lymph nodes and thymuses. Thymic infiltration by polymorphonuclear leucocytes is evident. However, endodermal reticular cells appear normal. Cells from control animals do not show evidence of incorporation of tritiated thymidine into thymocytes. However, a few labelled reticular lymphocytes are found in lymph nodes in the vicinity of medullary sinusoids and in subcapsular and trabecular areas of the spleen. Uptake of tritiated thymidine is usually found only in the most immature lymphocytes as well as in splenic megakaryocytes.

In contrast to these findings in control mice, the sections from thymosin-treated animals 24 hr after irradiation show a better preservation of the cell populations in both lymph nodes and spleen as compared to controls.

In both organs, little evidence of thymidine incorporation is evident at this time. The thymus is similar to controls, with regard to both cellular destruction and lack of thymidine incorporation.

At 48 hr after irradiation, lymph nodes of control animals are still small, showing extreme depletion or pyknosis of lymphocytes with little evidence of the presence of radioactivity (Fig. 12). Labelling of the white

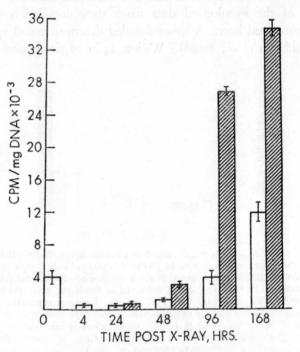

FIG. 11. Effect of injecting thymosin on the incorporation *in vivo* of [³H]thymidine into lymph node DNA of CBA/W mice exposed to 700 R whole-body X-ray irradiation. Conditions of irradiation: 700 R delivered as 138 R per min; 280 kV; 20 mA; HVL, 1·94 mm Cu; TSD, 47 cm. Each bar represents the average value for a group of eight mice. Other experimental details in text. □ Saline injection; ▨ thymosin injection, 4 mg/day. I=range of mean values.

pulp of the spleen is evident but confined to only a few immature cells. Depletion of thymic lymphocytes is still apparent with complete lack of labelled cells.

Lymph nodes from thymosin-treated mice show a marked increase in the number of immature cells 48 hr after irradiation. The thymidine label is slightly increased over the 24 hr group (Fig. 13). A similar increase in immature splenic lymphocytes with an increased number of labelled

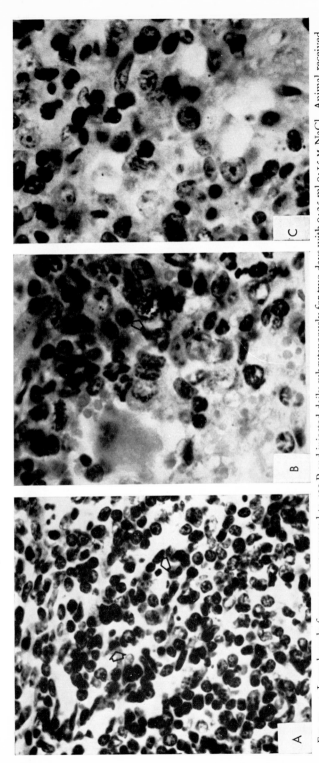

FIG. 12. A. Lymph node from a mouse exposed to 400 R and injected daily subcutaneously for two days with 0·25 ml 0·15 M-NaCl. Animal received 10 μc [³H]thymidine intraperitoneally 1½ hr before being sacrificed 48 hr after irradiation. Stain: haematoxylin–eosin. Autoradiographic emulsion. Note pyknotic lymphocytes and cellular fragments (see arrows).
B. Spleen section from same mouse as in A. Note depopulation of cells, pyknosis and presence of immature, labelled cell in centre right area (arrow).
C. Thymus cortex and medullary section from animal as in A. Note extreme cell depopulation, pyknotic lymphocytes and absence of labelled cells.

To face page 228

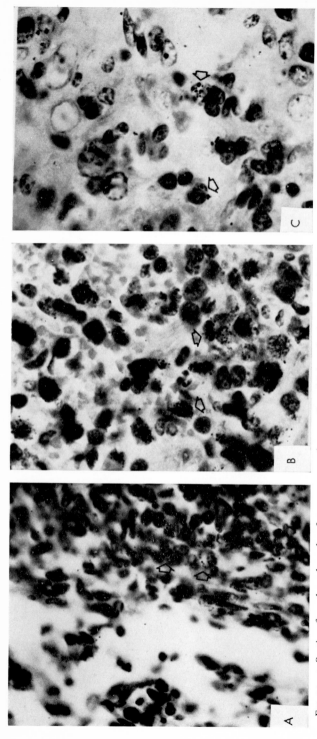

Fig. 13. A. Section from lymph node of mouse exposed to 400 R and injected subcutaneously daily for two days with 3 mg thymosin in 0·25 ml 0·15 M-NaCl. Other details as in Fig. 12. Note increase in labelling of cells along medullary cortical junction (arrows). B. Spleen section from same mouse as in A. Red pulp of subcapsular region with labelled immature cells. C. Thymus section from same mouse as in A. Medullary area. Only a few labelled cells are evident (arrows).

cells in the vicinity of the capsule and trabeculae is evident. An increase in labelled cells in the medullary portion of the thymus is also noted. Most striking differences between the morphology of the tissues of control and thymosin-treated mice begin to be apparent 72 hr after irradiation, after three injections of thymosin. At this time, control nodes still show depletion of lymphocytes and very few grain counts in free cells. However, labelling is rather extensive in immature lymphocytes surrounding the medullary sinuses. At this time tritiated thymidine appears in the cells of the red pulp of the spleen. A marked increase in the number of labelled cells with a concomitant increase in label per cell is evident; however, the label is limited to immature lymphocytes. Thymosin-treated animals show at 72 hr after irradiation marked splenic and lymph node mitotic activity with an increase in incorporation of tritiated thymidine. Radioautographs of lymph node sections of these mice show a marked increase in labelled immature lymphocytes. There is a marked reduction in cellular debris in the thymus but little evidence of cellular regeneration or labelling.

The available radioautographic evidence indicates that the cells responsible for the increased uptake of [³H]thymidine are the more primitive lymphoid elements.

The data obtained from the X-irradiation studies indicate that the regeneration of lymphoid structures after their involution by X-irradiation is accelerated by thymosin treatment. Of obvious interest is the question of whether this observed stimulation by thymosin is of functional significance; that is, can thymosin be useful in attenuating, either prophylactically or remedially, the deleterious effects of radiation on the structure and function of lymphoid tissue? Preliminary unpublished data suggest that the life of adult CBA/W mice exposed to lethal doses of whole-body X-irradiation may be prolonged significantly by daily administration of thymosin, beginning immediately after irradiation.

INFLUENCE OF THYMOSIN ON CELL-MEDIATED IMMUNITY IN NORMAL MICE

The influence of thymosin on cell-mediated immunity in normal animals has been explored in collaboration with Drs M. A. Hardy, J. Quint and D. State of our Department of Surgery. In a recent study (Hardy et al., 1968), we have used $B_{10}D_2$/Sn mice as recipients of a skin graft from C57BL/6 mice. Recipient animals were injected subcutaneously daily for seven days before receiving the allograft and injections were continued as long as the allograft was retained. These mice were then used again as second-set skin allograft recipients. The data obtained are illustrated in Figs. 14 and 15. The results indicate that administration of thymosin accelerates the rate of rejection of first and second-set skin allografts. We interpret the data to mean that thymosin increases the cell-mediated responsiveness of the recipient host animals. Thus it

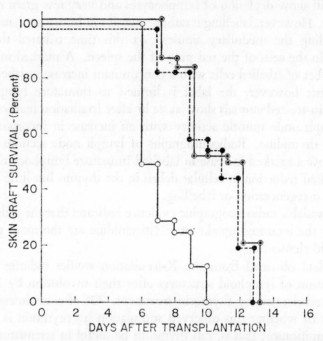

FIG. 14. Effect of thymosin on first-set skin grafts. Adult $B_{10}D_2/Sn$ mice were used as recipients of grafts, and C57BL/6 mice as donors. Prospective recipients were injected daily with 4·0 mg of thymosin (o — o), saline (● – – ●), or 4·0 mg of liver extract (⊙ — ⊙) beginning at seven days before skin transplants were made, and injections were continued daily for the duration of the experiment. Beginning at six days after transplantation, each animal was examined and evaluations were made of the percentage of the skin area transplanted that had survived (ordinate). Each point is the average value for a group of 15 animals. (From Hardy et al., 1968.)

appears that cell-mediated homograft responses, in contrast to the humoral immune response, can be stimulated in normal as well as in immunologically deficient animals by giving thymosin.

An antiserum to thymosin has been prepared in rabbits by standard procedures, using complete Freund's adjuvant. The Ouchterlony diffusion technique applied to this antiserum shows two main precipitin bands against the thymosin preparation used as antigen. One of these bands

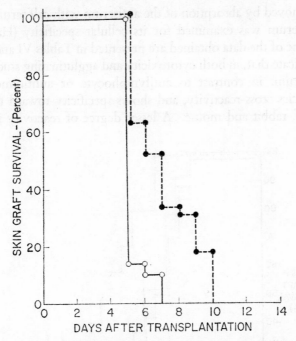

FIG. 15. Effect of thymosin on second-set skin grafts; $B_{10}D_2/Sn$ mice that had rejected a first-set skin graft from C57BL/6 mice were used as recipients of a second C57BL/6 skin graft. Treatment, evaluation, and number of animals same as in Fig. 14, except that only one control group, which received saline injections ($\bullet$ - - - $\bullet$), is presented; thymosin-treated animals ($\circ$ — $\circ$). (From Hardy et al., 1968.)

TABLE VI

CYTOTOXIC EFFECT OF A RABBIT ANTISERUM TO CALF THYMOSIN
ON LYMPHOID CELLS FROM THREE SPECIES

	Percentage dead cells after incubation with:		
Origin of cells	Saline	Normal rabbit serum	Antithymosin serum
Calf			
Thymus	12	9	96
Lymph node	15	12	30
Spleen	12	17	44
Mouse (A/J)			
Thymus	6	13	77
Lymph node	12	15	21
Spleen	20	21	18
Rabbit (New Zealand White)			
Thymus	16	21	49
Lymph node	20	15	18
Spleen	22	18	15

(From Hardy et al., 1969.)

can be removed by absorption of the antiserum with calf serum albumin. This antiserum was examined for its cellular specificity (Hardy *et al.*, 1969); some of the data obtained are presented in Tables VI and VII. The results indicate that, in both cytotoxicity and agglutinating studies *in vitro*, the antiserum, in contrast to antilymphocyte or antithymocyte sera, shows species cross-reactivity, and shows specificity toward thymocytes of the calf, rabbit and mouse. A lesser degree of reactivity is observed

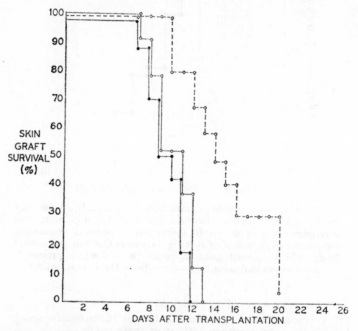

Fig. 16. Effect of a rabbit antithymosin serum on first-set C57BL/6 skin grafts on A/J mice. Other details as in Fig. 14, except that groups of recipient mice either were not treated (o — o) or received either 0·15 ml normal rabbit serum (● — ●) or 0·15 ml antithymosin serum (o – – – o) daily for seven days before the allogeneic transplant and daily thereafter. (From Hardy *et al.*, 1968.)

with cells from calf lymph nodes and spleen. These findings may be related to the possibility that thymosin, or a protein impurity still present in the thymosin preparation used as antigen, could be a thymus-specific antigen.

The thymosin antiserum has also been tested in our skin allograft studies (Hardy *et al.*, 1968); the data in Figs. 16 and 17 indicate that the administration of this antiserum to mice depresses their capacity to reject first and second-set skin allografts.

TABLE VII

AGGLUTINATION OF THYMIC, SPLENIC AND LYMPH NODE CELLS FROM THREE ANIMAL
SPECIES BY A RABBIT ANTISERUM TO CALF THYMOSIN
(From Hardy et al., 1969.)

Origin of cells	Saline	Normal rabbit serum	Antithymosin serum*
Calf			
Thymus	0	0	1024
Lymph node	0	0	512
Spleen	0	0	4
Mouse (A/J)			
Thymus	0	0	256
Lymph node	0	0	2
Spleen	0	0	32
Rabbit (New Zealand White)			
Thymus	0	0	32
Lymph node	0	0	8
Spleen	0	0	4

* Reciprocal of highest dilution causing agglutination.

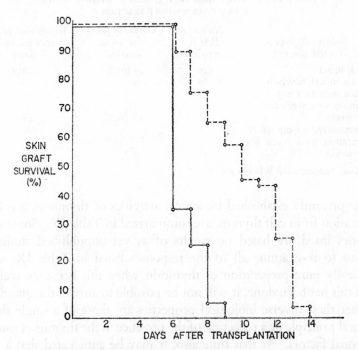

FIG. 17. Effect of a rabbit antithymosin serum on second-set skin
grafts. Other details as in Fig. 16 except that the recipient A/J mice
had rejected a first-set skin graft, and only two groups of animals are
presented, one injected with 0·15 ml antithymosin serum daily
(o – – – o) and the other injected with 0·15 ml normal rabbit serum
daily (o — o). (From Hardy et al., 1968.)

Another cell-mediated immune phenomenon which has been used to explore the role of the thymus in the immunological maturation of lymphoid tissue is the sensitive graft-versus-host reaction. In collaboration with Dr Lloyd Law of the National Cancer Institute, we have recently described (Law, Goldstein and White, 1968) the effect of thymosin in the graft-versus-host experimental system. Some of the data obtained are presented in Table VIII. It is evident that giving thymosin to neonatally thymectomized mice restores to the spleen cells of these animals their immunological competence, as reflected in the capacity of these cells to elicit a graft-versus-host reaction when injected into an allogeneic mouse strain. In contrast, an extract prepared from spleen or other tissue extracts (unpublished results) does not restore the immunological competence of spleen cells of the neonatally thymectomized animals.

TABLE VIII

EFFECT OF THYMOSIN (FRACTION 3) AND A SPLEEN EXTRACT
ON GRAFT-VERSUS-HOST REACTION

Spleen cells from C57BL donors	Number of BALB/c recipients	Number of graft-versus-host reactions	Average time of death after injection (days)
Control, intact	40	25 (63%)	18·3
Thymectomized, newborn	66	0	—
Thymectomized + 1 mg thymosin 3 × a week for four weeks	63	48 (76%)	18·3
Thymectomized + 1 mg spleen preparation 3 × a week for four weeks	41	0	—

(After Law, Goldstein and White, 1968.)

Our presently established biological activities of thymosin, a cell-free preparation from calf thymus, are summarized in Table IX. Some of the activities listed are based on results of as yet unpublished studies. It remains to re-examine all of the responses listed in Table IX with a chemically pure preparation of thymosin when this becomes available. Until this has been done, it will not be possible to answer the question of whether these diverse biological properties are those of a single thymic humoral principle or a reflection of the presence in the thymus of multiple hormonal factors. At that time also, it may be anticipated that a more rational approach could be initiated to the possible clinical applications of thymic preparations. For the present it may be concluded that the thymus is indeed an endocrine gland, and that its humoral roles are of prime significance in the regulation of host resistance.

TABLE IX

EFFECTS OF THYMOSIN ON LYMPHOID TISSUE STRUCTURE AND FUNCTION

A. *Enhances lymphocytopoiesis in:*
  1. Normal mice, rats and guinea pigs
  2. Adrenalectomized mice
  3. Germ-free mice
  4. X-irradiated mice
B. *In neonatally thymectomized mice:*
  1. Reduces wasting and mortality
  2. Increases body weight gain
  3. Increases absolute numbers of blood lymphocytes
  4. Increases lymphoid organ size
  5. Partially restores lymph node and spleen morphology
C. *Influence on humoral antibody response:*
  1. Slight to sheep erythrocytes in neonatally thymectomized mice (plaque assay)
  2. Slight to sheep erythrocytes in adult thymectomized, lethally irradiated, bone marrow restored mice (plaque assay)
  3. None to sheep erythrocytes in normal or X-irradiated mice (haemagglutinin titres)
D. *Influence on cell-mediated immune response:*
  1. In normal mice:
    (a) accelerates 1st and 2nd-set skin allograft rejection
    (b) an antithymosin serum delays 1st and 2nd-set skin allograft rejection
    (c) potentiates or negates effects of antilymphocyte serum
  2. In neonatally thymectomized mice:
    (a) restores ability to reject skin allografts
    (b) restores to spleen cells their capacity to elicit graft-versus-host reaction

SUMMARY

Major emphasis in this paper is given to summarizing recent and current investigations in our own laboratory concerned with endocrine studies of the thymus and its humoral role in the regulation of host resistance. A brief description is provided of the preparation, purification and properties of thymosin, a lymphocytopoietic factor present in the thymic tissue of a variety of species. Data are presented to support the conclusion that thymosin is a thymic specific factor which influences the development of host immunological competence. As a consequence of this activity, administration of thymosin is shown to restore partially the capacity of neonatally thymecto- mized mice to grow normally, to prevent serious depletion of their lymphoid cells, and to maintain their capacity to elicit a normal homograft response. In addition, the spleen cells of such thymosin-treated mice can induce a graft-versus-host reaction when injected into an allogeneic strain of mouse. The lymphocytopoietic activity of thymosin in normal CBA/W mice is also seen in animals whose lymphoid tissue has been exposed to an involuting dose of X-irradiation. Such thymosin-treated mice exhibit an accelerated rate of regeneration of lymphoid tissue, as compared to irradiated mice injected either with saline or bovine serum albumin, as reflected in the weight of lymphoid tissue, in the quantitative incorporation of tritium-labelled nucleosides into lymph node DNA, and in radio-autographic examination of tissue sections. This radioactivity is localized primarily in the immature lymphocytes of the lymph nodes and spleen.

Thymosin administration to recipient mice accelerates first and second-set skin allograft rejections. A thymosin antiserum, prepared in rabbits and administered to recipient mice, delays first and second-set skin allograft rejection.

The positive influence of thymosin on parameters of cell-mediated homograft responses is in contrast to the apparent lack of effect of thymosin administration on humoral immune responses.

Acknowledgements

The studies described in this paper represent the combined efforts of a number of individuals in addition to the authors. Particular mention should be given to Dr Sipra Banerjee, a post-doctoral fellow from the University of Calcutta, and Dr Yoshitsugu Asanuma, a visiting faculty member from the University of Osaka School of Medicine. Able technical assistance has been provided by Miss Norma Robert, Mrs Saraswati Lathi and Mr James Oliver. Collaborative studies with other laboratories are referred to in the text.

The data presented in this paper from the authors' laboratory derive from investigations supported by grants from the National Science Foundation (GB-6616X), the Damon Runyon Fund for Cancer Research (DRG-920B), the National Cancer Institute, National Institutes of Health, USPHS(CA-07470(5)), and the American Cancer Society (P-68-J). Grateful acknowledgement is also made of the helpful financial and technical support of Merck Sharp and Dohme Research Laboratories.

Allan L. Goldstein is the recipient of a Career Scientist Award of the Health Research Council of the City of New York under contract 1-519.

REFERENCES

ARCHER, O. K., and PIERCE, J. C. (1961). *Fedn Proc. Fedn Am. Socs exp. Biol.*, **20**, 26.
ARNASON, B. G., JANKOVIC, B. D., WAKSMAN, B. H., and WENNERSTEN, C. (1962). *J. exp. Med.*, **116**, 177.
ASANUMA, Y., GOLDSTEIN, A. L., and WHITE, A. (1969). Submitted for publication.
BRACCI, C. (1905). *Riv. Clin. pediat.*, **3**, 572.
CAMBLIN, J. G., and BRIDGES, J. B. (1964). *Transplantation*, **2**, 785.
CHASE, M. W. (1945). *Proc. Soc. exp. Biol. Med.*, **59**, 134.
CHASE, M. W. (1953). In *The Nature and Significance of the Antibody Response*, pp. 156–169, ed. Pappenheimer, A. M., Jr. New York: Columbia University Press.
COHN, E. J., STRONG, L. E., HUGHES, W. L., JR., MULFORD, D. J., ASHWORTH, J. N., MELIN, M., and TAYLOR, H. L. (1946). *J. Am. chem. Soc.*, **68**, 459.
DEFENDI, V., and METCALF, D. (ed.) (1964). *The Thymus*. Philadelphia: The Wistar Institute Press.
DOUGHERTY, T. F. (1952). *Physiol. Rev.*, **32**, 379.
DOUGHERTY, T. F., CHASE, J. H., and WHITE, A. (1944). *Proc. Soc. exp. Biol. Med.*, **57**, 295.

DOUGHERTY, T. F., and WHITE, A. (1945). *Am. J. Anat.*, **77**, 81.
DOUGHERTY, T. F., and WHITE, A. (1946). *Endocrinology*, **39**, 370.
DOUGHERTY, T. F., WHITE, A., and CHASE, J. H. (1945). *Proc. Soc. exp. Biol. Med.*, **59**, 172.
FICHTELIUS, K. E., LAURELL, G., and PHILLIPSSON, L. (1961). *Acta path. microbiol. scand.*, **51**, 81.
GALANTE, L., GUDMUNDSSON, T. V., MATTHEWS, E. W., TSE, A., WILLIAMS, E. D., WOODHOUSE, N. J. Y., and MacINTYRE, I. (1968). *Lancet*, **2**, 537.
GOLDSTEIN, A. L., ASANUMA, Y., BATTISTO, J. R., HARDY, M. A., QUINT, J., and WHITE, A. (1969*a*). Submitted for publication.
GOLDSTEIN, A. L., BANERJEE, S., SCHNEEBELI, G. L., DOUGHERTY, T. F., and WHITE, A. (1969*b*). *Radiat. Res.*, in press.
GOLDSTEIN, A. L., SLATER, F. D., and WHITE, A. (1966). *Proc. natn. Acad. Sci. U.S.A.*, **56**, 1010.
GOOD, R. A., DALMASSO, A. P., MARTINEZ, C., ARCHER, O. K., PIERCE, J. C., and PAPERMASTER, B. W. (1962). *J. exp. Med.*, **116**, 773.
GOOD, R. A., and GABRIELSON, A. E. (ed.) (1964). *The Thymus in Immunobiology*. New York: Hoeber.
GREGOIRE, C., and DUCHATEAU, G. (1956). *Archs Biol. Liège*, **67**, 269.
HARDY, M. A., QUINT, J., GOLDSTEIN, A. L., STATE, D., and WHITE, A. (1968). *Proc. natn. Acad. Sci. U.S.A.*, **61**, 875.
HARDY, M. A., QUINT, J., GOLDSTEIN, A. L., WHITE, A., STATE, D., and BATTISTO, J. R. (1969). *Proc. Soc. exp. Biol. Med.*, **130**, 214.
HESS, M. W. (1968). *Experimental Thymectomy. Possibilities and Limitations*. New York: Springer-Verlag.
JANKOVIC, B. D., WAKSMAN, B. H., and ARNASON, B. G. (1962). *J. exp. Med.*, **116**, 159.
JERNE, H. K., and NORDIN, A. A. (1963). *Science*, **140**, 405.
KLEIN, J. J., GOLDSTEIN, A. L., and WHITE, A. (1965). *Proc. natn. Acad. Sci. U.S.A.*, **53**, 812.
KLEIN, J. J., GOLDSTEIN, A. L., and WHITE, A. (1966). *Ann. N.Y. Acad. Sci.*, **135**, 485.
LAW, L. W., GOLDSTEIN, A. L., and WHITE, A. (1968). *Nature, Lond.*, **219**, 1391.
METCALF, D. (1956). *Br. J. Cancer*, **10**, 442.
METCALF, D. (1965). *Nature, Lond.*, **208**, 1336.
METCALF, D. (1966). *The Thymus. Its Role in Immune Responses, Leukaemia Development and Carcinogenesis*. New York: Springer-Verlag.
MILLER, J. F. A. P. (1961). *Lancet*, **2**, 748.
MILLER, J. F. A. P. (1965). *Nature, Lond.*, **208**, 1337.
MILLER, J. F. A. P. (1967). *Lancet*, **2**, 1299.
MILLER, J. F. A. P., and OSOBA, D. (1967). *Physiol. Rev.*, **47**, 437.
NYGAARD, O. F., and POTTER, R. L. (1960). *Radiat. Res.*, **12**, 120.
PARK, E. A., and McCLURE, R. D. (1919). *Am. J. Dis. Child.*, **18**, 317.
POTOP, I., BOERU, V., and MREANĂ, G. (1966). *Biochem. J.*, **101**, 454.
ROBERTS, S., and WHITE, A. (1949). *J. biol. Chem.*, **178**, 151.
SCHWARTZ, H., PRICE, M., and ODELL, C. A. (1953). *Metabolism*, **2**, 261.
TAYLOR, R. B. (1965). *Nature, Lond.*, **208**, 1334.
TRAININ, N., BURGER, M., and KAYE, A. M. (1967). *Biochem. Pharmac.*, **16**, 711.
WHITE, A. (1947–48). *Harvey Lect.*, **43**, 43.
WHITE, A., and DOUGHERTY, T. F. (1945). *Endocrinology*, **36**, 207.
WHITE, A., and DOUGHERTY, T. F. (1946). *Ann. N.Y. Acad. Sci.*, **46**, 859.
WHITE, A., and GOLDSTEIN, A. L. (1968). *Perspect. Biol. Med.*, **11**, 475.
WOLSTENHOLME, G. E. W., and PORTER, R. (ed.) (1966). *Ciba Fdn Symp. The Thymus. Experimental and Clinical Studies*. London: Churchill; Boston: Little, Brown.

INTERACTION BETWEEN THYMUS CELLS AND BONE MARROW CELLS IN RESPONSE TO ANTIGENIC STIMULATION

J. F. A. P. Miller and G. F. Mitchell

Walter and Eliza Hall Institute of Medical Research, Melbourne, Australia

An immunological role for the thymus was established in 1961 when it was observed that neonatal thymectomy in mice had profound effects on lymphopoiesis and immunogenesis (Miller, 1961, 1962a, b). The defects associated with neonatal thymectomy in many species of rodents can be summarized as follows (reviewed by Miller and Osoba, 1967, and Miller and Mitchell, 1969a): (1) a severe diminution in the number of long-lived small lymphocytes which circulate in blood and lymph and which specifically populate the paracortical areas of the lymph nodes and periarteriolar lymphocyte sheaths of the spleen, and (2) an impairment of the capacity to effect cell-mediated immunities such as delayed hypersensitivity and transplantation reactions. Plasma cells, lymphoid follicles, germinal centres, and immunoglobulin and humoral antibody production are generally unaffected. In birds, the bursa of Fabricius, an organ analogous to the thymus, is responsible for the development of the cells in lymphoid follicles and of plasma cells, and thus of cells which can produce humoral antibody and immunoglobulins (reviewed by Warner, 1967). The mammalian equivalent of the bursa has not been unequivocally identified but is thought to be situated in the lymphoepithelial regions of the intestine (Fichtelius, 1967). There are notable exceptions to the rule that thymectomy does not impair humoral antibody production: thus, the capacity of thymectomized mice to mount an antibody response to heterologous erythrocytes and serum proteins, and to a few other antigens, is severely impaired (Miller and Mitchell, 1969a; Taylor, 1969).

LYMPHOPOIESIS

Several processes are involved in the control of lymphopoiesis. Among these are the influences of the thymus and of antigen.

Lymphopoiesis within the thymus is intense and independent of antigenic stimulation. The intense cellular proliferation takes place in the cortex, not in the medulla. Radioautographic studies in mice have shown that small lymphocytes arise from the rapid division of primitive lymphoid cells (large and medium lymphocytes) and that the mean intrathymic lifespan of the majority (90–95 per cent) is 3 to 4 days. The lower rate of production of lymphocytes in spleen and lymph nodes reflects the lower percentage of primitive lymphoid cells in these tissues (Metcalf, 1966).

The proliferative stimulus to thymic lymphoid cells is intrinsically controlled and dependent upon the epithelial cytoreticulum. Thus, it has been established that the rate of inflow into the thymus of lymphoid precursor cells, and the proliferative rate of thymus lymphoid cells, are identical in implants grafted to normal or thymectomized hosts, to hosts bearing as many as 23 other thymus implants, or to hosts that had previously been subjected to resection of lymph nodes or spleen. Furthermore there is no compensatory regeneration of a thymus fragment after subtotal thymectomy (Metcalf, 1966). These findings imply that thymus tissue is not subject to thymus-specific feedback inhibition of proliferative activity.

The hallmark of thymus lymphopoiesis is that, unlike lymphopoiesis in lymph nodes and spleen, it is totally independent of antigenic stimulation. In the foetus, which is to a great extent shielded from antigenic stimuli, proliferative activity in thymic lymphoid cells is marked whereas it is low or absent in lymphoid cells elsewhere (Metcalf, 1966). Lymphopoiesis in lymph nodes and spleen is much lower in germfree than in conventional mice but thymic lymphopoiesis is unchanged (Wilson, Bealmear and Sobonya, 1965). An analogous situation exists in respect to both the thymus and bursa of Fabricius in birds (Thorbecke et al., 1957) and one would expect a high rate of antigen-independent lymphopoiesis in whatever organ is the mammalian equivalent of the bursa. Parenteral injection of antigens does not increase the mitotic activity of thymus lymphoid cells (Metcalf, 1966) but induces lymphopoiesis in lymph nodes and spleen (differentiation and proliferation of pyroninophilic cells and plasma cells, and production of germinal centres).

It can thus be concluded that the stimulus regulating differentiation and proliferation of thymus lymphoid cells is intrinsic, antigen-independent and linked to the cells of the epithelial cytoreticulum. The exact nature of this stimulus is unknown but there is experimental evidence from work on the reconstitution of thymectomized mice by thymus in diffusion chambers (Miller and Osoba, 1963), and by thymus extracts (Law, Goldstein and White, 1968), that thymus-specific humoral factors may play a role in the conversion of lymphoid precursor cells to lymphoid cells which can eventually take part in immune responses. Much more work is however required to establish the nature and mode of action of such a humoral thymus influence.

THYMUS ORIGIN OF RECIRCULATING LYMPHOCYTES

Experiments with chromosomally marked cells have established that there are, in bone marrow, haemopoietic stem cells that can differentiate to lymphoid cells within the thymus and lymph nodes (Micklem et al., 1966). Thymectomy of adult mice, subsequently subjected to total body irradiation and marrow protection, did not prevent lymphoid cell differentiation in certain regions of the lymphoid system, such as in lymphoid follicles (Balner and Dersjant, 1964). It did, however, prevent recovery of the normal number of recirculating small lymphocytes (Miller and Mitchell, 1967, 1969a). It seems therefore that there are at least two separate lines of lymphoid cell differentiation: one, thymus-dependent, and responsible for the origin of recirculating small lymphocytes; the other, thymus-independent, and responsible for the differentiation of cells in lymphoid follicles and of plasma cells. While it is evident that the cells in both pathways have their initial origin in bone marrow, we will, for the sake of simplicity, reserve the term "thymus-derived cells" for those lymphoid cells that can differentiate only in the presence of the thymus, and "marrow-derived cells" for those that can differentiate in its absence.

There is clear evidence that, in the rat, the recirculating small lymphocytes are immunologically competent cells that can initiate certain types of immune responses such as graft-versus-host reactions, host-versus-graft reactions and the haemolysin response to sheep erythrocytes (Gowans and McGregor, 1965). The origin of recirculating cells is still debated but the evidence from various types of experiments suggests that these cells may be thymus-derived:

(1) When thymus lymphocytes were labelled with tritiated thymidine, in situ, in the young rat, labelled cell migrants appeared exclusively in those regions of the lymphoid tissues which are populated specifically by recirculating small lymphocytes and depleted specifically by thymectomy, namely the paracortical areas of the lymph nodes and periarteriolar lymphocyte sheaths of the spleen (Weissman, 1967).

(2) The recirculating small lymphocytes can be drained by a thoracic duct fistula. This technique was exploited in order to estimate the size of the circulating lymphocyte pool. It was found that close to 100 million cells could be drained in 48 hours from a normal 6-week-old CBA mouse but only 3 to 4 million cells from a neonatally thymectomized mouse of the same age and strain (Miller, Mitchell and Weiss, 1967). In irradiated mice given bone marrow the number of lymphocytes mobilized within 48 hours rose from a total of about 5 million, 1-3 weeks after irradiation, to about 60 million 8 weeks later. If, however, the mouse had been thymectomized (even as an adult) before irradiation, this number did not rise significantly in 8 weeks above its post-irradiation level (Miller and Mitchell, 1967, 1969a). This severe depletion in recirculating small lymphocytes clearly accounts for the immunological defects of thymectomized mice.

(3) The fate of radioactively labelled or chromosomally marked thymus lymphocytes was determined in mice subjected to heavy doses of total-body irradiation (Miller and Mitchell, 1969b). The cells localized in the periarteriolar regions of the irradiated spleen and appeared there as small lymphocytes (cf. Figs. 1 and 2). If sheep erythrocytes were given at the same time, there appeared in these regions, within 24 to 48 hours, large pyroninophilic blast cells and dividing cells (Fig. 3). These were derived from the inoculated thymus cells, since cytological analyses revealed the chromosome marker of the donor (Nossal et al., 1968). The fate of the large pyroninophilic cells was determined in various experiments: (1) they could be labelled with tritiated thymidine and small lymphoid cells were found among their progeny (Figs. 4 and 5); (2) they were serially passaged through successive generations of heavily irradiated mice together with sheep erythrocytes but at no time were haemolysin-forming cells detected in any of these recipients (Miller and Mitchell, 1967); (3) finally, suspensions of cells from the irradiated spleens containing the labelled progeny of these pyroninophilic cells were injected intravenously into normal mice in which a thoracic duct fistula had been established (about 50 million cells per mouse). After 15 hours labelled small lymphocytes but no labelled large cells appeared in the thoracic duct lymph (Fig. 6) and constituted up to 0·6 per cent of the cells examined between 15 and 30 hours. When the draining animals were sacrificed 36 hours after injection, labelled small cells were observed in the paracortical areas of the lymph nodes and periarteriolar lymphocyte sheaths of the spleen, but no labelled cells appeared in the follicles. These results suggest that cells derived from the thymus can react to certain antigens (such as sheep

erythrocytes in mice) by transforming to large pyroninophilic cells which divide to give rise to a progeny of small lymphocytes that can readily recirculate but cannot transform to haemolysin-forming cells. The similarity in the fate of thymus lymphocytes following "activation" by sheep erythrocytes and the fate of thoracic duct lymphocytes in allogeneic hosts (Ford, Gowans and McCullagh, 1966), is striking.

ORIGIN OF HAEMOLYSIN-FORMING CELLS

Neonatally thymectomized mice exhibit both a deficiency of recirculating small lymphocytes and an impairment of the capacity to respond to sheep erythrocytes by producing haemolysin-forming cells (Miller, Mitchell and Weiss, 1967). This impaired immune response could easily be corrected by inoculating either thymus or thoracic duct lymphocytes but not bone marrow cells (Miller and Mitchell, 1968). It was tempting to speculate that stem cells in marrow must migrate through the thymus before they could give rise to recirculating small lymphocytes which, themselves, could be recruited by antigen into the spleen where they would differentiate to haemolysin-forming cells. As indicated above, however, thymus cells never gave rise to haemolysin-forming cells in irradiated hosts, suggesting either that conditions in the irradiated spleen were not conducive to the differentiation of thymus-derived cells to antibody-forming cells or that these cells had lost the potential to differentiate along these lines. In other experiments, it was demonstrated that while neither thymus nor marrow cells, given alone, could restore the capacity of irradiated mice to respond to sheep erythrocytes, haemolysin-forming cells were produced if both populations were given together (Claman, Chaperon and Triplett, 1966; Davies et al., 1967; Miller and Mitchell, 1967). What then was the role of marrow cells in this system— did they provide a favourable environment in the irradiated spleen for the differentiation of thymus cells to antibody-formers or were they, themselves, the haemolysin-forming cell precursors? Further experiments unequivocally established that marrow-derived cells, but not thymus-derived cells (be these thymus cells or thoracic duct lymphocytes), gave rise to haemolysin-forming cells:

(1) Neonatally thymectomized CBA mice could respond to sheep erythrocytes equally well whether syngeneic or allogeneic thymus or thoracic duct cells were given. Immuno-serological and chromosome marker techniques established the identity of the haemolysin-forming cells as that of the thymectomized host and not of the inoculated thymus

or thoracic duct cells (Mitchell and Miller, 1968a; Miller and Mitchell, 1968; Nossal et al., 1968).

(2) CBA or (CBA × C57BL)F$_1$ mice, thymectomized as adults, were subjected to heavy doses of total-body irradiation and protected with CBA bone marrow. Two weeks later they were given (CBA × C57BL)F$_1$ thoracic duct cells and sheep erythrocytes. Immuno-serological analyses of the haemolysin-forming cells revealed that they had the immunogenetic characteristics of the marrow-derived cells (CBA and not F$_1$). When, however, F$_1$ thoracic duct cells were given to heavily irradiated CBA mice along with sheep erythrocytes but without marrow cells, haemolysin-forming cells of F$_1$ origin were produced. The number of these was, however, considerably increased by a prior injection of marrow cells and, in this case, the majority (75 per cent or more) of the haemolysin-forming cells were derived from the marrow donor (Mitchell and Miller, 1968b; Nossal et al., 1968; and Table I).

TABLE I

IDENTITY OF HAEMOLYSIN-FORMING CELLS IN RECONSTITUTED IRRADIATED MICE

Recipient	Inoculum*	Peak number of haemolysin-forming cells per spleen	Percentage reduction of haemolysin-forming cells with: anti-CBA serum	anti-C57BL serum
Irradiated CBA	10^7 (CBA × C57BL)F$_1$ TDL + SRBC	822	85, 92, 93	85, 85, 95
Adult thymectomized irradiated CBA + CBA marrow	10^7 (CBA × C57BL)F$_1$ TDL + SRBC	19 800	95, 95, 96, 97	0, 0, 0, 5
Adult thymectomized irradiated (CBA × C57BL)F$_1$ + CBA marrow	10^7 (CBA × C57BL)F$_1$ TDL + SRBC	18 633	96	24

* TDL, thoracic duct lymphocytes; SRBC, sheep erythrocytes.

It is evident, therefore, that neither thymus lymphocytes nor the majority of thoracic duct cells can transform to haemolysin-forming cells. There are, however, in thoracic duct lymph, some cells which can differentiate to haemolysin-forming cells, and one wonders whether these may not belong to a population which is not thymus-derived, but which is mobilized from those regions of the lymphoid tissues populated by marrow-derived cells, such as, for instance, the follicles.

We can therefore conclude that thymus-derived cells are essential for the haemolysin response in order to enable the differentiation of marrow-derived cells into haemolysin-forming cells. How do they act? Is their influence specific or not? Which cell is involved in antigen recognition? Which is the target cell for the induction of tolerance and which cell is responsible for carrying the immunological memory?

ROLE OF THYMUS–DERIVED CELLS

The influence of thymus-derived cells in the haemolysin response can be specific or non-specific. If it is not specific, one can ascribe to thymus-derived cells one of the following roles:

(1) *Trephocytic.* By acting as a source of nucleosides, thymus-derived cells would facilitate the differentiation and proliferation of marrow-derived cells in response to certain antigens. It must, however, be recalled that lymphocytes are not the only cells capable of giving up their nucleosides to be reutilized by other cells: granulocytes (Robinson and Brecher, 1963) and other marrow cells (Metcalf, 1968) can do this. Yet marrow cells were never able to repair the immunological defects of thymectomized mice (Miller and Mitchell, 1968). The most cogent argument against a trephocytic role for thymus-derived cells is the fact that lymphocytes from donors specifically tolerant of sheep erythrocytes failed to restore to normal the reactivity to sheep red blood cells but could confer the capacity to respond normally to horse erythrocytes (see below).

(2) *Phagocytic.* On this theory thymus-derived cells would be the precursors of macrophage-like cells essential for processing the antigens of heterologous erythrocytes, in order to render these immunogenic to marrow-derived cells. This interpretation is not supported by any experimental evidence.

(3) *Pharmacological.* It is known that activation of lymphocytes by certain antigens in delayed hypersensitivity reactions triggers the production of factors that have a multitude of biological activities—for example, the migration inhibitory factor acting on macrophages and factors which induce blast transformation in other lymphocytes (David, 1968). Once elaborated, however, these factors have no immunological specificity. It may be that thymus-derived cells are activated by antigen to release a factor which facilitates the recruitment of marrow-derived cells and their differentiation to haemolysin-forming cells. Experimental evidence for this is, however, lacking.

(4) *Absorption of "carrier-antibodies".* Bretscher and Cohn (1968) have proposed that certain cells passively absorb "carrier-antibodies" which,

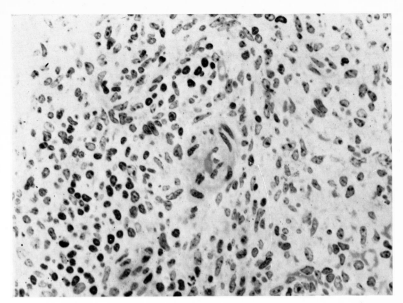

FIG. 1. Spleen of lethally irradiated CBA mouse injected intravenously 48 hours before with sheep erythrocytes. Note extreme depletion of small lymphocytes around arteriole. Methyl green pyronine. × 470

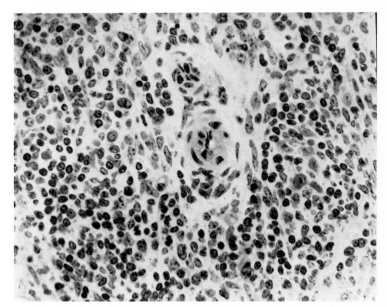

FIG. 2. Spleen of lethally irradiated CBA mouse injected intravenously 48 hours before with 50 million CBA thymus cells but not with sheep erythrocytes. Note appearance of small lymphocytes around arteriole. Methyl green pyronine. × 470

To face page 244

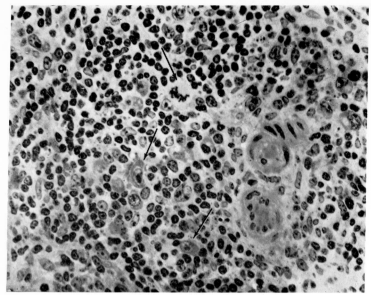

Fig. 3. Spleen of lethally irradiated CBA mouse injected intravenously 48 hours before with 50 million CBA thymus cells and sheep erythrocytes. Note appearance of small lymphocytes, large pyroninophilic blast cells (arrows), and dividing cell (arrow). Methyl green pyronine. × 470

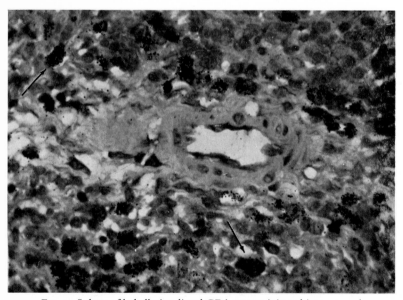

Fig. 4. Spleen of lethally irradiated CBA mouse injected intravenously 60 hours before with 50 million CBA thymus cells and sheep erythrocytes. Two injections of tritiated thymidine were given 12 and 24 hours before this section was taken. Note labelled lymphocytes and large pyroninophilic cells (arrows) in periarteriolar region. Methyl green pyronine. × 560

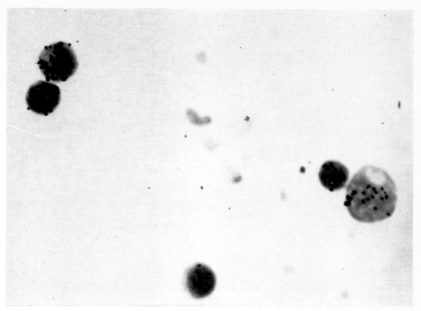

FIG. 5. Smear of spleen cell suspension from CBA mouse 5 days after lethal irradiation and injection of 50×10^6 CBA thymus cells and sheep erythrocytes. Five injections of tritiated thymidine were given twice daily from 24 hours on. Note presence of labelled lymphocytes including heavily labelled small lymphocytes. Methyl green pyronine. × 1080

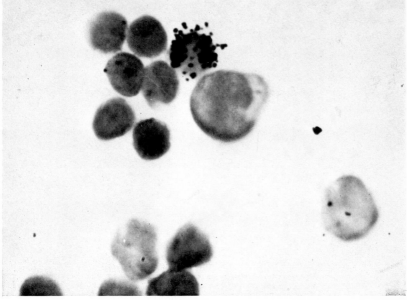

FIG. 6. Thoracic duct cells from a normal CBA mouse injected intra-venously 15 hours before with the spleen cell suspension shown in Fig. 5. Note appearance in the lymph of labelled small lymphocytes and absence of label in large lymphocytes. Giemsa. × 1080

by combining with the "carrier determinants" on complex antigenic molecules, present to "haptene-reactive" cells well-exposed and suitably oriented haptenic determinants. Union between the haptene receptor and the antigen–carrier–antibody complex distorts the membrane of the haptene-reactive cell and switches on anti-haptene antibody production. Thymus-derived cells might thus exert their role by passively absorbing large amounts of carrier-antibodies which, by complexing with antigenic determinants on heterologous erythrocytes, would trigger haemolysin production in marrow-derived antibody-forming cell precursors. Henry and Jerne (1968) showed that immunologically specific 19 s haemolysin-containing preparations exerted a positive feedback on haemolysin production. It is conceivable that such preparations contained carrier antibodies or that thymus-derived cells, themselves, absorbed 19 s haemolysin molecules which, in combination with the specific antigenic determinants, facilitated activation of marrow-derived haemolysin-forming cell precursors. According to this type of hypothesis, thymus-derived cells would not act in a specific way and thus need not have receptors with immunological specificity.

There are some experimental data which suggest that the activity of thymus-derived cells may be specific in the immunological sense. (1) It has been demonstrated that both immunological memory (Gowans and Uhr, 1966) and tolerance (McGregor, McCullagh and Gowans, 1967) are properties linked to thoracic duct lymphocytes. (2) The activation of thymus-derived cells by antigen appears to be specific. In one experiment, for instance, lethally irradiated CBA mice received either thymus cells alone (group 1), thymus cells and sheep erythrocytes (group 2) or thymus cells and horse erythrocytes (group 3). After one week, their spleens were transferred together with sheep erythrocytes and marrow cells to other irradiated CBA mice. A significant haemolysin response to sheep erythro-cytes occurred only in those receiving spleens transferred from mice of group 2 (Mitchell and Miller, 1968a). (3) Thoracic duct cells, harvested from normal mice or from donors specifically tolerant of sheep erythro-cytes, were injected together with sheep red cells, horse red cells, or both, into neonatally thymectomized recipients. The results (Table II) showed that cells from normal donors were as effective as cells from tolerant donors in enabling a response to horse erythrocytes. By contrast, lympho-cytes from non-tolerant donors were superior to cells from tolerant donors in activating the response to sheep erythrocytes.

These results imply that tolerance is a property linked to thymus-derived cells and therefore that the thymus produces unspecific cells of

TABLE II

HAEMOLYSIN-FORMING CELLS IN NEONATALLY THYMECTOMIZED MICE RECEIVING THORACIC
DUCT CELLS

			Peak number of haemolysin-forming cells per spleen	
Inoculum*	Antigen challenge*	Number of mice	anti-SRBC	anti-HRBC
No cells	SRBC	10	2555	—
	HRBC	6	—	2812
	Both	7	3814	1071
Non-tolerant TDL	SRBC	9	65 067	—
	HRBC	5	—	36 020
	Both	9	57 789	33 877
SRBC-tolerant TDL	SRBC	10	11 690	—
	HRBC	4	—	32 715
	Both	17	21 106	30 241

* TDL, thoracic duct lymphocytes; SRBC, sheep erythrocytes; HRBC, horse erythrocytes.

the antigen-reactive type, each being sensitive to certain antigenic determinants and to the induction of specific tolerance. The data given in the experiments just mentioned do not support the various hypotheses which would give to thymus-derived cells a passive or non-specific role, whether trephocytic, antigen-processing, pharmacological or the absorption of "carrier-antibodies".

If it is accepted that the role of thymus-derived cells is specific, one can suggest two hypotheses in an attempt to explain the nature of the interaction which takes place between thymus-derived and marrow-derived cells.

(1) *Information transfer.* Since marrow-derived cells are the haemolysin-formers, one will have to determine whether they, too, are unispecific cells. So far, we have been unable to do this. If it turns out that they are not unispecific, it may be that the genetic control of antibody synthesis has to be shared by two cell types: one, marrow-derived, differentiated to the stage where it can express the information essential for the synthesis of the constant half of the immunoglobulin molecules (which determines class specificity); the other, thymus-derived, differentiated to the stage where it can express the information necessary for the synthesis of only the variable half of the molecule (in which resides the specificity of the antibody-combining site). Fusion of the two halves would occur in the marrow-derived cell, after its interaction with antigen-activated thymus-derived cells, at the level of DNA or possibly m-RNA, as can be envisaged by various models, episomal or viral, such as the one suggested some years ago by Smithies (1965).

(2) *Antigen recognition and focusing.* If, on the other hand, as seems more likely, both thymus-derived and marrow-derived cells are unispecific cells of the antigen-reactive type, it may be that antigenic determinants which elicit the haemolysin response have to be focused on marrow-derived cells by a mechanism which involves prior recognition of other antigenic determinants by thymus-derived cells. An analogous situation has been suggested by Mitchison (1967) in order to explain the immune response to haptene–protein conjugates where one cell is believed to recognize the haptene but other cells are essential to recognize carrier determinants on the protein molecule. Carrier effects have been implicated in the immune response to heterologous erythrocytes. Thus, in the chicken, for instance, erythrocytes incompatible for the A blood group (a weak determinant) provoke only a weak anti-A antibody response during the course of normal immunization. By contrast, cells incompatible for the strong B blood group induce a strong anti-B response. If, however, the A determinants are introduced on to B-incompatible cells, the anti-A response increases remarkably (Schierman and McBride, 1967). The presence of a strong determinant acts, as it were, as an adjuvant or carrier to facilitate the response to a second, weak determinant.

It is perhaps difficult to imagine how two rare unispecific cells can find each other and meet during the response to certain complex antigens. There are, however, a number of factors which would ensure this contact: (1) the recirculating, thymus-derived lymphocytes migrate in large numbers through gaps in the marginal sinus in the spleen and populate the periarteriolar lymphocyte sheaths (Goldschneider and McGregor, 1968); (2) it is in the vicinity of the marginal sinus and in the periarteriolar lymphocyte sheaths that the very first haemolysin-forming cells appear (Fitch, Steiskal and Rowley, 1969); (3) in this region too, thymus-derived cells are activated by antigen to produce large pyroninophilic cells which divide to give rise to more small lymphocytes (see above, p. 242); (4) it is likely that there is on the erythrocyte a multiplicity of antigenic determinants which could activate a large number of clones of unispecific thymus-derived cells, each of which would be involved in focusing on to marrow-derived cells those antigenic determinants which can elicit the haemolysin response. An alternative possibility would be that thymus-derived cells actually export their antibody for short distances on to the surface of cells such as dendritic reticulum cells and that the interaction between antigen and marrow-derived cells occurs there.

The question may be asked whether thymus–bone marrow interactions occur in other antigenic systems. In the case of "cellular immunities" (delayed hypersensitivity and allograft immunity), the specific cells which initiate the reaction are likely to be thymus-derived. We do not, however, know whether the effector cells themselves are these same thymus-derived cells or whether they are derived from a separate cell line originating in bone marrow. It has been established that the majority of the cells which constitute the cellular infiltrate in delayed hypersensitivity lesions are non-specific cells derived from bone marrow (Lubaroff and Waksman, 1968). In "humoral immunities" it is likely that an interaction between thymus-derived and marrow-derived cells takes place, (1) if the antigen can provoke the proliferation of thymus-derived cells, (2) if thymectomy impairs the antibody response to that antigen, and (3) if thymus-derived cells can restore to irradiated mice protected with bone marrow the capacity to respond by producing specific humoral antibody.

The existence of interactions between two distinct lymphoid cell lines in the immune response raises many fundamental questions about the mechanism of the response and opens up yet another intriguing chapter in immunology.

SUMMARY

Among the processes involved in the control of lymphopoiesis are the influences of the thymus and of antigen. The thymus recruits stem cells from the bone marrow and channels their differentiation exclusively along lymphoid pathways. Lymphopoiesis in the thymus cortex is intense and independent of antigenic stimulation. It is intrinsically controlled, being linked in some way to the epithelial cytoreticulum. Some thymic lymphocytes migrate out and eventually join the pool of recirculating small lymphocytes which mediate the "cellular immunities" such as delayed hypersensitivity and transplantation immunity. Thymectomy is associated with a fall in the population of these cells and an impairment of the capacity to undertake cell-mediated immune reactions.

Extra-thymic lymphopoiesis is antigen-dependent. Recent work indicates that thymus-derived lymphocytes can be "activated" in the spleen by certain antigens such as heterologous erythrocytes in mice, to produce large pyroninophilic cells which divide to give rise eventually to a progeny of small lymphocytes which behave as typical recirculating cells. These do not transform to antibody-forming cells. Precursors of antibody-formers are derived initially from bone marrow (not via thymus) and differentiate to antibody-forming cells only after interaction with antigen-activated, thymus-derived cells. These findings were established in thymectomized and irradiated mice reconstituted with

THYMUS–BONE MARROW INTERACTION 249

syngeneic or semi-allogeneic thymus or thoracic duct cells, and challenged
with sheep erythrocytes. The identity of the haemolysin-forming cells
was determined by means of immuno-serological and chromosome
marker techniques.

Thymus-derived cells from donors specifically tolerant to sheep red
cells failed to restore the capacity to react to the specific antigen but could
confer excellent adoptive immune response to horse erythrocytes. This
result does not support those hypotheses which postulate a passive or
non-specific role for thymus-derived cells, such as a trephocytic, phagocytic (antigen processing) or pharmacological role, or the absorption of
"carrier antibodies". It seems more likely that thymus-derived cells are
unispecific cells of the antigen-reactive type. It has not yet been determined whether marrow-derived cells, too, are unispecific cells. If they
are not, then the possibility of transfer of genetic information from thymus-derived cells will have to be examined. If, on the other hand, both
thymus-derived and marrow-derived cells are unispecific, it may be that
those antigenic determinants which elicit the haemolysin response from
marrow-derived, antibody-forming cell precursors have to be focused
on to these precursors by some mechanism which requires recognition of
other antigenic determinants by thymus-derived cells.

REFERENCES

BALNER, H., and DERSJANT, H. (1964). Nature, Lond., 204, 941.
BRETSCHER, P. A., and COHN, M. (1968). Nature, Lond., 220, 444.
CLAMAN, H. N., CHAPERON, E. A., and TRIPLETT, R. F. (1966). Proc. Soc. exp. Biol. Med.,
    122, 1167.
DAVID, J. R. (1968). Fedn Proc. Fedn Am. Socs exp. Biol., 27, 6.
DAVIES, A. J. S., LEUCHARS, E., WALLIS, V., MARCHANT, R., and ELLIOTT, E. V. (1967).
    Transplantation, 5, 222.
FICHTELIUS, K. E. (1967). Expl Cell Res., 46, 231.
FITCH, F. W., STEISKAL, R., and ROWLEY, D. A. (1969). In Lymphatic Tissues and Germinal
    Centers in Immune Response. Adv. exp. Med. Biol., vol. 3, ed. Cottier, H., and Congdon,
    C. C. New York: Plenum.
FORD, W. L., GOWANS, J. L., and McCULLAGH, P. J. (1966). Ciba Fdn Symp. The Thymus:
    Experimental and Clinical Studies, p. 58. London: Churchill.
GOLDSCHNEIDER, I., and McGREGOR, D. D. (1968). J. exp. Med., 127, 155.
GOWANS, J. L., and McGREGOR, D. D. (1965). Prog. Allergy, 9, 1.
GOWANS, J. L., and UHR, J. W. (1966). J. exp. Med., 124, 1017.
HENRY, C., and JERNE, N. K. (1968). J. exp. Med., 128, 133.
LAW, L. W., GOLDSTEIN, A. L., and WHITE, A. (1968). Nature, Lond., 219, 1391.
LUBAROFF, D. M., and WAKSMAN, B. H. (1968). J. exp. Med., 128, 1437.
McGREGOR, D. D., McCULLAGH, P. J., and GOWANS, J. L. (1967). Proc. R. Soc. B, 168, 229.
METCALF, D. (1966). The Thymus: Its Role in Immune Responses, Leukemia Development
    and Carcinogenesis. New York: Springer.
METCALF, D. (1968). J. cell. Physiol., 72, 9.

MICKLEM, H. S., FORD, C. E., EVANS, E. P., and GRAY, J. (1966). *Proc. R. Soc. B*, **165**, 78.
MILLER, J. F. A. P. (1961). *Lancet*, **2**, 748.
MILLER, J. F. A. P. (1962a). *Ciba Fdn Symp. Tumour Viruses of Murine Origin*, p. 262. London: Churchill.
MILLER, J. F. A. P. (1962b). *Ciba Fdn Symp. Transplantation*, p. 384. London: Churchill.
MILLER, J. F. A. P., and MITCHELL, G. F. (1967). *Nature, Lond.*, **216**, 659.
MILLER, J. F. A. P., and MITCHELL, G. F. (1968). *J. exp. Med.*, **128**, 801.
MILLER, J. F. A. P., and MITCHELL, G. F. (1969a). *Transplant. Rev.*, **1**, 3.
MILLER, J. F. A. P., and MITCHELL, G. F. (1969b). In *Lymphatic Tissues and Germinal Centers in Immune Response. Adv. exp. Med. Biol.*, vol 3, ed. Cottier, H., and Congdon, C. C. New York: Plenum.
MILLER, J. F. A. P., MITCHELL, G. F., and WEISS, N. S. (1967). *Nature, Lond.*, **214**, 992.
MILLER, J. F. A. P., and OSOBA, D. (1963). *Ciba Fdn Study Grp The Immunologically Competent Cell: its Nature and Origin*, p. 62. London: Churchill.
MILLER, J. F. A. P., and OSOBA, D. (1967). *Physiol. Rev.*, **47**, 437.
MITCHELL, G. F., and MILLER, J. F. A. P. (1968a). *Proc. natn. Acad. Sci. U.S.A.*, **59**, 296.
MITCHELL, G. F., and MILLER, J. F. A. P. (1968b). *J. exp. Med.*, **128**, 821.
MITCHISON, N. A. (1967). *Cold Spring Harb. Symp. quant. Biol.*, **32**, 431.
NOSSAL, G. J. V., CUNNINGHAM, A., MITCHELL, G. F., and MILLER, J. F. A. P. (1968). *J. exp. Med.*, **128**, 839.
ROBINSON, S. H., and BRECHER, G. (1963). *Science*, **142**, 392.
SCHIERMAN, L. W., and McBRIDE, R. A. (1967). *Science*, **156**, 658.
SMITHIES, O. (1965). *Science*, **149**, 151.
TAYLOR, R. B. (1969). *Transpl. Rev.*, **1**, 114.
THORBECKE, G. J., GORDON, H. A., WOSTMAN, B., WAGNER, M., and REYNEIRS, J. A. (1957). *J. infect. Dis.*, **101**, 237.
WARNER, N. L. (1967). *Folia biol., Praha*, **13**, 1.
WEISSMAN, I. L. (1967). *J. exp. Med.*, **126**, 291.
WILSON, R., BEALMEAR, M., and SOBONYA, R. (1965). *Proc. Soc. exp. Biol. Med.*, **118**, 97.

DISCUSSION

Edelman: Is it your feeling, Dr Miller, that the thymus plays a fundamental role in the origin of antibody diversity, and therefore that it is in a central position in all immune responses?

Miller: I assume that the same generator of diversity operates in randomizing patterns displayed on the surface of either thymus-derived or non-thymus-derived lymphocytes, but that differentiation within the thymus is such as to preclude its cells from exporting their antibody (in the sense of humoral antibody as we know it). Hence, they can only take part in cell-mediated immunities such as delayed hypersensitivity or, in some instances, collaborate in some way with antibody-forming cell precursors to augment their response to certain antigens.

Edelman: Do you feel then that humoral antibodies and delayed hypersensitivity "antibodies" are really different species? Is it in fact true that for humoral antibody production the thymus is absolutely required in ontogenesis?

Miller: Presumably humoral and cell-bound antibodies display the same range of specificities but differ in the secondary biological properties of the molecule. I would say that the thymus is not essential for humoral antibody production in most cases.

Edelman: That is a very fundamental point. My second point is a technical one: have you used allotypic markers rather than chromosomal markers?

Miller: We have been working with congenic strains of mice which are syngeneic but bear allotypic markers, and have determined the γG_2a response to sheep red blood cells, because the allotypic marker is on the γG_2a molecule in the mouse. In the few results so far obtained, the allotypic marker is that of the neonatally thymectomized host or of the marrow donor: it is not that of the thymus donor. The results therefore agree with those obtained with isoantigenic or chromosomal marker experiments.

Monod: To speculate, let us say that there are immunoglobulin genes, with a constant region (C) gene somewhere and a number of variable region (V) genes elsewhere. The first assumption is that there are cells which differentiate in such a way that the V region in them tends to be excluded out of the chromosome into an episome. This is nothing very extraordinary; we know of cases of this sort in bacteria. I would further add that this is true of lymphocytes in particular, but it is more or less true depending upon which class of lymphocyte you are looking at. The properties of thymus cells would be such that in these cells this particular genetic region will exist mostly in the episomal form. So we have an interpretation of what your interaction means: it is a transfer of information, but information concerning only the V region. Therefore this is not contradicted by the fact that the allotype comes from the host.

Miller: I know; but we would like to devise an experiment to check on this.

Monod: The bone marrow cell may be one where the differentiation is such that the episome tends to be inserted back very quickly into the system.

A last speculative point is that we are pretty sure that there must be some signal for the beginning of transcription of any gene; and therefore we may ask why the C gene is not expressed in a bone marrow cell that has not translocated the V region into it. It is not unreasonable to say that the region of insertion of the V gene into the C gene, which has to be a small homologous region, is such that it creates the promoter that is needed for the protein to be synthesized.

Miller: This is why we are interested to find an allotypic marker on the variable half of the chain.

Cohn: I have considered the episome hypothesis (Cohn, 1968). The usual assumption (Rajewsky *et al.*, 1969) that the "thymus–bone marrow" cooperation involves episome transfer of the variable region controlling the combining site is faced with the fact that the postulated episome transfer requires recognition of antigen by both cells, donor and recipient. The proof of this has come from the demonstration that neither tolerant thymus nor tolerant bone marrow will show cooperative effects with immunized bone marrow or thymus (Habicht, Chiller and Weigle, 1969). This dual recognition of antigen was one reason that Bretscher and I (1968) invented carrier-antibody. If transfer is antigen-dependent then the bone marrow cell has to express receptor-antibody and the thymus-derived cell, carrier-antibody. Since the specificity resides in the bone marrow cell before the postulated episome transfer occurs, the hypothesis in this form is untenable. It is possible to propose, as do Rajewsky and his co-workers (1969), that the bone marrow cell expresses one specificity which after episome transfer is converted to another. However, this ignores the reason why the episome hypothesis was invented in the first place by shifting the problem to that of the origin of the specificity of the bone marrow receptor before episome transfer.

Edelman: Dr Miller, you say that in delayed sensitivity reactions the thymus has an important role. Have you any experimental data or any suspicions about how it might work with delayed sensitivity cells?

Miller: I would predict that thymus-derived lymphocytes are the only specific cells; that they recognize antigen specifically, are activated by antigen and, during the process of being activated, produce certain factors, pharmacological or otherwise, which seduce *other* cell types into the area, and that the lesion is actually produced by these non-specific cells which are not thymus-derived.

Edelman: In that case it is a *single* cell phenomenon with respect to specificity in delayed hypersensitivity?

Miller: I would say yes, but it has not been satisfactorily proved yet.

Fortier: Dr Miller, can one relate the role of the thymus in channelling the stem cells from the bone marrow along lymphoid pathways, as inferred from your experiments, to the immunological effects ascribed by Professor White to his recently isolated humoral principle of thymic origin?

Miller: We have hypothesized (Miller, 1965) from work with marker chromosomes that stem cells migrate into the thymus where their differentiation is channelled exclusively to lymphoid pathways. Some thymus lymphocytes emigrate and eventually function as part of the recirculating pool of lymphocytes. Now, is it not possible that a humoral factor, which is produced by the thymus epithelial cells, normally acts within the thymus environment to channel this differentiation? And that under artificial conditions, when cells cannot migrate into or from the thymus, or when the thymus is absent, differentiation of stem cells could be induced outside the thymus, by giving this humoral factor? Another possibility is that some thymus-derived lymphocytes may have escaped from the thymus before birth, and hence before neonatal thymectomy, and that the agent which Professor White has isolated (thymosin) is inducing the proliferation of these few thymus-derived lymphocytes, thus augmenting the immunological capacity of the animal.

White: Our studies with thymosin, the lymphocytopoietic fraction we have isolated from calf thymic tissue, have led us to the present working hypothesis, based upon our evidence (see pp. 210–237) that this humoral factor appears to restore to immunologically incompetent animals some degree of their host-cell-mediated immune functions without, as yet, significant repair of synthesis of humoral antibody. We would like to propose that the maturation and proliferation of immunologically competent cells involved in cell-mediated immune responses, such as allograft rejection and the graft-versus-host reaction, require only the endocrine influence of the thymus. That is to say, a cell-free fraction of the thymus, for example thymosin, can function in lieu of the thymus in such systems. In contrast, immunological maturation of cells involved in humoral antibody production may require both thymosin and the *in situ* environment of the thymus. In this latter case, the thymus functions as what Dr Miller has described as a "finishing school"; that is, it completes or makes possible the completion from bone-marrow-derived cells of lymphoid cells which are not capable of recognizing antigen. These cells are concerned with humoral antibody production, either at the level of recognition of antigen or at a later stage in the maturation process. The presence of a viable thymic locus appears to be required for the proper development of certain types of primitive stem cells into antibody-producing cells, as shown by Dr Miller's studies. We are suggesting leaving unsettled at present the question of whether in this last type of immune response, the production of 19s antibody, the function of the thymus must be supplemented by an action of a humoral agent. The data we have presented (pp. 224–229) in our radiation studies suggest that thymosin does stimulate the proliferation of the more immature lymphoid elements of

9*

spleen and lymph nodes. It remains for further studies to elucidate the factors that determine the channelling of proliferating, immature stem cells into pathways leading to mature lymphocytes versus pathways leading to antigen-reactive, or reacting, cells concerned with host immunological competence.

May I ask, Dr Miller, whether spleen cells will function in lieu of bone marrow cells in conjunction with thymocytes in your experimental system?

Miller: The spleen is a composite organ in which you have haemopoietic cells including stem cells, marrow-derived lymphoid cells and thymus-derived recirculating-type lymphocytes. If we used spleen, interaction could thus be taking place between the two lymphoid cell types within the same population.

White: I ask only because of your emphasis on *bone marrow.*

Miller: We use bone marrow because it lacks the thymus-derived population.

Fortier: Have you speculated, Dr Miller, on the meaning of the extreme lability of the thymus in response to glucocorticoids, and of the species differences in the rate of thymus involution?

Miller: I hesitate to speculate in front of Professor White, who has studied corticosteroids intensively, but I think that the environment in which cells are found might determine some of their phenotypic characteristics, such as susceptibility to corticosteroids.

White: The high mitotic index of cells in the thymus implies a high rate of turnover of the cell population in this organ. This may be the basis of our older evidence (Dougherty and White, 1945) that a primary effect of cortisone and other glucocorticoids is to inhibit mitosis of lymphoid cells and that the thymus is more sensitive, temporally, to corticoids than are peripheral lymph nodes.

REFERENCES

BRETSCHER, P. A., and COHN, M. (1968). *Nature, Lond.,* 220, 444–448.
COHN, M. (1968). In *Differentiation and Immunology,* pp. 1–28, ed. Warren, K. B. New York: Academic Press.
DOUGHERTY, T. F., and WHITE, A. (1945). *Am. J. Anat.,* 77, 81.
HABICHT, G. S., CHILLER, J. M., and WEIGLE, W. O. (1969). In *Developmental Aspects of Antibody Formation and Structure,* ed. Šterzl, J. Prague: Publishing House of Czechoslovak Academy of Sciences.
MILLER, J. F. A. P. (1965). *Br. med. Bull.,* 21, 111.
RAJEWSKY, K., SHIRRMACHER, V., NASE, S., and JERNE, N. K. (1969). *J. exp. Med.,* 129, 1131–1143.

ANTICIPATORY MECHANISMS OF INDIVIDUALS

MELVIN COHN

The Salk Institute for Biological Studies, San Diego, California

I THE GENERAL CASE

WHAT are the common denominators of systems, present in an individual, that enable him to react *specifically* to a vast number of stimuli, unlikely to have been selective forces in evolution? For example, you can make antibody to D-tyrosine or learn to whistle Yankee Doodle.

I shall discuss two aspects:

The interaction between an unexpected stimulus and the recognition element responsible for specificity.

The origin of the recognition elements for unexpected stimuli.

A. What are the ground rules?

Genes, not stimuli, determine the structure of the recognition elements which interact with the unexpected stimuli, and, as such, are subject to an evolution driven by variation and selection.

As a consequence of this,

(1) the functionally poised, specific recognition element *precedes* the stimulus, and

(2) the recognition element upon successful interaction with a given stimulus is established as a long-lived product.

The stimulus *does not induce* the formation of the recognition element; *it selects for it.* It is for this reason that I have called immunity and learning *anticipatory* mechanisms (Cohn, 1967a).

B. What are the recognition elements?

Since the selective force of the stimulus acts on cells, the recognition elements must be expressed in such a way that this is possible.

255

In the immune system, interaction with *properly presented* antigen triggers the antigen-sensitive cell to both multiply and differentiate to a cell which secretes antibody with a specificity directed against the triggering antigen. The presumption is that the recognition of any antigenic determinant must reside in antibody itself (Jerne, 1955). Since the specificity of the recognition element must be the same as that of the induced antibody—that is, both recognize the triggering antigen—and since combining specificity is a function of the primary sequence of amino acids, I would argue that the primary sequence determining combining specificity must be the same for the recognition element and the induced secreted antibody.

In the learning system, I can only make a reasonable guess. The recognition element could be a module of circuitry made up of cells which specifically communicate with each other via synapses formed by complementary membrane-bound proteins (Szilard, 1964; Cohn, 1968a). On interaction with a *properly presented* learning stimulus, this module is fixed permanently into the circuitry of the memory system.

C. What determines the breadth of responsiveness?

The immune and learning systems have vast but not unlimited capabilities to respond, because, eventually, the limitation becomes the number and kinds of genes which can participate in the response. Obviously the more genes you have, the greater the range of stimuli you can recognize. This is what gives the individual foresight. However, the total number of germ-line genes is limited because those coding for the recognition of very rare stimuli are effectively silent and would be expected to be lost by mutational drift before selection could save them. The way that hypotheses cope with this dilemma is to reduce the number of germ-line genes to a point where each one is expressed sufficiently often for selection to operate effectively. This can be done either by having the products of a few germ-line genes interact in all permutations and combinations to generate a large number of recognition elements, and/or by having new recognition elements generated as a consequence of somatic variation throughout life. The extreme form of the somatic model is that *spontaneous* somatic mutation is the source of diversity in the individual's responsiveness (Cohn, 1968a). This raises questions of whether the rate of appearance of spontaneous random mutants could be high enough to generate the known numbers of functional variants. As a consequence, models of controlled somatic variation (Edelman and Gally, 1967) have been proposed in which the solution to this problem must be weighed against the likelihood of the specialized mechanism of variation. The precise relationship between germ-line and somatic genes is one of the unresolved questions in the understanding of anticipatory mechanisms.

Whatever the origin of the diversity (germ-line or somatic), two points should be stressed. First, somatic selection must operate to amplify and store in an orderly way those recognition elements which correspond to actually encountered inductive or learning stimuli. In fact, the key to any understanding of anticipatory mechanisms will be a precise description of the selective pressure. This is what gives the individual hindsight. Second, the capacity of the individual to respond to a large number of unexpected stimuli is a function of the size of the pool of primed recognition elements present throughout the duration of the stimulus. Of course, there must be a class of stimuli which are invisible to the recognition system simply because it operates via the functioning of a limited number of gene products—that is, proteins which are responsible for antibody activity or the membrane-membrane interactions which constitute learning circuitry. If the presently invisible stimuli were to become the selective forces of the future, the organism which had increased its germ-line genes, by duplication and divergence of them, would have an advantage because it would be capable of generating a new family of somatic derivatives whose products could now interact with the previously invisible stimulus.

D. How does an anticipatory mechanism avoid self-destruction?

This problem is more easily posed for the immune system where antibody formation to self-components would be lethal. In the learning system the module of learning circuitry should not get its cues from random noise or random internal previously learned or innate circuits.

Since the recognition element precedes the stimulus, a paralytic regulatory mechanism must operate to prevent harmful responses to self-stimuli or noise. Either these self-stimuli are not registered for anatomical reasons or there is an active mechanism to prevent a response. This latter mechanism (seen most clearly in the immune system) is presumed to use the same recognition elements as the one for induction of it. As a consequence, I shall postulate that the difference between response and no response resides in the number of "determinants" on a stimulus which is detected. In other words, induction of antibody formation or memory storage will require the *associated recognition* of two or more distinct determinants or input signals.

II   THE IMMUNE SYSTEM

A. Regulation: induction and paralysis of the immune system

The starting point of the model (Bretscher and Cohn, 1968; Cohn, 1969) is an antigen-sensitive cell that expresses a unique recognition element (receptor antibody), identical in specificity to the antibody it would make if induced. There is a constant birth of new antigen-sensitive cells throughout the life of the animal. This cell, depending upon how antigen is presented to it, is either stimulated to divide and make more antibody or killed, resulting in tolerance. The proposed pathway of differentiation leading to tolerance or antibody formation is illustrated below (see Landy and Braun, 1969 for discussion):

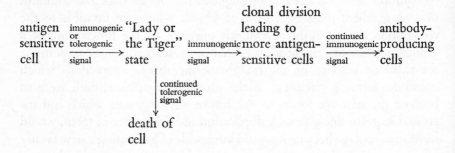

The arguments that an antigen-sensitive cell, expressing a unique receptor, can go one way or the other on reaction with properly presented antigen, have been discussed (Dresser and Mitchison, 1968; Bretscher and Cohn, 1968).

The tolerogenic signal is pictured to occur by a reaction between receptor antibodies and antigenic determinants of the tolerogen on a one-to-one basis (Fig. 1). In order that the cell sense this signal the receptor antibody is postulated to exist in two configurations, unbound (U) and bound (B) (Fig. 1). The antigen-sensitive cell expresses its receptor antibody essentially in the unbound conformation because the equilibrium between U and B is far in favour of U (U $\rightleftharpoons$ B). The reaction with antigen selects for the B conformation, the tolerogenic signal.

Whereas the tolerogenic signal is postulated to require the specific recognition of only one determinant, the immunogenic signal involves the recognition of at least two determinants on the immunogenic molecule. According to our ground rules, the recognition of these two determinants must be mediated by two antibodies, one of which, as in the case of a tolerogenic signal, is the receptor of the antigen-sensitive cell and the other, carrier-antibody, provides, directly or indirectly, the additional information for immunogenesis. I shall present a minimal model and then discuss possible alternatives.

The minimum requirements for an immunogenic signal (Fig. 2) are:

(1) The recognition of the immunogen by both receptor-antibody (anti-a) and carrier-antibody (anti-c).

(2) The conformational transition from unbound (U) to bound (B) occurring in both receptor and carrier-antibody.

(3) An additional signal, symbolized as a lightning-arrow in Fig. 2, which is activated by the conformational transition $U \rightarrow B$ in the carrier-

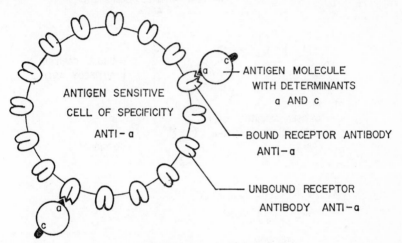

FIG. 1. The tolerogenic signal.

antibody. This signal, which I shall call the GO-signal, tells the antigen-sensitive cell (anti-a) to go down the pathway of division and differentiation leading to antibody formation.

I shall discuss each of these requirements in the light of probable molecular level events later on.

Determinants a and c are equivalent. They are distinguished here to illustrate the immunogenic signal. However, determinant a can act as does determinant c in the induction of anti-c. Carrier-antibody, which might be a special immunoglobulin class, is induced by exactly the same mechanism as that used for any other immunoglobulin.

What are examples of the phenomena which this minimum theory explains?

(1) The induction of antibody formation is described as requiring a macromolecule (Benacerraf, Paul and Green, 1969). The theory says that

the antigen must have two or more sufficiently well-separated foreign determinants to permit the immunogenic interaction (Fig. 2). "Macro"-molecules in general have this property whereas "small" molecules do not. However, it is evident, that for any given case, one should be able to determine the minimum size which permits two determinants to be expressed in an immunogenic configuration (Fig. 2). This has been done for the α,DNP-[L-lysine]$_n$ polymers for which it has been found that $n \geqslant 7$ is essential for immunogenicity (Stulbarg and Schlossman, 1968; David and Schlossman, 1968).

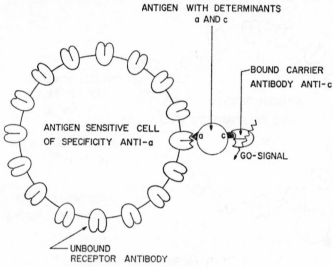

FIG. 2. The immunogenic signal (minimum model).

(2) In order to induce the formation of antibody to the haptene-determinant of a haptene-carrier antigen, the carrier moiety must be immunogenic. This implies that both haptene and carrier determinants must be specifically recognized (see review of Benacerraf, Paul and Green, 1969; Bretscher and Cohn, 1968; Mitchison, 1967; Rajewsky and Rottlander, 1967; Rajewsky et al., 1969). In general, these two determinants would be different but they need not be. However, in practice, the larger the number of total foreign determinants on an antigen and the greater the number that are different, the more rapid will be the onset of induction. In fact one of the key properties of macromolecules used as immunogens (see point 1) is that they possess many different foreign determinants compared to small molecules. Having fewer identical determinants on an antigen tends to lead to tolerance because the probability of

an encounter leading to a binary inducing interaction with both receptor and carrier-antibody will be low compared to that occurring between antigen and receptor antibody alone, leading to tolerance (Fig. 1). The fact that formation of antibody to any given determinant (in this example a haptene) requires that another determinant on the same molecule (in this example the carrier) be specifically and in an associated way recognized, is the central observation upon which this theory depends.

(3) An antibody appearing *de novo* would be expected to act as an antigen to induce the formation of anti-antibody and thus set off a chain induction which chokes the system. How is this avoided?

One can immunize an animal against its Fab portion (containing the combining site) if the Fc portion also contains a foreign antigenic determinant (Lieberman and Potter, 1966; Cohn, 1969; Cohn, Notani and Rice, 1969). A unique immunoglobulin derived from a myeloma in one inbred mouse strain (BALB/c) is injected into another mouse strain (A/J) differing by an antigenic determinant (a genetic marker) in the Fc portion of the molecule. The antibody response in A/J is directed against not only the Fc but also the Fab portion. Since one can neither immunize the BALB/c strain with its own myeloma protein nor the A/J strain with the Fab fragment (separated from Fc), it is clear that a carrier effect is being described as well as a possible breaking of tolerance to the unique Fab in the A/J mouse. I would assume that any given unique immunoglobulin would not differ from the total immunoglobulin pool by more than a few foreign Fab determinants, many of which cannot act as carrier for the others because of steric reasons; therefore, tolerance is favoured (see point 2 above) except of course when foreign determinants are present in Fc.

(4) How did the phenomena of induction and tolerance arise simultaneously in evolution (Cohn, 1967*b*)? One structure, antibody itself, is used to recognize the antigen, whether it is delivering a tolerogenic or immunogenic signal. Therefore, the regulatory device selected for had to include the tolerogenic signal in the immunogenic signal. A cell with receptor antibody which cannot undergo an unbound-to-bound transition (leading to tolerance) cannot be induced.

The initial events leading to tolerance-killing or antibody-induction are postulated to be identical, as symbolized by the "Lady or the Tiger" state (see above, p. 258). A cell carrying a tolerogenic signal is not immediately killed but can be put on an induction-pathway if the immunogenic signal is received before "death". The point of the "Lady or the Tiger" state is to assure that any cell which is inducible can be rendered tolerant. Under conditions of induction, following the injection of antigen, the average time to kill a cell by a tolerogenic encounter must be longer than the average time it takes for an inductive encounter between antigen-sensitive cell, antigen and cell-bound carrier-antibody.

The selective pressure for this is easy to visualize. An animal whose germ-line genes coded for a molecule which could not undergo the tolerogenic signal would be eliminated by auto-immunity and one which could not be induced would also be selected against.

(5) Why is tolerance to self-components stable? Consider a self-component, A, with determinants a_1–a_{20}. An antigen-sensitive cell anti-a_7 arises *de novo*. Determinant a_7 will now be recognized as foreign. Self-component A will not be immunogenic because it possesses only one foreign determinant, a_7. However it will be tolerogenic so that the antigen-sensitive cell, anti-a_7, will be killed. Now suppose that two antigen-sensitive cells of different specificity arise simultaneously, anti-a_7 and anti-a_{12} (a producer of carrier-antibody). Stability of tolerance can still be assured, if the time taken to paralyse the two antigen-sensitive cells is considerably shorter than the time required for the antigen-sensitive cell, anti-a_{12}, of the carrier-antibody class to build up an effective concentration of carrier-antibody.

(6) In the case of potential immunogens, induction of antibody and of tolerance shows a characteristic dose relationship (Mitchison, 1968b). The antigen-dose which induces antibody formation is intermediate between a low and high-dose zone, each of which provokes tolerance. There is competition between the induction of low or high-zone tolerance and of intermediate-zone antibody formation.

Consider the two competing processes, tolerance and induction, tolerance signalled solely by the interaction between antigen (Ag) and receptor-antibody (AbR), and induction triggered by the associated interaction between antigen and both receptor and carrier-antibody (AbC). At low concentrations the complexes Ag·AbR and Ag·AbC will be favoured, the former leading to tolerance, the latter without consequence unless an associated encounter with an antigen-sensitive cell occurs. This inductive encounter will be rare at low dose and become more frequent as the concentration of antigen increases to that of the intermediate-dose zone, because the level of both Ag·AbR and Ag·AbC becomes sufficiently high to make the probability of an encounter with free AbR or AbC leading to antibody formation measurable. This means that the rate-limiting step is the killing of the cell by tolerance, a process which becomes constant in rate before all of the free AbR on its surface has reacted. The rate of the inducing reaction continues to rise exponentially because of the formation of AbR· Ag· AbC and it overtakes the tolerance reaction, the rate of which is now unchanged by increasing dose of antigen. The inducing reaction takes precedence during the minimum period (the "Lady or the Tiger" state) before which a cell carrying a tolerogenic signal (Ag· AbR) is killed. In the intermediate-dose zone AbR and AbC are not saturated with antigen, thus allowing the inducing reaction

$$\left. \begin{array}{c} Ag \cdot AbR + AbC \\ \\ Ag \cdot AbC + AbR \end{array} \right\rangle AbR \cdot Ag \cdot AbC$$

to occur. At the tolerogenic high-dose zone both Ag· AbR and Ag· AbC are saturated so that an inducing interaction (AbR· Ag· AbC) is impossible and the cell is killed by the Ag· AbR signal.

Thus low and high-zone tolerance occur by the same mechanism. In the high zone many of the receptors on the antigen-sensitive cell are in the tolerogenic configuration (Fig. 1) whereas in low zone few are bound.

The Mitchison phenomenon illustrates how clever is Mother Nature. She made the probability of rendering an antigen-sensitive cell tolerant much greater than that of inducing it. This is a check-valve on the immune system which constantly threatens to annihilate its host by reacting to it. Detroit might well notice that a car is safer if it is easier to stop than to start.

(7) Tolerance can be broken rapidly by injecting cross-reacting immunogens (Weigle, 1969). Consider two antigens A and B. The animal is tolerant to antigen A which has determinants a_1–a_{20}. Tolerance can be broken by antigen B which has determinants a_1–a_{10} which cross-react with a_1–a_{10} and determinants b_1–b_{10} which are foreign.

If a new antigen-sensitive cell with specificity anti-a_7 appears, it may be induced to make antibody anti-$\tilde{a}_7$ by antigen B as this antigen has foreign determinants b_1–b_{10} which act as carrier sites. Thus tolerance is broken.

It may be argued that the reaction between anti-$\tilde{a}_7$ and a_7 is too weak for A to have originally paralysed the cell anti-$\tilde{a}_7$. If this were so, the simultaneous injection of A and B should have no effect on the breaking of tolerance by B. However, this is not what is found. The injection of A and B prevents the breaking of tolerance by B. This is expected because A cannot act as an immunogen, only as a tolerogen.

(8) Under special conditions two antigens will competitively inhibit the response to either one (Brody, Walker and Siskind, 1967). Consider

two unrelated haptenes, H_1 and H_2 and two unrelated carriers, X and Y. Four antigens can be made by linking them: H_1-X, H_1-Y, H_2-X, H_2-Y. The immunization of an animal with any one of these antigens results in a maximal anti-H_1 or anti-H_2 response. The immunization with both H_1 and H_2 linked to different carriers, e.g. H_1-X+H_2-Y, or H_1-Y+H_2-X, also results in maximal anti-H_1 and anti-H_2 responses. There is no competition under these conditions. Competition is seen only when the two haptenes H_1 and H_2 are injected each associated with the same carrier, eg. H_1-X+H_2-X or H_1-Y+H_2-Y. A marked decrease in anti-H_1 and anti-H_2 response is found.

Since the antigen-sensitive cells, anti-H_1 and anti-H_2, have receptors with no overlapping specificities the competition cannot be at the level of the antigen-sensitive cell. This *indicates* that there is competition for carrier-antibody under conditions where the amount of carrier-antibody is limiting. Other experiments support this conclusion.

An animal immunized with H_1-X, so that it could produce high levels of anti-H_1 on injection of H_1-X, hardly responds when injected with H_1-Y (Mitchison, 1967). This suggests that, for H_1-Y, the poor anti-H_1 response is due to a limiting amount of carrier-antibody to Y. This experiment has been extended (Rajewsky et al., 1969) by immunizing with H_1-X and Y. The fact that this animal now shows a good anti-H_1 response to H_1-Y confirms the interpretation. Furthermore, only if an irradiated recipient is reconstructed (Mitchison, 1969b) with two spleen cell populations, one from a donor immunized with H_1-X and the other with Y, will it show a good anti-H_1 response upon challenge with H_1-Y. These cooperative effects between cell populations show that carrier-antibody is induced on immunization and is limiting in the naïve animal. If carrier-antibody were in excess in the primary induction, the animal would have difficulty in maintaining tolerance to self-components, as discussed previously.

(9) Why does the initial antibody response to antigen show in general a progressive increase in affinity with time (Siskind and Benacerraf, 1969)? This phenomenon has never had a satisfactory explanation. Before the present formulation, the simple interaction between processed antigen and the responsive cell was believed to lead to induction. Clearly then the first antibody to appear should have been that of highest, *not lowest* affinity. Since the above arguments indicate that carrier-antibody limits the initial antibody response in the naïve animal, antigen-sensitive cells of highest affinity would be eliminated by a tolerogenic signal in the initial contact with antigen. As the carrier-antibody level increases, these

cells of highest affinity are selected for as they appear. Of course, subsequent phenomena such as feedback by the induced circulating antibody make the selective pressure even stronger for the induction of cells of increasing affinity (Mitchison, 1968a). It is predictable that a primary immunization with X followed by a secondary challenge with H-X would permit initially a primary induction of high-affinity anti-H cells not seen when H-X is injected into a naïve animal.

(10) How does the theory explain *original antigenic sin* (Fazekas de St. Groth, 1967)? This phenomenon is observed when an animal is primed with H-X and then challenged with h-X. H and h are haptenes which cross-react. The initial antibody produced in the secondary response has a higher affinity for H than for h (Dubert, 1959; reviewed by Mitchison, 1968a). This result implies that the primary immunization with H-X increases not only the level of carrier-antibody but also that of high-affinity anti-H antigen-sensitive cells. The secondary injection of h-X stimulates these anti-H cells because carrier-antibody levels do not limit their induction (see 8 above).

Thus far I have shown that the minimal model accounts for a wide variety of phenomena. Now I would like to deal with alternative forms of the model by discussing the postulated conformational transitions and the GO-signal.

Conformational changes in antibody upon reaction with antigen provide the only precise molecular level mechanism for regulation by antigen thus far proposed. The minimum model requires only that the transition from U→B be sensed. Actually two conformations of antibody appear to exist, a bound and a stretched form. The latter, stretched, configuration might be part of the immunogenic signal (Bretscher and Cohn, 1968). In order to evaluate this proposal, let us look at what is known about such events.

Neither receptor nor carrier-antibody is yet amenable to physicochemical studies. Therefore, I will argue from data concerning serum antibody, assuming, as was done previously (Bretscher and Cohn, 1968), that the extrapolation is a reasonable one.

The following observations indicate conformational changes in antibody on interaction with haptene and with antigen.

(1) Haptene-bound antibody has a decreased frictional coefficient (Warner, Schumaker and Karush, 1969) and an increased resistance to proteolytic digestion (Grossberg, Markus and Pressman, 1965) when compared to unbound antibody. The molecule in the bound conformation

(B), leading to the tolerogenic signal, has a more compact or collapsed configuration.

(2) The stretched conformation is revealed by immunochemical methods when high molecular weight antigens are complexed with antibody (Najjar and Fisher, 1955; Najjar, 1963; Henney, Stanworth and Gell, 1965; Henney and Stanworth, 1966; Kelus and Gell, 1968), or by optical rotation (Ishizaka and Campbell, 1959), or by electron microscopy (Feinstein and Rowe, 1965; Valentine, 1967). The elicitation as well as the induction of immediate and delayed hypersensitivity appears to require the stretched conformation (Feinstein and Rowe, 1965; Levine and

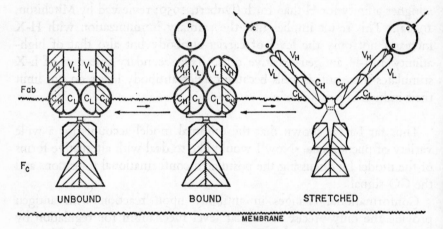

FIG. 3. Conformational transitions in receptor-antibody corresponding to the tolerogenic signal (bound) and immunogenic signal (stretched).

Redmond, 1968; Lawrence and Landy, 1969). The molecule in the stretched configuration reveals new antigenic determinants not present in the unbound or bound forms. In order to go from an unbound (U) to a stretched (S) configuration, the receptor-antibody must pass through the bound (B) form (see Fig. 3).

Two kinds of preferential subunit-interactions might contribute to the conformations.

(1) There appears to be a higher affinity between identical light chains— that is, those derived from a given immunoglobulin—than between two non-identical light chains chosen at random (Stevenson and Straus, 1968). A similar situation for heavy chains (H) has not been described. At present, there is no need to assume that a preferential H–H interaction plays a role.

(2) There is in general a higher affinity between a light and heavy chain derived from one immunoglobulin molecule than between a light and heavy chain chosen at random (Edelman *et al.*, 1963; Roholt, Onoue and Pressman, 1964; Grey and Mannik, 1965; Mannik, 1967).

The complexity of protein structure allows for several models to account for the conformational changes, U, B and S. The simplest and clearly first-approximation model divides the molecule into four nearly equal regions each with a specialized behaviour (Fig. 3). The molecule is usually looked upon as consisting of an Fab portion containing the combining site and the Fc portion essential for mediating the diverse interactions of immunoglobulins, e.g. complement fixation, placental passage, cytophilic and skin reactivity. I propose to further subdivide the molecule in order to account for the proposed conformational transitions. For this reason the Fab region is subdivided into a variable portion (V_{Fab}) in which V_H and V_L interactions dominate and a constant portion (C_{Fab}) in which C_L and C_H interactions occur. The Fc region, for our argument, can be left as a single block but the symmetry of the sequence which can be seen as having originated from four repeated gene duplications (Doolittle, Singer and Metzger, 1966; Lennox and Cohn, 1967; Hill *et al.*, 1967; Edelman, 1967; Singer and Doolittle, 1968; Edelman, 1970) leads one to divide it into two regions also, the amino-terminal half in which the carbohydrate is situated and the carboxyl-terminal half. Of course the prediction is that these two Fc regions will possess distinct functional roles with respect to the reactions listed above.

The antibody combining site is presumably minimally (or not) expressed in the unbound conformation. The L–L interaction via the V_{Fab} part of the molecule might be visualized as maximal and the L–H interaction as minimal (Fig. 3). The unbound conformation is in equilibrium with the bound conformation in which the appearance of the antibody combining site in V_{Fab} is expressed by a decrease (possibly to zero) in the L–L interaction and an increase (possibly to a maximum) in the L–H interaction responsible for the combining specificity. Stretching not only reduces the L–L interaction to zero and increases the L–H interaction to a maximum (Cohn, 1969) but also reveals a new configuration which has been assumed to be part of the immunogenic signal (Bretscher and Cohn, 1968; Cohn, 1969). The available data (Feinstein and Rowe, 1965; Gell and Kelus, 1967) indicate that stretching causes configurational changes in the Fc portion of the molecule, as represented in Fig. 3.

At this point it is instructive to give the background to the previous suggestion (Bretscher and Cohn, 1968) that the transition from B to S is part of the immunogenic signal. The reaction pictured as in Fig. 4 was that receptor-antibody, anti-a, is stretched between two identical determinant groups, *a*, on two immunogenic molecules themselves cemented together by a specific interaction between another determinant, *c*, on the carrier and stretched carrier-antibody, anti-c. The stretched conformation (S) as the *sole* immunogenic signal is insufficient since large antigens such as red blood cells can stretch the receptor-antibody and would induce, therefore, in the absence of carrier-antibody. In other words self-antigens on cells, such as erythrocytes, would induce auto-

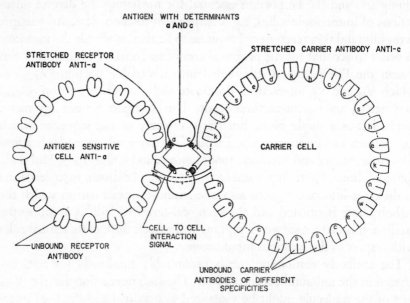

FIG. 4. The immunogenic signal (a reasoned choice among the alternatives).

antibody (Cohn, 1969). In our original model (Bretscher and Cohn, 1968) we proposed that polymeric antigens did not reach antigen-sensitive cells intact but were broken down by phagocytic cells to a size which required carrier-antibody. However, it is improbable that all auto-polymeric antigens are processed such that they never reach the antigen-sensitive cell intact. Therefore, even if the immunogenic signal were to require the stretched configuration, the GO-signal must be postulated in order to make the reaction with carrier-antibody obligatory. This is accomplished if the antigen-sensitive cell and the GO-signal are both activated by a conformational change, B or S, occurring in the receptor and carrier-antibody respectively. Furthermore, this assumption avoids the induction of an antigen-sensitive cell with a specificity directed against a carrier-antibody.

The stretched conformation was introduced (Bretscher and Cohn, 1968) as a possible component of the immunogenic signal for the following reasons:

(1) It accounts for the fact that aggregates tend to be immunogenic whereas non-aggregates (monomers) tend to be tolerogenic (reviewed by Siskind and Benacerraf, 1969). Aggregates and large polymeric antigens (red cells) are known to stretch the antibody molecule (reviewed in Bretscher and Cohn, 1968).

(2) It makes the finding that, in general, immunogens are macromolecular interpretable in terms of the requirement for a stretching interaction with receptor and carrier-antibody (Fig. 4) which would necessitate rigidly held together, well-separated determinants (discussed in Bretscher and Cohn, 1968).

(3) It predicts the passage from low-zone tolerance to intermediate-zone induction. Under the minimum model (p. 262, point 6) this transition as antigen dose increases is due to the rate of the tolerance reaction reaching a plateau while that of the induction reaction increases until it predominates. In other words the minimum model accounts for the switch-over from low to intermediate-dose phenomena but does not predict it without additional assumptions concerning the rate-limited tolerance reaction. The introduction of the stretched configurational signal predicts the transition from low-zone tolerance to intermediate-zone induction because at low concentrations of antigen the tolerogenic interaction Ag·AbR predominates while at higher concentrations the probability of two antigen molecules being aggregated together as Ag_2·AbR and Ag_2·AbC increases to a point where induction due to the immunogenic signal (Fig. 4) is measurable.

These arguments suggest that the stretched configuration is part of the immunogenic signal.

I should add, now, that the concept of carrier-antibody as an *obligatory* part of the immunogenic signal does not go uncontested. In fact, most immunologists led by Mitchison (1969a, b) and Miller (Miller and Mitchell, 1970) consider carrier-antibody as an antigen-trapping or concentrating device only. The treatment of the carrier-antibody system solely as a transport mechanism implies that its role is to provide a high enough concentration of immunogen at the receptor site of the antigen-sensitive cell. Any mechanism, even an imaginary non-specific one, could in principle fulfil this requirement. Furthermore, not only is the distinction between tolerance and induction left to another process but different mechanisms are needed to account for the difference between low-zone and high-zone tolerance. Therefore, I prefer to regard recognition of the immunogen by carrier-antibody as an *essential* and *specific* requirement of the immunogenic signal, rather than merely a transport or concentrating device.

Now let us consider two possible GO-signals, direct cell-to-cell communication and an antigen-sensitive cell activator.

The cell-to-cell communication signal takes two forms. It can be assumed that the carrier-cell either transports on its surface many different passively acquired cytophilic carrier-antibodies, among them anti-c (Fig. 4), or is an antigen-sensitive cell which synthesizes carrier-antibody, anti-c, which is not secreted. In both cases the immunogen possessing two foreign determinants, *a* and *c*, cements an antigen-sensitive cell, anti-a, to a carrier-cell, anti-c. Each cell reads the conformational transition in its respective antibody recognition element as a result of which there is communication (the GO-signal) between them, such as transfer of a chemical signal or membrane–membrane interaction, leading to the division and differentiation of the antigen-sensitive cell, anti-a.

The activator GO-signal is included here to emphasize that there is only weak evidence to support the idea of direct cell-to-cell communication. There are two models. In the first, the conformational transition in the carrier-antibody leads to its reaction with a soluble circulating activator which when cemented to an antigen-sensitive cell (Fig. 2) stimulates the cell to divide and differentiate. In the second, a "macrophage" transporting cytophilic carrier-antibody on its surface interacts with an immunogen, as a consequence of which the immunogen is metabolized to an antigen–activator complex. This complex is either released or remains on the cell surface where it eventually interacts specifically via its antigenic determinant with the receptor-antibody on the antigen-sensitive cell. The activator then acts to trigger the cell (for one specific model see Braun, 1969; Nossal, 1969). There is a massive literature claiming this activator to be ribonucleic acid (for one example see Gottlieb, 1969). Of course, some postulate concerning the metabolic step specifically linking any immunogen with a given activator has yet to be made. I shall leave activator models at that and return to cell-to-cell communication models by discussing the immunoglobulin class of carrier-antibody and where it comes from.

It is likely that carrier-antibody, whether it is associated with a multi-potential or unipotential cell, is of a special immunoglobulin class. The cooperative effects between thymus and bone marrow derived cells (Miller, 1969) suggest strongly that carrier-antibody is made by thymus-derived cells (Mitchison, 1969b; Good, 1969; Cohn, 1969). Furthermore, it has been shown (Mitchison, 1969b) that two spleen populations, one immunized against a haptene (H) on a carrier (X), ie. H-X, and the other against an unrelated carrier (Y), cooperate to produce an anti-H response to the haptene–carrier complex, H-Y. These results show that the contribution of the immunized thymus is exported to the spleen. This contribution is postulated to be antigen-sensitive cells of the special carrier-antibody class (discussed in Mitchison, 1969b; Cohn, 1969). These cells are destined to undergo one of two reactions, depending on whether the GO-signal involves a carrier-cell which transports many different carrier-antibodies or is the above thymus-derived cell itself. In the former case, the antigen-sensitive cell of the carrier-antibody class is induced to secrete carrier-antibody cytophilic for the carrier-cell (Fig. 4). Possibly the so-called humoral thymic factor (hormone) is carrier-antibody (White and Goldstein, 1970). In the latter case, the antigen-sensitive cell of the carrier-antibody class interacts via immunogen with antigen-sensitive cells of any immunoglobulin class to induce them.

Consider first the multipotential carrier-cell model (Fig. 4). The carrier-antibody it passively acquires must (1) be cytophilic for the carrier-cell, (2) be in very low free concentration in the serum, and (3) have a rather short half-life, in order to account for the kinetics of tolerance and induction (Bretscher and Cohn, 1968). From what is known of the various immunoglobulin classes this model then *requires* that carrier-antibody be of a special class. The multipotential carrier-cell has, as possible candidates, the macrophage and dendritic cells of the germinal centres (Nossal, 1969; Humphrey, 1969).

An analysis of the model that the carrier-cell is a unipotential thymus-derived antigen-sensitive cell requires an assumption about whether the carrier-cell itself is induced by cell-to-cell interactions only within that class. If the carrier-cells were the thymus-derived cells which are known to mediate tissue-graft and delayed hypersensitivity reactions (Mitchison, 1969a), I would presume that the carrier-cell is not induced as is the other antigen-sensitive cell upon an interaction between them. Since the induced carrier-cell would secrete toxic effectors, such as lymphotoxin (reviewed in Lawrence and Landy, 1969), immunization leading to antibody formation would flood the animal with toxin. This problem is reduced but not eliminated if a carrier-cell is induced only by interaction with another carrier-cell. Therefore, although I think that this suggestion is unlikely, it leads to the picture that the carrier-class is unique as a cell but leaves open the question of whether its receptor is of a special class. Even if the carrier-cell and toxic effector-secreting cells are not equated as above, the conclusion is the same.

A look at the evolution of the immune system does not allow further dissection of these ideas because there are no species known in which cell-mediated and humoral antibody phenomena are separated. The hag-fish shows no immune response of any kind and the lamprey shows both categories of response. However, the first antibody-producing animals are only known to express one class of immunoglobulin, believed to be IgM, which conceivably could act as its own carrier-antibody. The enhance-ment of sheep erythrocyte-induced antibody synthesis in the IgM class by antibody of that same class (Henry and Jerne, 1968) might be used as an argument that IgM has carrier role for itself, reminiscent of an ancestral role. In fact, Mitchison (1969b), in considering this enhancement pheno-menon, makes the suggestion that there are two classes of carrier-antibody, IgM responsible for the induction of IgM only, and IgX, an unknown class responsible for the induction of IgG only. I, personally, feel that other explanations of the enhancement of IgM antibody synthesis by passively given IgM antibody are more likely—for example, enhanced breakdown of red blood cells to immunogenic fragments.

I shall mention only in passing two problems raised by the multi-potential carrier-cell but not the unipotential carrier-cell model. These are (1) the need to prime the immunogenic reaction during ontogeny with maternal carrier-antibody, and (2) the requirement that experienced (memory) antigen-sensitive cells of the carrier-antibody class be induced by residual antigen (leakage of antibody) to maintain a high enough level of carrier-antibody in order that recall responsiveness exist (Bretscher and Cohn, 1968). The prediction that maternal carrier-antibody primes the initial responsiveness of the embryo to antigen is of special interest because it greatly simplifies the assumptions necessary in order to account for the selective component in the building of a library of diversified antigen-sensitive cells (see later).

If it is correct that the thymus contains and exports antigen-sensitive cells which express the carrier-antibody class, the following picture emerges. The effect of thymectomy will depend upon how many of these cells are functional in other tissues at the moment of the operation. A neonate mouse would, for thymus-dependent antigens, have few antigen-sensitive cells of the carrier-antibody class in other tissues. An adult mouse would have many which can be killed by X-irradiation. Since bone marrow has few cells of this class, restoration of the irradiated thymectomized adult mouse to antibody synthesis requires in addition to bone marrow, thymus-derived cells, or, for that matter, any source of antigen-sensitive cells of the carrier-antibody class, such as thoracic duct lymphocytes (Miller, 1969).

Why might antigen-sensitive cells of the carrier-antibody class be associated with an organ like the thymus? The amount of carrier-antibody for a naïve response is limiting. This is no surprise as the amount of carrier-antibody must be regulated at a very low level to minimize auto-immune accidents. If the probability of inducing an antigen-sensitive cell of the carrier class were proportional to the ratio of these cells to the total number, then the total antibody response would be slow, limited by the level of carrier-antibody which must be built up. Very rapid feedback inhibition by circulating non-carrier antibody (Möller, 1969) might quench the response. If the only class of antigen-sensitive cells which the thymus contains is that of carrier-antibody and if the carrier-cell and immunogen (Kölsch, 1968) are present there also, then the rapid induction of carrier-antibody relative to the other classes is assured even though carrier-antibody is at first limiting.

The thymus turns over, by a birth and death cycle, a large number of cells. It exports a few and imports a sufficient number so that it is con-stantly repopulated with peripheral cells (Metcalf, 1967). The following picture emerges.

(1) The export–import cycle may be a way of guaranteeing an antigen-sensitive population which is fully tolerant to self-antigens be-cause it has been exposed to them during the *séjour* in the extra-thymic tissues.

(2) The export–import plus the birth–death cycle may be part of the mechanism for generating and selecting for diversity (see later).

I wish to stress two very important practical consequences of this theory for the recognition of and response to antigen.

First, non-immunogenic substances should be able to induce tolerance. This prediction of the theory contradicts the present-day generally accep-ted belief that in order for a substance to be tolerogenic, it must be immu-nogenic. This point has never been adequately tested. The phenomenon of breaking tolerance tells us why this is so (Bretscher and Cohn, 1968; Cohn, 1969). Consider an animal suspected of being tolerant to a haptene, H, as a result of an injection of H alone. In order to assay for tolerance to H, an immunogen consisting of H linked to a carrier X, H-X, must be injected. This is equivalent to breaking tolerance to H by the injection of a cross-reacting antigen, H-X, as I discussed previously.

If this prediction is correct, one should be able to isolate non-immunogenic tolerogenic fragments, for example from transplantation antigens, which would allow an adult to be rendered specifically tolerant to transplants. If immunogenicity and tolerogenicity are inseparable, the present formulation is incorrect and the new theory will have as its key postulate a molecular level mechanism which links them.

Secondly, if an individual cannot make carrier-antibody, he cannot make antibody to anything; he is only rendered tolerant on contact with antigen. This means that the specific suppression of the carrier-antibody class by such methods as are used to suppress allelic immunoglobulin classes (Herzenberg, McDevitt and Herzenberg, 1968), would render the individual unable to reject grafts or show any hypersensitivity reactions, auto-immune, immediate or delayed. If the inhibition of carrier-antibody synthesis were reversible, tolerance to a specific determinant could be made permanent if this determinant were present during and following the recovery period. Such a method would work in adults.

B. The origin of the diversity of antibodies

What is the relationship between the germ-line and somatic gene?

We imagine two ways in which the structural genes for recognition elements could be carried by an individual. Either all of the genes are carried in the germ-line and expressed randomly in somatic cells (germ-line models), or few genes are carried in the germ-line and varied in somatic cells where the derivative genes are expressed (somatic models). The sequence of events which follows interaction with a stimulus is the same under either hypothesis.

Many indirect arguments have been advanced as to why germ-line models are unlikely (Lennox and Cohn, 1967; Cohn, 1968a). I will not review these arguments simply because not one of them is decisive.

I have interpreted the sequence data as meaning that *spontaneous* mutation is the generator of diversity (Dreyer and Gray, 1968; Cohn, 1968a). The major question arises in determining experimentally whether this occurred during the period of germ-line evolution (Dreyer and Gray, 1968) or somatic evolution (Cohn, 1968a). Since the spontaneous mutational frequency leading to an amino acid replacement has been shown to be very high (Kimura, 1968; Ohno et al., 1969; Ohno, 1969), in fact of the order of 10^{-4}/cistron/division (B. Ames, personal communication), somatic mutational models need add no special mechanisms to account for the large numbers of variants which can be generated. When I say that the origin of the diversity is *spontaneous* somatic mutation, I do not wish to imply that subsequent events (which are known for germ-line evolution) such as intragenic recombination cannot add to and amplify the diversity. I will leave the question at that by assuming that diversity results from spontaneous mutation, because I wish to deal only with the extreme case of selection in the evolution of diversified cells.

I had previously pointed out that two models of selection seemed to be relevant, subunit selection and antigenic selection (Cohn, 1968a). I tried unsuccessfully to analyse each process independently of the other, avoiding in particular what I considered to be a "hedge", namely sequential selection, first by fitting of subunits then by antigen. The latter was obvious, the former seemed *ad hoc*. However, with the discovery of the specific L–L interaction (Stevenson and Straus, 1968), and the development of a precise model for regulation by antigen (Bretscher and Cohn, 1968), the postulate of sequential selection, first by subunits than by antigen, becomes attractive but by no means the only model.

In order to illustrate the problems involved in building a library of cells of diverse specificity in a clone (an individual), we start with a stem-cell population which generates variants by somatic mutation. These variants are the virgin antigen-sensitive cells. I would imagine that these cells will be acted upon by the following selection pressures.

(1) Selection for L–L interaction (before antigen).

The tolerogenic signal must be included in the inductive signal in order to avoid the generation of a cell which can be induced but not be paralysed (Bretscher and Cohn, 1968). I will assume a somatic model in which virgin antigen-sensitive cells expressing germ-line L and H genes are continuously generated with their receptors in the bound conformation. I have pictured this configuration to result from a weak L–L interaction (Fig. 3). This is equivalent to saying that the expression of the germ-line gene yields a light chain in which the L–L affinity is so low that the cell expressing it is killed by the same mechanism as leads to death in tolerance. Mutations in the light chain gene which increase the L–L interaction above a threshold value are therefore strongly selected for, as they survive. The first selection then is for a receptor in the unbound configuration. This explains the finding that identical light chains preferentially interact (Stevenson and Straus, 1968).

Under a germ-line model, L–L subunit selection would have gone on during germ-line evolution so that the antigen-sensitive cell would be born with its receptor in the unbound configuration. This means that the germ-line gene product would be a light chain in which the L–L affinity is high, and the cell expressing it would not be killed by the tolerance signal. In this situation only selection by antigen is relevant. If, under a somatic model, it is assumed that the germ-line genes code for a high L–L interaction (unbound configuration of the receptor), then the observed light chain variability would be the result of antigenic selection, as originally proposed (Cohn, 1968a). The difficulties of this theory that selection is uniquely by antigen are overcome by the assumption that the germ-line receptor is expressed in a tolerogenic configuration, simply because subunit selection gives a starting variety of specificities upon which antigen can subsequently select.

This subunit selection step assures that the tolerogenic signal is included in the inductive signal.

(2) Selection for L–H interaction (after antigen).

Under a somatic model, the cells which survive subunit selection express a receptor-antibody in which the light chain is varied but the heavy chain (except for rare mutants) is the unvaried expression of the germ-line gene. This virgin antigen-sensitive cell population, as well as all derivatives from it, is now selected upon by the tolerogenic signal from self-components. The survivors represent a limited range of specificities because only the light chain contributes to the diversity. Two factors are involved in the antigenic selection step: (1) the presence of a complementary combining site, and (2) the ability to undergo a configurational transition from the unbound to the bound or stretched form. Both of these factors are assumed to entail selection by immunogen for preferential L–H interactions, thus explaining this curious finding.

A minor population will express a larger spectrum of specificities because the heavy chains will have one *effective* mutation in them. Any virgin antigen-sensitive cell which undergoes an inducing interaction produces, on the pathway to dead-end plasma cells which secrete antibody, many more antigen-sensitive cells which are experienced memory cells.

In the case of the carrier-antibody class of antigen-sensitive cell in the thymus, the cells surviving L–L subunit selection are virgin antigen-sensitive cells which divide but do not leave the thymus unless an immunogenic stimulus is presented to them. This accounts for the high turnover of cells which never leave the thymus.

The experienced cell is postulated to leave the thymus but not divide in the peripheral tissues unless induced by an immunogenic stimulus. Thus cells of this class expressing one mutation in the heavy chain accumulate in the extra-thymic tissues and repopulate the thymus. In the thymus, they again divide and do not leave unless again stimulated by an immunogen. Since those expressing two effective mutations in the heavy chain have a higher probability of being induced because their range of specificities is greater, the process is repeated for this class and the cycle of repopulation of the thymus builds up the heavy-chain variability seen in the serum. Of course, light-chain mutations would also be selected for provided they did not decrease the L–L interaction to below the tolerogenic threshold. The selection is for greater and greater diversity as each sequential mutation in the variable region of each of the heavy-chain classes results not only in a wider range of combining specificities, but also an ever-increasing potentiality to generate increasing diversity with subsequent mutations. Thus a diverse large library of long-lived responsive cells of the carrier-antibody class is selected for sequentially.

Under a germ-line theory preferential L–H interactions can only derive from somatic selection by antigen.

Now I must turn to the question of how the selection by immunogens is primed. Since an immunogenic stimulus requires carrier-antibody, selection by immunogens cannot be initiated unless a selected population of antigen-sensitive cells of the carrier-antibody class pre-exists. This is a paradoxical vicious circle. The problem of priming the selection is independent both of the origin of the diversity, germ-line or somatic, and of the nature of the carrier-cell, multipotential or unipotential. For this reason maternal carrier-antibody was postulated to fill the role of primer (Bretscher and Cohn, 1968; Cohn, 1969). Clearly, this question has not been adequately investigated particularly in lower forms, fish and amphibians, where carrier-antibody might be detected in the egg. Since maternal carrier-antibody will not in general be directed against the self-components of the offspring the problem of auto-immunity is not serious. Nevertheless, polymorphisms in antigenic self-determinants could distinguish mother and offspring, leading to a potential difficulty. If the half-life of maternal carrier-antibody is short compared to the time it takes to generate an antigen-sensitive cell which is anti-self component, this problem is minimized. The maternal carrier-antibody permits selection to start. However, even with maternal primer, a special mechanism seems to be necessary in order to continue the selective pressure and it is for this reason that I would guess that antigen-sensitive cells of the carrier-antibody class are associated with the thymus, which exports experienced cells of that class only, by a mechanism I have already discussed.

Where does the diversity in the other immunoglobulin classes come from?

There are at least three ways to deal with this question.

(1) There are other organs in which antigen-sensitive cells of the other classes are generated randomly, by mechanisms similar to those described for the thymus. Carrier-cells, transporting carrier-antibody, migrate to these other organs to permit the second immunogen-dependent step in the selection. Cells of each class are then cycled through their respective enclaves because they home there and undergo successive selection steps by immunogenic stimuli. The thymus, being the enclave for the carrier-antibody class, is the only organ in which induction does not require cooperation from outside cells. Were it not for priming by maternal carrier-antibody the other classes would lag enormously far behind the thymic carrier-antibody class in the numbers of diversified cells.

(2) Large numbers of short-lived virgin antigen-sensitive cells of all classes are generated randomly in extra-thymic tissues. If they do not encounter an immunogenic stimulus, they die. If they are induced, they differentiate to long-lived memory cells which accumulate. These memory cells are selected upon by successive immunogenic stimuli which provoke division and appearance of mutants. This very direct and simplest view of the mechanism of selection has as its only drawback the problem of numbers of cells and the distribution of mutants.

(3) The carrier-antibody class after it leaves the thymus can, when it reaches other organs, be switched on to express another class, depending on the signal it gets. For example, the gut might switch on IgA production. It is reasonable to assume that this event affects only the heavy chain class because the variable regions of kappa and lambda light chain classes are not interchangeable (discussed in Cohn, 1968b) whereas the heavy chains of all classes share the same variable regions (reviewed in Lennox and Cohn, 1967; Cohn, 1967b; Cohn, 1968b). Therefore, selection, increasing the number of diversified cells in the carrier-antibody class, automatically affects all classes and the identical range of specificities would be found to arise simultaneously in all classes. In the successive selections for diversity by repopulation of the thymus either only the carrier-antibody class of cells homes there or any cell which arrives there is switched back to the carrier-antibody class.

In view of the fact that spontaneous somatic mutation affects all genes, selection by stimuli only acts if the gene is expressed in such a way that this is possible. Since somatic mutation would in general disrupt an organism, usually the differentiated cell expressing its unique product does not divide and selection is not therefore operative (Cohn, 1967c; Cohn, 1968c). When somatic evolution is operative the burden of providing a functional unit is largely a question of selection. What is striking about the structure of immunoglobulin subunits is that the variation is limited essentially but not uniquely (Hood and Ein, 1968; Milstein, Clegg and Jarvis, 1968; Postingl, Hess and Hilschmann, 1968) to the amino-terminal sequence of about 110 amino acids. The selection has to account for this since mutation must be occurring throughout the gene. Mutations affecting the function of the constant region either in its ability to form an intact immunoglobulin or to interact as a receptor with the other cell components eliminate that variant. However, this cannot be true for all amino acid replacements in the constant region. Why are such replacements rarely found?

We are considering here somatic selection only. Germ-line evolution ensures subunits that have constant region sequences which are functional to begin with. Somatic mutations affecting the ability of a molecule to carry out its function as an immunoglobulin, for example to fix complement, would not be selected upon any more than a somatic mutation inactivating a given reticulocyte's haemoglobin. Amino acid replacements in the constant region which affect L–L, L–H or H–H interactions either block induction and/or synthesis of that molecule or are neutral. It is because neutral mutations are not selected upon that they are seen only very rarely.

Since somatic selection acts only on the ability of the molecule to undergo a tolerogenic and inducing transition, it must be concluded that variation of the germ-line product outside of the first 100 or so amino-terminal amino acids rarely affects these properties in a positive way. This argument says that most amino acid replacements in the constant region of the light or heavy chain contribute neither to an increased L–L interaction, postulated to be essential to the survival of a virgin antigen-sensitive cell, nor to the L–H interaction necessary to form an antibody molecule which can be induced. With some exceptions, only replacements in the amino-terminal V region improve the L–L fit needed to survive subunit selection and permit an L–H interaction compatible with inductive selection. These considerations are a further justification for treating the V portion of the molecule as a relatively independent unit (Fig. 3). A simple minded guess would be that V_L and V_H taken from a given immunoglobulin will interact to give a functional combining site in the complete absence of C_L and C_H.

I have assumed that tolerance results in the death of the antigen-sensitive cell. I could imagine that tolerance simply silenced the cell. However, the danger of accumulating large numbers of antigen-sensitive cells which are anti-self is sufficient to warrant the assumption that tolerance kills the cell. The postulate that tolerance results in death leads to difficulties with the theory that variability is generated by *spontaneous* somatic mutation (Cohn, 1968*b*) because tolerance can be rapidly broken, in a matter of weeks (Weigle, 1969). This implies that *de novo* antigen-sensitive cells of the specificities to which the animal is tolerant are being generated at an unexpectedly high rate. One way out of this difficulty, suggested to me by Dr Peter Bretscher, would be to have *induced* cells undergo a step in differentiation in which they return to a stem-cell state which expresses no receptor. In a random way these stem cells are turned on to become antigen-sensitive cells expressing their original specificity. The range of antibody specificities represented in the pool of stem cells is determined by the past immunogenic history of the individual.

Since I was led to consider, as a possible hypothesis, one in which all diversity is generated in cells expressing the carrier-antibody class and switches to other immunoglobulin classes later, it might be postulated that receptor-antibody is of the carrier-antibody class in all cells no matter which class they will make on induction. The receptor-antibody and the antibody made on induction would have the identical V portion of the molecule but different C portions. Although such an assumption greatly simplifies the structural and conformational considerations for induction, it seems unlikely because it forces one to assume that allotypic or class suppression of immunoglobulin synthesis (Herzenberg, Minna and Herzenberg, 1967) by appropriate anti-heavy chain sera is due to the constant killing of only those cells which have been induced by an immunogen so that they start expressing the non-receptor antibody class they were programmed for.

The last point concerns the assumption that a germ-line model operates in generating diversity. The maximum range of specificities would be expressed explosively to produce a population of antigen-sensitive cells which are selected upon by tolerogenic and immunogenic stimuli. The events which one must postulate to be operating in the soma would be much simpler than those discussed above. However, this simplicity is illusory because the complexity returns when the events necessary to explain the germ-line evolution are considered. If a plausible scheme accounting for the known diversity is spelled out under a germ-line model, it involves as many assumptions as the somatic one discussed here (Cohn, 1968*a*).

III THE LEARNING SYSTEM

As in the case of the immune system, two steps are to be defined: the origin of the diversity and the mechanism of selection by learning stimuli of the corresponding elements which are stored in the library of memory and recall. The extent to which selection plays a role in building the library of responses depends upon the degree of programming of the circuitry before the selection. At one extreme, if the diversity is all in the germ-line the degree of programming is potentially maximal and selection bears less of the burden in determining the "how and what" of learning. At the other extreme, if there is random somatic variation, such as the mutation of a few germ-line genes, selection bears more of the brunt in establishing memory and recall because of the low degree of pre-programming. In both cases, however, selection by learning stimuli is the key to what the brain stores, because a great capacity to learn is only possible for structures in which learning occurs during construction (Offner, 1965). I shall illustrate with models of both extremes. Of course controlled (not random) somatic variation, for example recombination, can be envisaged, but these intermediate situations need not concern us now. Thus there are two problems: the logic of the circuitry and the selection of circuits.

In spite of the fact that the immune system is far better understood than learning, psychologists have known for some forty years what immunologists have just realized. Stated in the most general terms, in order for an unexpected signal to be sensed as inducing, it must be tightly associated with or coupled to another signal under conditions where both are recognized. In the induction of antibody synthesis one of these signals (called a determinant) is recognized by the antigen-sensitive cell which senses the signal as an inducing one because the associated signal is recognized by the carrier-antibody cell. In the induction of a learning pathway, a stimulus is presented which consists of two signals coupled in time, one of which is recognized by the learning-sensitive-circuit and the other by already learned or innate circuitry. What does this phenomenon look like?

An *idealized* Pavlov-dog might behave as follows (Pavlov, 1926). It would salivate when a biscuit is put into its mouth. This is recognition and response mediated by *innate* neural circuits. If a bell were rung just before a biscuit was given, the dog would eventually salivate at the sound of the bell alone. If subsequently a light were flashed every time the bell was sounded it would salivate after the light flash alone. If the bell were rung at random, for example after the biscuit was fed, it would become very difficult to couple subsequently the bell–biscuit stimulus in a way that would teach the dog. If the bell were rung and the dog painfully shocked, the salivating response of the taught dog would be extinguished and it would be more difficult to re-teach the bell–biscuit association. If the biscuit were given but no bell rung, or if the bell were rung and no biscuit given, the taught dog would gradually forget but at any time be much easier to re-teach. Similarly the dog may forget the "extinguishing" pain–bell stimulus and the original salivating response to the bell would return slowly and spontaneously. If the biscuit were made distasteful and fed to the taught dog until it no longer salivated, it might still salivate when the bell was rung.

I have synthesized this example of learning to illustrate the effects of the cooperative interaction due to the recognition of the two signals which comprise a learning stimulus.

A. What is the logic of the circuitry?

In order for the dog to distinguish the bell as part of a learning stimulus from the enormous number of stimuli constantly entering its senses, the bell must be associated with the biscuit which is recognized by an established pathway, whether it be previously learned or innate. Therefore the learning like the immunogenic stimulus must present at least two determinants which are recognized, in this case the bell and the biscuit. I will call the former the learning (bell) and the latter the carrier (biscuit) determinant.

The linked encounter between a learning-determinant with its corresponding virgin recognition element and the carrier-determinant with an established or experienced recognition element induces memory storage. For the initial learning events, only innate circuitry can act as the carrier-recognition element and there is a severe limit on the variety of signals which can be learned. As each new recognition element is wired into the circuitry by learning stimuli, the variety of carrier recognition elements greatly increases because one "learned circuit" can convert many "naïve circuits" into learned ones. The range of stimuli which can become learning signals increases autocatalytically. Thus learned circuitry is wired by associative pairing with previously learned circuitry. This relationship should hold from the smallest circuit that can learn to the maze of circuits added during the learning lifetime of man. *Associative pairing is the logic of the circuitry.*

B. Upon what structure in the brain does selection by learning stimuli act?

Of the many models, it seems most likely to me that new functional synapses must be formed in order for new learning circuits to be established (Eccles, 1966; Brindley, 1967). Since our ground rule is that the recognition element must precede the stimulus, "learning" synapses must be modifiable (discussed by Brindley, 1967) as a consequence of the associated input signals. In other words, synapses which are temporarily functional must be made permanently functional as a result of the selective action of learning stimuli.

I might illustrate the genesis of circuits by considering the two extreme situations.

(1) During ontogeny, a vast, highly programmed network of fixed but short-lived synaptic connexions is generated. These synapses are being turned on and off like the blinking neon lights of Broadway. These short-lived connexions (which I surmise are the basis for short-term memory) are made randomly functional until learning stimuli select those to stabilize as long-lived, in the orderly sequential process described above.

(2) During ontogeny, the innate circuits are generated as well as a large pool of long-lived non-dividing memory neurons capable of making short-term connexions with each other and with innate circuits. The

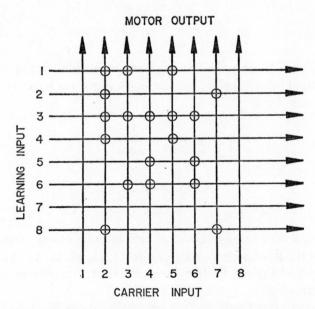

FIG. 5. The Willshaw, Buneman and Longuet-Higgins model for associative memory and recall in a *maximally* pre-programmed circuitry.

wandering memory neurons make short-lived connexions randomly with any complementary neuron. These short-term connexions are made long-lived by learning stimuli. The variant on this hypothesis is that there is a pool of neuroblast precursors that generate short-lived memory neurons which become long-lived as a result of learning stimuli. In other words, new memory neurons are added to the learning pathways.

Let us consider a very provocative example of a maximally pre-programmed network that has been analysed by Willshaw, Buneman and Longuet-Higgins (1969). This model (Fig. 5) arranges the learning and the carrier neurons in rectangular coordinates illustrated here by placing the learning neurons along the ordinate and the carrier neurons along the abscissa. The grid is only illustrative. The essential point is that a given memory neuron makes connexion with many carrier neurons and *vice versa*. The carrier neurons activate a motor output.

The associated firing of a learning with a carrier neuron activates the synapse pictured in Fig. 5 where they cross by a circle. Consider memory storage of five learning events due to the following associated firings:

MEMORY STORAGE

Learning input	Carrier input
13	23
28	27
36	34
14	25
35	46

If recall involves only the learning input we should recover the correct motor output which, before the learning event, was activated uniquely by the carrier input:

RECALL

Learning input	Motor output
13	23(5)
28	27
36	34(6)
14	25
35	46

The output is determined by the threshold firing of two synapses. The output for the five above learning events (13, 28, 36, 14, 35) is reasonably correct with two errors illustrated in parentheses. The system works well if not overloaded.

Now let us consider the second, minimally pre-programmed situation by returning to Pavlov's dog.

In Fig. 6A, two circuits, carrier and learning, are represented. The biscuit stimulates the carrier circuit, which is an established connexion with the salivary gland via an interneuron. At first the bell stimulates many unrelated pathways connected by memory neurons in short-term connexion, as well as the learning circuit we are interested in, that leading from auditory neurons to salivation. All of these pathways are fired by the bell. The repetition of the bell–biscuit association results in repeated firing of the memory neuron connecting the input from the bell to salivation only. The other pathways then drop out, since I assume that consolidation by repeated firing is necessary to convert the bell-memory neuron synapse into a long-lived one. Once fixed, the newly learned pathway can either be added to (Fig. 6B) by associating with the bell, for example, a light flash which now stimulates salivation, or extinguished (Fig. 6C) by associating pain. The new pathway, in this case via a light

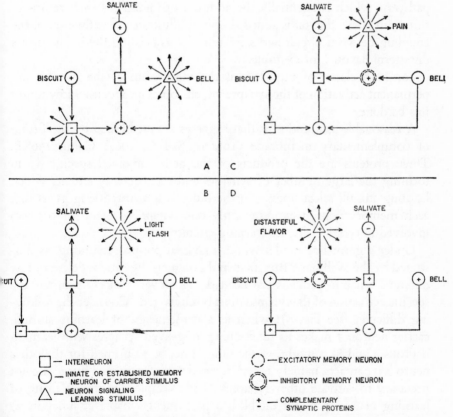

FIG. 6. Associative learning in *minimally* pre-programmed circuitry.

10*

flash, involves the addition of other memory neurons which can now use the already learned (bell) circuit as a carrier circuit. Extinction of the bell–salivation circuit involves the addition of an inhibitory memory neuron which is selected for by the associated bell–pain circuits.

Forgetting occurs when no learning stimulus is given over a long period of time. For example, the bell is never rung or the painful shock is no longer applied. In this latter case, the extinguishing pain–bell stimulus (Fig. 6c) spontaneously disappears and the original bell–salivation response reappears. Forgetting could be a measure of the half-life of a learned connexion and/or it could be an active process identical to extinction at the molecular level. If they are identical, then forgetting, like extinction, must result from an associated learning input. The difference would be that in the case of extinction we know what this input is, for example pain–bell (Fig. 6c), whereas in the case of forgetting, unknown associated signals must be adding connexions to the bell–salivation or bell–pain pathway, which is essentially the addition of "noise". I shall return to this point later. Fig. 6D is added simply to illustrate the extinction of the coupling between biscuit and salivation, by associating the biscuit and a distasteful flavour, for example.

For both classes of model, it is the mechanism of the specific and permanent activation of the synapse which comes into focus. How might this be done?

A reasonable assumption is that synapses are established by the fitting of complementary membrane proteins (Szilard, 1964; Cohn, 1968a). These proteins are the products of the genes involved specifically in forming the large number of synapses and consequently circuits which learning stimuli select upon. Furthermore it is a simplifying guess that each memory neuron expresses only one synaptic protein (or pattern) involved in creating diverse learning circuitry.

Under a germ-line model with a maximal pre-programming, as illustrated by the Willshaw, Buneman and Longuet-Higgins model, very few different synaptic proteins are needed. However a mechanism for the specific activation of the synapse must be envisaged. Consider the following difficulty (see Fig. 5): when as a consequence of learning stimuli, carrier neuron 7 makes its functioning long-lived synapses with learning neurons 2 and 8, it might, at the same time, be so primed for the other neurons it contacts, namely 1, 3, 4, 5, 6 and 7, that associated firings are not necessary to activate other synapses. For example, the simple firing of learning neuron 7 might establish a permanent synapse at junction 77 without carrier neuron 7 having been associated specifically with it. If it is assumed that complementary synaptic proteins are revealed upon associated firing and these make the contact function, the problem is to avoid expressing all of those made by one neuron when it is established by associated firing, so that the subsequent non-associated firing of any other neuron will establish a learned connexion. This difficulty can be avoided if one firing neuron making one unique synaptic protein expresses it only at the point where the associated firing neuron expresses its unique anti-synaptic protein. The membrane of the neuron is then discontinuous in structure so that the other contacts have neither of the complementary synaptic proteins expressed in a long-lived state. The picture is that the firing of a neuron causes all of its contacts to express the synaptic protein but only in those contacts in which the complementary protein is being simultaneously expressed is the synapse eventually locked into place. The half-life of an unfixed synaptic protein must be short compared to the time it takes stray or random inputs to fix it and long compared to the time it takes repeated associated learning inputs to fix it.

Now let us look at the problem of minimum pre-programming in which new memory neurons are added to the learning pathways.

In order to construct such a model, I shall assume that cells coming from a pool of memory neurons make functional synaptic connexions with already established, learned and innate pathways by means of connexions which are not modifiable. In addition they contact the *de novo* learning pathway by a modifiable synapse which is short-lived, but functional before the learning stimulus is delivered. In other words, the memory neuron makes at least three connexions (Fig. 6), two post-synaptic to the carrier (biscuit) and learning (bell) circuits and one pre-synaptic to the motor interneuron (salivate). The carrier and motor interneuron synapses could be large classes of invariant complementary connexions while the learning synapse might be due to two complementary unique protein patterns, one on the presynaptic endings of the learning pathways (represented by a diamond in Fig. 6) and the other on the memory neuron (represented by a dotted circle in Fig. 6). It is this latter synapse only which is modified by learning. This does not mean that a memory neuron can make only one specific learning synapse. It can introduce itself into many possible established circuits because the fit between the complementary synaptic proteins can be more or less good.

Although for illustrative reasons I shall discuss the *minimum* pre-programmed model much of the argument applies to the maximum pre-programmed model also. Two factors determine both how well a synaptic connexion functions and its life-span: first the affinity (association constant) between the two synaptic proteins which interact, and second the number of interacting molecules which are participating. The combination of affinity and number defines the avidity of the synaptic connexion. The avidity has a threshold value below which there is simply no functional synapse but above which some modulation of the efficiency of synaptic transmission is effected until the second threshold is reached. Above this the avidity between the membranes in synapse is so high that transmission is optimal—that is, limited by other factors. I shall assume as a first approximation that, if a short-term synapse is at all functional, transmission is all or none and that the increase in avidity only affects the life-span of the connexion.

Since a measurable learned response involves many circuits of the kind discussed above, the life-span of these connexions (memory of the response) is affected by both the avidity of synaptic proteins and the number of circuits involved in that particular memory. The actual measurement of the half-life of memory (forgetting) is probably dominated by the number of circuits and the degree of redundancy. For example, learning a nonsense syllable such as "rgm" would activate few synapses because there are not many associations you can make with it. However a war experience which involves innumerable associations with death, friend, fear, pain, hatred is recalled and thought about throughout the life of an individual. Given a constant probability of synaptic degeneration, the war experience would be remembered longer than the nonsense syllable.

The learning event consists of a consolidation period during which a memory neuron incorporated into a circuit becomes long-lived by further differentiation as a result of its repeated functioning consequent on the associated firing of two circuits, carrier and learning. The short-term association of memory neurons is the basis of short-term memory which is converted to a long-term memory circuit during the consolidation period. The continued presence of the associated learning stimulus is necessary for the continued differentiation to a long-lived memory neuron. This requirement for use during a critical period may be explained by assuming that the recognition elements binding the synapses specifically together are expressed in greater and greater numbers as both the memory and learning cells are fired more and more. The memory cell is fired by the carrier circuit and the learning cell by the associated sensory input. Before the memory–learning cell connexion is established only a few molecules of the synaptic protein are exposed so that a synapse, however short-lived, can be made. Each time the memory–learning cell pair is fired, the two synaptic membranes rearrange further to expose more and more molecules of recognition protein, something like knocking over sequentially, parallel lines of dominoes (Changeux and Podelski, 1968). Thus, the synapse is stabilized and when a large enough number of molecules (of high enough affinity) are interacting, the synapse is long-lived.

Forgetting can occur if each neuronal connexion has a certain probability of degenerating which is most likely a function of its avidity. This assumption by Brindley (1967) that forgetting involves the loss of the established memory neuron connexion by random decay with a given half-life can only be a partial solution, in the light of the psychological experiments of Ceraso (1967) that forgetting can be induced by associated interfering inputs. This experiment indicates that forgetting can also occur by erasing, which means that new inhibitory neurons have been introduced or a given learned pathway has been connected to another one which amplified the number of associations so as to introduce noise. This makes forgetting and extinction identical at the level of neuronal circuitry. If forgetting is active erasure by inhibition, it is a specific event; if forgetting is random decay, it is equivalent to ablation, which is a non-specific event.

A memory neuron might have on it a large number of synapses from different learning and carrier circuits; any two that are fired together might fix this neuron permanently. The consequence is that some circuits might be non-specifically incorporated in the learning event. Even if the type of solution which I discussed as part of the Willshaw, Buneman and Longuet-Higgins model were not operative, non-specificity of connexions is minimized if the circuits are packets separated anatomically to receive different inputs and activate different outputs. The synaptic proteins can be subdivided into classes to do this. The degree to which such pre-programming occurs affects the selection envisaged but in a rather minor way. Unrelated learning associations are also minimized by the fact that the number of learning and carrier circuits is great compared to the number which can be passively associated non-specifically. In most cases, the result will be innocuous or even rarely useful. If the non-specific association is incompatible with function, a feedback carrier circuit could select for an inhibitory memory neuron to erase the association. Incompatibility is recognized by the learning system because a new signal is generated which interferes with the functioning of an established circuit. For example, incompatibility is formally equivalent to a learning circuit in which pain or a distasteful flavour (Figs. 6c and 6d) extinguishes the bell-induced salivation response. If the bell-induction were incompatible with function such that pain or discomfort were induced, the bell-association would be extinguished. Originally, inhibitory neurons might have been invented by evolution to erase incompatible associations. In the simplest of nerve nets as seen in ctenophores, if the muscles involved in feeding are excited, those involved in swimming are inhibited, these two activities being incompatible (Horridge, 1968). Of course, in present-day complex species inhibitory neurons provide circuitry not possible with excitatory neurons only (von Euler, Skoglund and Soderberg, 1968). What is conspicuously lacking in any model is how excitatory or inhibitory memory neurons get selected for in a given case, although it is easy to imagine that certain learning stimuli of the painful or distasteful category would activate only inhibitory neurons by virtue of a limitation imposed by a classification of their synaptic proteins.

To recapitulate, whatever model of learning one considers, the functioning long-lived wiring of memory neuron circuits must be derived from somatic selection. It seems most reasonable that learning involves selection for long-lived active synapses by the associated interaction of a learning and carrier circuit. The virgin circuitry consists of memory cells which are connected via modifiable synapses to the *de novo* learning pathways and via fixed synapses to established cells from the innate or previously learned carrier pathways. Recording in the memory involves the specific long-lived activation of a short-lived synapse by the *associated* firing of the two kinds of cells. The accumulation of experience occurs as a consequence of the selective forces of the learning stimuli acting sequentially on memory cells in the soma.

C. *Comments on methodology*

What an individual is confronted with, in an actual learning experience, is a stimulus made up of many determinants which are looked at by many recognition elements each of which sees a different determinant. In other words, any learning stimulus, complex enough to evoke a learned behavioural response, activates many recognition elements. Since, with present methodology, the units comprising the stimulus are poorly defined and the assay of correctness of the recall of the response is crude, there appears to be an enormous redundancy in the learning mechanism. A similar situation exists for antibody formation because evolution had to solve a similar problem. This characteristic of anticipatory mechanisms tells us why experiments using ablation of, or injury to, large regions of the brain do not tell us about the logic of the circuitry. These techniques do tell us which regions of the brain are concerned with say visual or auditory recording because they look at the major and mostly innate circuits. The conclusion that memory patterns for a given multi-determinant stimulus are topologically spread over a large region of the brain is probably correct. However, ablation of any part of it must result in specific blurring of the details of a learned pattern. The fact that it is not seen is only a comment on the poor assays for responsiveness. This is why Lashley (1950) failed to find the engram. Since we do not know how to measure responsiveness to the determinants involved in the stimulus, it may take only 1 per cent of the recognition elements to react in order for a recall response to be scored as normal or to be of sufficient adaptive value to the individual. The heterogeneity of the response (or apparent redundancy) greatly reduces the malfunction and error in the response, of particular importance when it has survival value. Apparent redundancy is a consequence of poor experimental resolution of a given response into its units. If the fine structure of the response could be analysed many differences would be seen. At the molecular or cellular level there is probably no redundancy. The difficulty with ablation is not technical since even single neurons could be removed; it lies with the assay of the response. The same problem is raised when genetic blocks (mutants) are used as specific surgeon's knives. In principle, ablation or genetic surgery may result in *specific* losses of memory, for example the name of your dog, not the name of your cat, the number 10, not 20, the price of a haircut, not a shave; the chances of finding it are low. If all the information to make recognition elements is carried in germ-line genes on a one gene–one element basis, then specific genetic blocks in learning are predictable but unlikely to be found, whereas if derivative somatic genes code for the elements (or if germ-line gene products are used in permutations and combinations; Szilard, 1964), no specific losses could be found, only large-scale effects on apparently unrelated learned events, resembling those seen when massive ablation is carried out.

A learned response is usually acquired gradually, not all or none. Since the circuits described here are all or none, the assumption would be that we are examining the accelerating phase of acquisition of a learned response, the bringing into play of more circuits—the heterogeneity. The response appears gradually because different circuits recognize part of the stimulus and they add up as they are consolidated at different rates. So-called one-trial learning in lower animals (assuming they have no "consciousness") must take place by an entirely different mechanism. In man, one-trial learning exists because "consciousness" can be used as a reverberating circuit to present the stimulus repeatedly to the learning mechanism. "Consciousness" is a way of storing the stimulus.

This problem which I have posed in biological terms as the heterogeneity of the response is often described by analogy with the hologram (Longuet-Higgins, 1968; Gabor, 1968; Greguss, 1968; Pribram, 1969). There are two aspects to this analogy. The first is the recording of the photographed object (stimulus) such that partial destruction of the film still allows one to recover the total image, be it somewhat blurred. This is because every element in the original image is distributed over the entire photographic plate. The second is the *associative* recording of the incident and the scattered beam (from the object being photographed) on the film and the recovery of the image by illuminating the plate with the incident beam only. The first aspect has been discussed above as the heterogeneity of the response, and the hologram analogy is not illuminating. The second aspect, which is the theme of this analysis, is very important as it provides an associative model for analysing the logic of the circuitry. How far we shall be able to translate the physical analogues into actual circuits of neurons remains to be seen. I have already discussed one example, the Willshaw, Buneman and Longuet-Higgins model, which shows how powerful this analogy can be.

Now we come to a difference between the immune and learning mechanism due to the fact that circuitry is a property only of the latter. In the immune system, because one cell expresses one recognition element and because the *associative* inductive stimuli are *not* stored in a tightly coupled way, the secondary or recall response results from the same *associative* interactions as the primary response. The only difference is that more antigen-sensitive cells and a higher level of carrier-antibodies are involved. In the learning system the *associative* inductive stimulus is stored as a linkage between the carrier and learning circuits. Therefore recall of the associated stimulus can be evoked *non-associatively* by a signal coming via the learning circuit only. The associative aspect of the learning event is not required for the evocation of recall because the circuitry stores the association. In other words, the learning system records its associated input as a specific linkage whereas the immune system does not.

An example is "the tip of the tongue" phenomenon (Brown and McNeill, 1966) in which one tries to recall a familiar word by recalling words of similar form and meaning. It becomes clear that words are recorded in and recovered from the brain by an associative network. Clearly, we do not scan our total memory to recover a relationship. As a simple approximation, it is brought forth along the pathway along which it was recorded (Proust, 1913).

I would expect that the mechanism which pilots an innate neuron to its connexions during ontogeny or evolution will be the same as that which guides a memory neuron. The permanent fixation of a neuron into a circuit is most likely use-dependent in all cases. What we must know is whether all neurons, innate or memory, require associated inputs of the inducing and carrier type to fix them. This is an obvious problem of the regulation during ontogeny of germ-line-determined innate neuronal patterns. However, the problem of "ontogeny" of learning begins at this point. If randomly expressed memory neurons are consolidated in circuits by associated inputs, induced and carrier, then the questions we shall ask at the genetic and biochemical level are the same as those which have been answered in the immune system.

IV A VIEW FROM THE BRIDGE

The nervous system stores in a circuitry of cells the associated input of a *de novo* learning signal and an established carrier signal such that recall requires the learning signal only. The immune system stores in unconnected memory cells the haptene and carrier inputs such that the recall of the anti-haptene response requires the presence of experienced cells of anti-carrier specificity. The coupling during recall is not in the circuitry but in the stimulus. In principle, this difference allows the nervous system to store a good deal of the potentiality to diversify in pre-programmed circuitry. In a situation of *maximal* pre-programming of circuits (see Fig. 5), learning stimuli select on intracellular structures and activate one or another synapse for which a given cell is fixed. In the situation of *minimal* pre-programming (see Fig. 6) the learning and immune system approach each other, particularly if the memory neuron expresses only *one* synaptic protein pattern. Learning stimuli, like immunogenic stimuli, select on cells. The first task is to determine how much of learning (or immunity) is actually pre-programmed.

Why is the associated specific recognition of two signals necessary in order for a stimulus to be immunogenic or learning? Anticipatory mechanisms depend upon cells committing themselves to a vast number of unique interactions before these commitments can be tested for their compatibility with the life of the individual. In the immune system, this incompatibility involves the production of cells producing antibody to self, or auto-immunity. In the learning system, this incompatibility would arise by recording as a learning stimulus all of the noise which enters the senses. The reason why two such biologically different systems as immunity and learning require their inductive stimuli to be *associative* is to avoid these incompatibilities.

Since I assume that new functional synapses are made during learning, it is important to ask whether adding synapses implies adding cells which were generated explosively during one moment of ontogeny or are produced constantly from a pool of precursors during the life of the individual. For immunity the latter is clear and it is probable that spontaneous somatic mutation generates diversity which is selected upon by the stimulus. For learning, the analogous idea that new cells are added to the circuits is heretical. As has been done for the immune system, an insight might be gained by X-raying animals at doses which preferentially kill dividing cells. If the neuroblast stem cells giving rise to memory neurons were essential to *de novo* learning, a properly X-rayed animal would be able to recall but not learn (Strobel, Clark and MacDonald, 1968; discussed in Cohn, 1968a).

Today, even the assumption that new functional synapses are made during learning is strongly doubted and the literature contains many models for memory which leave synaptic connexions unchanged, such as varying firing frequency of cells, modulation of amplitude of spontaneous firing frequency, and induced synthesis of a vast array of transmitters (one cell–one transmitter). The reason I have focused on a synapse which is pictured to be formed by a complementary fit between two proteins, one on each neuron, is that the idea is simple and does no violence to the accumulated knowledge of molecular biology. As a result of my focus, molecular level models involving known aspects of gene behaviour and regulation by cell-to-cell communication are possible. This may predict the *new look*, but today the emperor has no clothes.

SUMMARY

Animals have developed systems for dealing with a wide variety of unexpected stimuli, for example, antibody to D-tyrosine or learning to whistle Yankee Doodle. These systems follow the ground rules:

(a) the functionally poised specific recognition element *precedes* the stimulus, and

(b) the recognition element upon successful interaction with a given stimulus becomes the product.

The stimulus *does not induce* the formation of the recognition element; *it selects for it*. Consequently, the anticipatory mechanism is an evolutionary one involving both variation and selection within each individual. The logic of the selection is based on the associative recognition of two or more distinct determinants or input signals which comprise an unexpected immune or learning stimulus. Using this idea, the origin of the diversity (variation), as well as the selective pressures operating in the immune and learning systems, have been analysed.

Antibodies and learning circuitry are theories made by the individual as to which stimuli exist in his environment and like all theories when subjected to test are either eliminated or amplified.

Acknowledgements

This work was supported by the National Institutes of Health, grant no. A105875.
The real fun in trying to understand this difficult subject has been the discussions with Dr Peter Bretscher and Dr Barry Komisaruk who contributed freely their ideas and enthusiasm. I have incorporated their insights both advertently and inadvertently. The flights of fancy which irritate most scientists but few poets are apologetically mine.

REFERENCES

BENACERRAF, B., PAUL, W. E., and GREEN, I. (1969). *Ann. N. Y. Acad. Sci.*, in press.
BRAUN, W. (1969). In *Immunological Tolerance*, pp. 187–191, ed. Landy, M., and Braun, W. New York: Academic Press.
BRETSCHER, P. A., and COHN, M. (1968). *Nature, Lond.*, **220**, 444–448.
BRINDLEY, G. S. (1967). *Proc. R. Soc. B*, **168**, 361–376.
BRODY, N. I., WALKER, J. G., and SISKIND, G. W. (1967). *J. exp. Med.*, **126**, 81–92.
BROWN, R., and MCNEILL, D. (1966). *J. Verbal Learng Verbal Behav.*, **5**, 325–337.
CERASO, J. (1967). *Scient. Am.*, **217**, 117–122.
CHANGEUX, J. P., and PODELSKI, T. R. (1968). *Proc. natn. Acad. Sci. U.S.A.*, **59**, 944–950.
COHN, M. (1967*a*). *Bull. All-India Inst. Med. Sci.*, **1**, 8–16.
COHN, M. (1967*b*). In *Gamma Globulins: Structure and Control of Biosynthesis*, Nobel Symposium 3, pp. 615–640, ed. Killander, J. New York: Wiley-Interscience.
COHN, M. (1967*c*). *Cold Spring Harb. Symp. quant. Biol.*, **32**, 211–221.
COHN, M. (1968*a*). In *Nucleic Acids in Immunology*, pp. 671–715, ed. Plescia, O. J., and Braun, W. New York: Springer-Verlag.
COHN, M. (1968*b*). In *Nucleic Acids in Immunology*, pp. 537–539, ed. Plescia, O. J., and Braun, W. New York: Springer-Verlag.
COHN, M. (1968*c*). In *Differentiation and Immunology*, pp. 1–28, ed. Warren, K. B. New York: Academic Press.
COHN, M. (1969). In *Immunological Tolerance*, pp. 283–335, ed. Landy, M., and Braun, W. New York: Academic Press.
COHN, M., NOTANI, G., and RICE, S. A. (1969). *Immunochemistry*, **6**, 111–123.
DAVID, J. R., and SCHLOSSMAN, S. F. (1968). *J. exp. Med.*, **128**, 1451–1459.
DOOLITTLE, R. F., SINGER, S. J., and METZGER, H. (1966). *Science*, **154**, 1561–1562.
DRESSER, D. W., and MITCHISON, N. A. (1968). *Adv. Immun.*, **8**, 129–174.
DREYER, W. J., and GRAY, W. R. (1968). In *Nucleic Acids in Immunology*, pp. 614–643, ed. Plescia, O. J., and Braun, W. New York: Springer-Verlag.
DUBERT, J. M. (1959). *Annls Inst. Pasteur, Paris*, **97**, 679–683.
ECCLES, J. C. (1966). In *Brain and Conscious Experience*, pp. 314–344, ed. Eccles, J. C. New York: Springer-Verlag.

296   MELVIN COHN

EDELMAN, G. M. (1967). In *Gamma Globulins: Structure and Control of Biosynthesis*, Nobel Symposium 3, pp. 89-108, ed. Killander, J. New York: Wiley-Interscience.

EDELMAN, G. M. (1970). This volume, pp. 304-317.

EDELMAN, G. M., and GALLY, J. A. (1967). *Proc. natn. Acad. Sci. U.S.A.*, **57**, 353-358.

EDELMAN, G. M., OLINS, J. A., GALLY, J. A., and ZINDER, N. D. (1963). *Proc. natn. Acad. Sci. U.S.A.*, **50**, 753-761.

EULER, C. VON, SKOGLUND, S., and SODERBERG, U. (ed.) (1968). *Structure and Function of Inhibitory Neuronal Mechanisms.* New York: Pergamon Press.

FAZEKAS DE ST. GROTH, S. (1967). *Cold Spring Harb. Symp. quant. Biol.*, **32**, 525-536.

FEINSTEIN, A., and ROWE, A. J. (1965). *Nature, Lond.*, **205**, 147-149.

GABOR, D. (1968). *Nature, Lond.*, **217**, 1288-1289.

GELL, P. G. H., and KELUS, A. S. (1967). *Adv. Immun.*, **6**, 461-477.

GOOD, R. (1969). In *Immunological Tolerance*, pp. 332-333, ed. Landy, M., and Braun, W. New York: Academic Press.

GOTTLIEB, A. A. (1969). In *Immunological Tolerance*, pp. 182-187, ed. Landy, M., and Braun, W. New York: Academic Press.

GREGUSS, P. (1968). *Nature, Lond.*, **219**, 482.

GREY, H. M., and MANNIK, M. (1965). *J. exp. Med.*, **122**, 619-632.

GROSSBERG, A. L., MARKUS, G., and PRESSMAN, D. (1965). *Proc. natn. Acad. Sci. U.S.A.*, **54**, 942-945.

HENNEY, C. S., and STANWORTH, D. R. (1966). *Nature, Lond.*, **210**, 1071-1072.

HENNEY, C. S., STANWORTH, D. R., and GELL, P. G. H. (1965). *Nature, Lond.*, **205**, 1079-1081.

HENRY, C., and JERNE, N. K. (1968). *J. exp. Med.*, **128**, 133-152.

HERZENBERG, L. A., McDEVITT, H. O., and HERZENBERG, L. A. (1968). *A. Rev. Genet.*, **2**, 209-244.

HERZENBERG, L. A., MINNA, J. D., and HERZENBERG, L. A. (1967). *Cold Spring Harb. Symp. quant. Biol.*, **32**, 181-186.

HILL, R. L., LEBOVITZ, H. E., FELLOWS, R. E., and DELANEY, R. (1967). In *Gamma Globulins: Structure and Control of Biosynthesis*, Nobel Symposium 3, pp. 109-126, ed. Killander, J. New York: Wiley-Interscience.

HOOD, L., and EIN, D. (1968). *Nature, Lond.*, **220**, 764-767.

HORRIDGE, G. A. (1968). *Interneurons*, pp. 436. San Francisco: Freeman.

HUMPHREY, J. (1969). In *Immunological Tolerance*, pp. 173-181, ed. Landy, M., and Braun, W. New York: Academic Press.

ISHIZAKA, K., and CAMPBELL, D. H. (1959). *J. Immun.*, **83**, 318-326.

JERNE, N. K. (1955). *Proc. natn. Acad. Sci. U.S.A.*, **41**, 849-857.

KELUS, A. S., and GELL, P. G. H. (1968). *J. exp. Med.*, **127**, 215-234.

KIMURA, M. (1968). *Nature, Lond.*, **217**, 624-626.

KÖLSCH, E. (1968). *Experientia*, **24**, 951-953.

LANDY, M., and BRAUN, W. (ed.) (1969). *Immunological Tolerance*. New York: Academic Press.

LASHLEY, K. S. (1950). *Symp. Soc. exp. Biol.*, **4**, 452-482.

LAWRENCE, H. S., and LANDY, M. (ed.) (1969). *Cellular Immunity*. New York: Academic Press (in press).

LENNOX, E., and COHN, M. (1967). *A. Rev. Biochem.*, **36**, 365-406.

LEVINE, B. B., and REDMOND, A. P. (1968). *J. clin. Invest.*, **47**, 556-567.

LIEBERMAN, R., and POTTER, M. (1966). *J. molec. Biol.*, **18**, 516-528.

LONGUET-HIGGINS, H. C. (1968). *Nature, Lond.*, **217**, 104-105.

MANNIK, M. (1967). *Biochemistry, N.Y.*, **6**, 134-142.

METCALF, D. (1967). *Cold Spring Harb. Symp. quant. Biol.*, **32**, 583-590.

MILLER, J. F. A. P. (1969). In *Immunological Tolerance*, pp. 125–140, ed. Landy, M., and Braun, W. New York: Academic Press.
MILLER, J. F. A. P., and MITCHELL, G. F. (1970). This volume, pp. 238–250.
MILSTEIN, C., CLEGG, J. B., and JARVIS, J. M. (1968). *Biochem. J.*, **110**, 631–654.
MITCHISON, N. A. (1967). *Cold Spring Harb. Symp. quant. Biol.*, **32**, 431–439.
MITCHISON, N. A. (1968a). In *Differentiation and Immunology*, pp. 29–42, ed. Warren, K. D. New York: Academic Press.
MITCHISON, N. A. (1968b). In *Regulation of the Antibody Response*, pp. 54–67, ed. Cinader, B. Springfield, Ill.: Thomas.
MITCHISON, N. A. (1969a). In *Cellular Immunity*, ed. Lawrence, H. S., and Landy, M. New York: Academic Press (in press).
MITCHISON, N. A. (1969b). In *Immunological Tolerance*, pp. 115–125, ed. Landy, M., and Braun, W. New York: Academic Press.
MÖLLER, G. (1969). In *Immunological Tolerance*, pp. 217–236, ed. Landy, M., and Braun, W. New York: Academic Press.
NAJJAR, V. A. (1963). *Physiol. Rev.*, **43**, 243–262.
NAJJAR, V. A., and FISHER, J. (1955). *Science*, **122**, 1272–1273.
NOSSAL, G. J. V. (1969). In *Immunological Tolerance*, pp. 55–80, ed. Landy, M., and Braun, W. New York: Academic Press.
OFFNER, F. (1965). *Biophys. J.*, **5**, 195–200.
OHNO, S. (1969). *Canad. J. Genet. Cytol.*, **11**, 457–468.
OHNO, S., STENIUS, C., CHRISTIAN, L., and SHIPMANN, G. (1969). *Biochem. Genet.*, in press.
PAVLOV, I. P. (1926). *Conditioned Reflexes*, pp. 430 [English translation and editor, Anrep, G. W. (1960)]. New York: Dover Publications.
POSTINGL, V. H., HESS, M., and HILSCHMANN, N. (1968). *Hoppe-Seyler's Z. physiol. Chem.*, **349**, 867–871.
PRIBRAM, K. H. (1969). *Scient. Am.*, **220**, 73–86.
PROUST, M. (1913). *A la Recherche du Temps Perdu*. (Re-edited Clarac, P., and Ferré, A.). Paris: Bibliothèque NRF de la Pléiade, Gallimard. Vol. I–III.
RAJEWSKY, K. (1969). In *Immunological Tolerance*, p. 148, ed. Landy, M., and Braun, W. New York: Academic Press.
RAJEWSKY, K., and ROTTLANDER, E. (1967). *Cold Spring Harb. Symp. quant. Biol.*, **32**, 547–554.
RAJEWSKY, K., SHIRRMACHER, V., NASE, S., and JERNE, N. K. (1969). *J. exp. Med.*, **129**, 1131–1143.
ROHOLT, O., ONOUE, K., and PRESSMAN, D. (1964). *Proc. natn. Acad. Sci. U.S.A.*, **51**, 173–178.
SINGER, S. J., and DOOLITTLE, R. F. (1968). In *Structural Chemistry and Molecular Biology*, pp. 172–183, ed. Rich, A., and Davidson, N. San Francisco: Freeman.
SISKIND, G. W., and BENACERRAF, B. (1969). *Adv. Immun.*, in press.
STEVENSON, G. T., and STRAUS, D. (1968). *Biochem. J.*, **108**, 373–382.
STROBEL, M. G., CLARK, G. M., and MACDONALD, G. E. (1968). *J. comp. physiol. Psychol.*, **65**, 314–319.
STULBARG, M., and SCHLOSSMAN, S. F. (1968). *J. Immun.*, **101**, 764–769.
SZILARD, L. (1964). *Proc. natn. Acad. Sci. U.S.A.*, **51**, 1092–1099.
VALENTINE, R. C. (1967). In *Gamma Globulins: Structure and Control of Biosynthesis*, Nobel Symposium 3, pp. 251–258, ed. Killander, J. New York: Wiley-Interscience.
WARNER, S., SCHUMAKER, V., and KARUSH, F. (1969). Personal communication.
WEIGLE, W. O. (1969). Personal communication.
WHITE, A., and GOLDSTEIN, A. L. (1970). This volume, pp. 210–237.
WILLSHAW, D. J., BUNEMAN, O. P., and LONGUET-HIGGINS, H. C. (1969). *Nature, Lond.*, **222**, 960–962.

DISCUSSION

Miller: The idea that a conformational change signals the difference between the pathways to tolerance and immunity is an interesting one. What I would question is the validity of your generalization that, in the absence of carrier-antibody, you cannot make antibodies of any sort to anything. How did the first carrier-antibody get there?

Cohn: Many differentiated systems require primers. For example, synthesis of liver glycogen requires glycogen (discussed in Cohn, 1968, 1969; Bretscher and Cohn, 1968). I would guess that the origin of the first carrier-antibody is maternal.

Miller: The idea of maternal transfer is not very appealing, because it has been shown (Silverstein, Prendergast and Kraner, 1963) that in the foetal lamb, which has sub-minimal amounts of IgM and none of the other immunoglobulins, you can infuse large quantities of rabbit anti-sheep immunoglobulins and yet the foetus responds perfectly well to all the antigens present in the foreign serum and to any other antigen you care to add, and it can also reject foreign skin grafts. This seems to argue against the idea that a carrier-antibody is present in the tissues or serum of the foetus.

Cohn: This result taken at face value argues against any model, since the anti-sheep immunoglobulin should block also immunoglobulin-mediated cell–cell interactions. I can only argue that there *is* a special carrier-antibody class.

Miller: Nevertheless the rabbit anti-sheep immunoglobulin ought to neutralize it. Secondly, passive antibody ought to mimic the booster effects of Rajewsky and Rottlander (1967) and of Mitchison (1967) as well as the tolerance effect of Benacerraf, Green and Paul (1967). On the contrary, passive antibody injected together with the specific antigen, according to the regime of Stuart, Saitoch and Fitch (1968), instead of creating an inducing complex does just the opposite: it switches off the immune response.

Cohn: Antiserum to any given antigen would be expected to have high levels of inhibiting antibody in all immunoglobulin classes and a trace amount of enhancing carrier-antibody in the special class. In order to reveal the latter in the face of the former a wide variety of concentrations and time schedules must be tried. This has not been done. Since serum antibody can at best have traces of carrier-antibody compared with that in the other classes, inhibition by passively given serum antibody is expected. Serum antibody in the non-carrier classes coats the antigen, removing it from any possibility of interaction with the antigen-sensitive cell.

Miller: Why do you have to make it a *general* law that there must be carrier-antibody in order to induce? Why can't certain polymeric structures or aggregated material substitute for carrier-antibody in some instances? It has recently been shown (W. Weigle, personal communication) that deaggregated normal rabbit gamma globulin is an extremely good tolerogen; if, however, the gamma globulin is prepared from a rabbit antilymphocytic serum and then deaggregated, it is no longer a tolerogen but an excellent immunogen. The antilymphocytic gamma globulin, by virtue of its capacity of sticking on to lymphocytes, might be presented to the immunologically competent cell in such a way that it will induce rather than paralyse.

Cohn: It was in an attempt to account for the immunogenic properties of aggregates, as contrasted to the tolerogenic tendencies of disaggregated substances, that the stretched form of the receptor was invoked (Bretscher and Cohn, 1968). Aggregates stretch antibodies. If you argue that aggregates are immunogenic and disaggregated substances tolerogenic *per se* without any requirement for carrier-antibody, then you are throwing the question of how to distinguish tolerance from induction upon another mechanism. For example, why don't you make antibody to your own red blood cells, which are aggregates? Furthermore, you ignore the fact that the carrier effect is specific.

Miller: I am not discussing whether the carrier effect is specific or not; we know that it is specific. I am just saying that there may be other ways of creating an inducing complex besides putting specific carrier-antibody on to the cell. The reason why you need a specific carrier in some instances may be that there is a weak determinant which, by itself, will not induce a cell into antibody formation; this determinant has to be concentrated, as it were, on to the antigen-sensitive cell by means of some mechanism which involves recognition of other determinants on the carrier molecule. So it is just one way of making an inducing complex. There may be other ways which can be used to induce cells, such as the use of aggregated material and non-specific inducers like phytohaemagglutinin.

Cohn: You are a leader in the school of thought that the role of carrier-antibody is solely that of an antigen-trapping or concentrating mechanism. If all that a carrier-antibody does is to trap or concentrate antigen, if you increase the concentration of antigen sufficiently you should be able to induce in the absence of carrier-antibody. In other words there would be another mechanism of induction. I am saying that you can increase the antigen concentration as much as you want but you will still require the carrier for induction.

Miller: But Mitchison (1967) showed that you can induce without a specific carrier effect. He primed mice with NIP-ovalbumin and transferred their spleen cells to irradiated recipients challenged with either NIP-ovalbumin or NIP-bovine serum albumin (NIP-BSA). In order to get a good response to the heterologous NIP-conjugate, he simply had to increase the antigen concentration 1000 times. Hence, just increasing the concentration of the haptene conjugated to a heterologous non-specific carrier allows an antibody response.

Cohn: You cannot decide whether carrier-antibody is uniquely an antigen-concentrating mechanism or a specific requirement of the signal by this experiment. In the idealized situation the antigen-concentrating model predicts that NIP-ovalbumin would at high enough concentrations induce anti-NIP in a NIP-BSA-primed cell population, whereas if carrier-antibody is required for the immunogenic signal, NIP-ovalbumin should not induce because there is no carrier-antibody directed against ovalbumin. However, this is not an idealized situation. Immunization with NIP-BSA induces increased numbers of antigen-sensitive cells specific for NIP in all classes including the carrier-antibody class. The secondary challenge with NIP-ovalbumin, as compared to NIP-BSA, reduces the number of determinants which can act as carriers for NIP. In fact, it may reduce the number virtually to NIP acting as carrier determinant for itself. At low immunogenic concentrations the NIP-BSA-primed cells give a marginal primary response to NIP because carrier-antibody is limiting. However as the antigen concentration is increased the probability of an encounter with anti-NIP carrier-antibody increases and NIP acts as a carrier determinant for itself, greatly accelerating the immunogenic response. Therefore, Mitchison's experiment does not distinguish the two models.

Monod: Dr Cohn, you want to account for the capacity of a given antigen-sensitive cell to be committed to either tolerance or induction. In order to do that you assume logically that there is some sort of co-operative interaction. When you have very low concentrations of antigen there is no cooperation and that induces tolerance, and this is experimental to the extent that low concentrations of antigen favour tolerance. When you have higher concentrations of antigen there is some cooperative effect which you wish to say accounts in itself for induction because it is co-operative. To put it more simply, the monomolecular interaction between the antigen-sensitive cell and an antigen leads to tolerance, and a multimolecular interaction leads to induction. If that is the core of your idea, I don't see that you need to introduce the stretching, because the cooperation could be between different receptor antibodies in the bound conformation on the surface of the interacting cells. I don't see that you are gaining anything in your theory by putting that precision into it, and furthermore I do not understand why you require a specific interaction with the carrier *cell*.

Cohn: You are perfectly correct; the minimal model suffices at present as far as the recognition of antigen is concerned. The carrier-cell interaction is introduced as one possible GO-signal in addition to the conformational change in the receptor in order to understand why you do not respond to your own cellular antigens. In other words the cell-to-cell interaction is one way of making carrier-antibody an obligatory component of the immunogenic signal.

Monod: You have another argument for the specific interaction between the two types of cells which is independent of anything else, namely the experiments that indicate that there is carrier specificity. I agree that if these experiments do say that there is carrier specificity it is very hard to avoid postulating an interaction between two cells.

Miller: Perhaps the antigen-sensitive cell can sense a difference between a micromolecular structure and a macromolecular structure on its surface —the former making it tolerant and the latter inducing it.

Cohn: But why don't you make antibody to your own red blood cells?

Edelman: I would like to mention a fundamental theoretical difficulty that precedes any of these considerations. In a selective immune system it suffices to hit just a few of the relevant cells with the antigen in order to induce antibody production. You don't have to hit all the cells that could interact with that antigen, if some of the cells mature and divide sufficiently rapidly. Whereas the fundamental difficulty in tolerance— and it arises because we lack a quantitative theory—is that to make an animal tolerant the antigen must encounter some large proportion of all the cells previously committed to that particular antigenic determinant. Paradoxically, you do it at a lower dose of antigen. This is an unanswerable difficulty at present but it is fundamental, because if it is not answered selective theories of antibody formation are in jeopardy. Either the selective theory is wrong or there must be some special mechanism whereby amplification also occurs in tolerance. I mention this because your model does not simplify the difficulty of selectivity; it complicates it by the fact that you require independent interaction with two different kinds of specificities in a selective system. You have to have a prior ensemble of cells for your carrier and a prior ensemble for your antigen. The problem is the probability of independent encounter: you have to knock out both the c and the a sites. Do you think you can do that without amplification?

Cohn: Yes!

Bhargava: Can you calculate the minimum number of cells that must be hit to get tolerance, assuming that one molecule per cell is sufficient?

Cohn: No.

Harris: Two points. For your model you require a carrier cell, but J. L. Gowans takes the view that sometimes you don't (Ellis, Gowans and Howard, 1969). Two-cell interactions may well commonly occur, but, in Gowans' view, they may not be a mandatory requirement for antibody formation.

My second point. Is there any experimental evidence—I wear logical necessity very lightly—that tolerance leads to cell death? Do we know that tolerant cells actually die?

Cohn: To your first point, Gowans' lymphocyte population must contain carrier-cells. To your second point: no, we don't.

Miller: Lymphocytes from *non-primed* animals could not be primed *in vitro* in the absence of macrophages (Ford, Gowans and McCullagh, 1966). Professor Gowans can do it with primed cells but *not* with lymphocytes from non-primed animals.

Anand: There is evidence that if you experimentally influence certain central nervous regions like the hypothalamus and thus change the body's homeostatic mechanisms, the response of the body in antibody formation is affected. We found that by stimulation or destruction of certain hypothalamic regions the activity of the reticuloendothelial system was markedly changed, as determined by the carbon clearance test (Thakur, 1968). Some Russian workers have also observed (personal communication) that after localized destruction in hypothalamic areas, antibody formation is altered. In other words if you change the homeostatic conditions you are introducing a third variable in the antigen-sensitive cells and you may not have exactly the same response of tolerance or antibody formation.

Cohn: Yes; clearly to get antigen-sensitive cells by differentiation from precursors, a whole series of hormone interactions is required.

REFERENCES

BENACERRAF, B., GREEN, I., and PAUL, W. E. (1967). *Cold Spring Harb. Symp. quant. Biol.*, **32**, 569.

BRETSCHER, P. A., and COHN, M. (1968). Nature, Lond., **220**, 444–448.
COHN, M. (1968). In *Differentiation and Immunology*, pp. 1–28, ed. Warren, K. B. New York: Academic Press.
COHN, M. (1969). In *Immunological Tolerance*, pp. 283–335, ed. Landy, M., and Braun, W. New York: Academic Press.
ELLIS, S. T., GOWANS, J. L., and HOWARD, J. C. (1969). *Antibiotica Chemother.*, **15**, 40–55.
FORD, W. L., GOWANS, J. L., and McCULLAGH, P. J. (1966). *Ciba Fdn Symp. The Thymus. Experimental and Clinical Studies*, pp. 58–79. London: Churchill.
MITCHISON, N. A. (1967). *Cold Spring Harb. Symp. quant. Biol.*, **32**, 431.
RAJEWSKY, K., and ROTTLANDER, E. (1967). *Cold Spring Harb. Symp. quant. Biol.*, **32**, 547.
SILVERSTEIN, A. M., PRENDERGAST, R. A., and KRANER, K. L. (1963). *Science*, **142**, 1172.
STUART, F. P., SAITOCH, T., and FITCH, F. W. (1968). *Science*, **160**, 1463.
THAKUR, P. K. (1968). M.D. Thesis, All-India Institute of Medical Sciences, New Delhi.

ANTIBODY STRUCTURE: A MOLECULAR BASIS FOR SPECIFICITY AND CONTROL IN THE IMMUNE RESPONSE

GERALD M. EDELMAN

The Rockefeller University, New York

DURING the last decade there has been a quiet revolution in theories concerned with the origin of specificity and control in the immune response. Classical instructive theories of antibody production stated that information on the three-dimensional structure of the antigen is necessary at the time of synthesis of the antibody molecule in order to assure that the combining sites of antibodies are complementary to the antigenic determinant. As a result of work on the structure of antibodies as well as on the cellular dynamics of the immune response (see Killander, 1967; Cairns, 1967; Edelman and Gall, 1969), instructive theories have been supplanted by selective theories (Jerne, 1955, 1966; Burnet, 1959). Selective theories suppose that the information for making antibodies with specificities for a large range of different antigenic structures *already* resides in the organism. The antigen serves only to stimulate (and thereby select) those antibody-forming cells which synthesize antibodies having appropriately complementary binding sites.

This idea, which is now rapidly acquiring the status of a dogma, has two profound corollaries. First, the number of different complementary antigen-binding sites must greatly exceed the number of different antigens, because these sites must pre-exist. Second, there must be an exceedingly efficient system for amplifying the production of specific complementary antibodies following selection or stimulation by a given antigenic determinant.

A diagram summarizing the minimum requirements for a selective system is shown in Fig. 1. There must be a means for diversifying the antigen-combining sites and this diversification must be expressed in cells each committed to a given specificity. Trapping of an antigen by interaction with antibodies on some of the cells must result in an enormous amplification of the synthesis of their antibody molecules. We now know that the diversity occurs in the amino acid sequence of the different antibody molecules (Edelman *et al.*, 1961; Koshland and Englberger, 1963; Haber, 1964). We also have evidence that the system is clonal; that is, each cell synthesizes a single variety of antibody and after encounter with the appropriate complementary antigen is stimulated to replicate

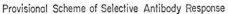

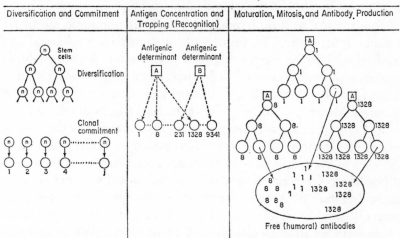

FIG. 1. Chain of events in a selective antibody response. The lower case "n" inside stem cells indicates that they are uncommitted and pluripotential. For clarity, diversification and commitment are presented as separate events, although a detailed model of differentiation from stem cells is not intended. Arabic numbers outside each cell represent the unique immunoglobulin produced after clonal commitment. These "receptor" antibodies can interact with various antigenic determinants. The degeneracy of the immune response is indicated by recognition of a single antigenic determinant (e.g. A or B) by several different cells which produce different antibodies. Interaction of a committed cell with antigen stimulates maturation, mitosis, and humoral antibody production.

and form progeny cells which synthesize that particular antibody at increased rates (see Cairns, 1967, for various features of this system).

After release from the cells the antibodies can bind the antigen used to stimulate their production and then mediate a number of immune reactions. These include complement fixation, skin fixation, and opsonization, and may be termed effector functions (EF). It is clear that effector functions do not depend directly on the specificity of the antigen recognition function (ARF).

The purpose of my remarks in this paper is to consider these two groups of molecular functions and the mechanisms of the selective response in terms of the molecular structure of antibodies. Antibodies or immuno-globulins represent a special molecular recognition system which appeared in the vertebrates—that is, relatively late in evolution. It is perhaps not surprising that a very special protein structure has evolved to serve as the essential molecule of immunity. The details of this structure provide substantial support for selective theories and prompt several new hypo-theses on mechanisms of the selective immune response.

Before turning to these hypotheses I shall describe some salient facts on antibody structure. To begin with, it should be emphasized again that all of the information for the specificity of antibodies is contained in the amino acid sequences of their constituent polypeptide chains. It follows from the selection dogma that there must be very many different sequences. Moreover, the bulk of the evidence suggests that the informa-tion is in both the light and the heavy polypeptide chains of the molecule (Singer and Doolittle, 1966). These facts pose operational problems, for antibodies to even a single haptenic antigen are heterogeneous, and it is not possible to determine an unequivocal amino acid sequence on such a complex mixture of proteins. Fortunately for research purposes, certain plasma cell tumours (myelomas) of mouse and man produce large amounts of homogeneous immunoglobulins. Each myeloma protein appears to have a unique sequence although its general structure resembles that of heterogeneous immunoglobulin. There is, however, no general way of determining the activity or specificity of myeloma proteins for antigens. We are therefore faced with a peculiar dilemma in structure–activity determination: either we must study a myeloma globulin the activity of which is unknown or we must attempt to analyse the amino acid sequences of an indeterminately large mixture of specific antibodies. Some efforts are being made to circumvent this dilemma by searching for homogeneous antibodies (Haber *et al.*, 1967), and for myeloma proteins with the capacity to bind antigens (Eisen *et al.*, 1967).

Regardless of the structure–activity problem, there is ample evidence that pure myeloma globulins are typical immunoglobulins. In fact, they are the main source of information on the detailed nature of antibody structure and diversity. The gross structure of a human γG immuno-globulin (Eu) from a patient with multiple myeloma is presented in Fig. 2. The tentative amino acid sequence of the entire protein has recently been determined (Edelman *et al.*, 1969). Each molecule consists of two identical heavy or γ chains (molecular weight 55 000) and two identical light or ϰ chains (molecular weight 23 000). The NH_2-termini of the heavy chains are blocked; there is evidence that the terminal residue is pyrollidone-carboxylic acid. The heavy and light chains interact via weak forces and a single disulphide bond, and one half of this interchain bond is the half-cystine at the COOH-terminus of the light chain. Similarly, each half-molecule is linked to its counterpart by weak forces and by two closely neighbouring disulphide bonds connecting the heavy chains. Just next to this region, the molecule may be cleaved by proteolytic enzymes such as trypsin and papain to produce two Fab fragments consisting of a light

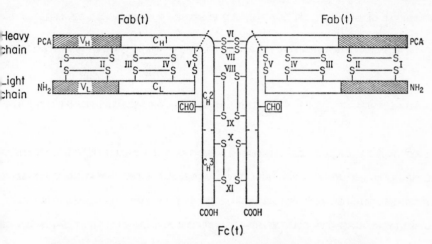

FIG. 2. Structure of a human γG1 immunoglobulin molecule from myeloma patient Eu. Site of cleavage by trypsin to form Fab(t) and Fc(t) fragments is indicated by dotted lines. Half-cystines contributing to disulphide bonds are designated by Roman numerals. V_H and V_L, variable regions of heavy and light chains. C_L, constant region of light chain. C_H1, C_H2, and C_H3, homology regions in constant region of heavy chain. PCA, pyrollidonecarboxylic acid. CHO, carbohydrate.

chain and the amino terminal part of the heavy chain, and one Fc fragment which is a dimer made up of the carboxyl terminal portions of each heavy chain. The Fc fragment contains carbohydrate which is linked by covalent bonds to each heavy chain. Studies on antibody molecules with known specificity indicate that the active site is present in the Fab fragment (Porter, 1959) and is shared by light and heavy chains (Singer and Doolittle, 1966).

The amino acid sequence of the light or $\varkappa$ chain of myeloma protein Eu is shown in Fig. 3. Free homogeneous light chains are excreted in multiple myeloma in the form of Bence-Jones proteins (Edelman and Gally, 1962), each with a unique primary structure. Because of their

KAPPA CHAINS SUBGROUP I

1   10   20
ASP–ILE–GLN–MET–THR–GLN–SER–PRO–SER–THR–LEU–SER–ALA–SER–VAL–GLY–ASP–ARG–VAL–TH
     VAL LEU LEU   THR SER   VAL   LEU ARG   ILE
                                             PHE

                                                30   40
ILE–THR–CYS–ARG–ALA–SER–GLN–SER–ILE–ASN–THR–TRP–LEU–ALA–TRP–TYR–GLN–LYS–PRO–
    ALA   GLN   ASP   SER SER TYR   ASN   GLY
                                       LYS ILE PHE
                                            LYS
                                            ASN

                                          50   6
GLY–LYS–ALA–PRO–LYS–LEU–LEU–MET–TYR–LYS–ALA–SER–SER–LEU–GLU–SER–GLY–VAL–PRO–SE
LYS   ILE   ILE   ASP   ASN   THR
                                                          LYS   ALA

                                            70   80
ARG–PHE–ILE–GLY–SER–GLY–SER–GLY–THR–GLU–PHE–THR–LEU–THR–ILE–SER–SER–LEU–GLN–PRO
                SER   THR   PHE   ASP   PHE   GLY

                                          90
ASP–ASP–PHE–ALA–THR–TYR–TYR–CYS–GLN–GLN–TYR–ASN–SER–ASP–SER–LYS–MET–PHE–GLY–G
GLU   ILE   PHE ASP THR LEU PRO ARG THR   G
                                         GLU ASN   LEU   P
                                         ASP   PRO
                                                           TYR

                                       110   1
GLY–THR–LYS–VAL–GLU–VAL–LYS–GLY–THR–VAL–ALA–ALA–PRO–SER–VAL–PHE–ILE–PHE–PRO–P
             LEU ASP ILE   ARG
                 LYS PHE
                     LEU

                                           130   1
SER–ASP–GLU–GLN–LEU–LYS–SER–GLY–THR–ALA–SER–VAL–VAL–CYS–LEU–LEU–ASN–ASN–PHE–T

                                         150
PRO–ARG–GLU–ALA–LYS–VAL–GLN–TRP–LYS–VAL–ASP–ASN–ALA–LEU–GLN–SER–GLY–ASN–SER–G

                                         170   18
GLU–SER–VAL–THR–GLU–GLN–ASP–SER–LYS–ASP–SER–THR–TYR–SER–LEU–SER–SER–THR–LEU–TH

                                         190   2
LEU–SER–LYS–ALA–ASP–TYR–GLU–LYS–HIS–LYS–VAL–TYR–ALA–CYS–GLU–VAL–THR–HIS–GLN–GL
                                            LEU

                                       210
LEU–SER–SER–PRO–VAL–THR–LYS–SER–PHE–ASN–ARG–GLY–GLU–CYS

FIG. 3. Amino acid sequence of the light chain of myeloma protein Eu. Amino acids which vary in κ chains of the same subgroup were determined by other investigators and are written under the variable positions. The available data on the amino acid sequences of κ chains and the appropriate references are summarized in Edelman and Gall (1969). The V_L region extends from residue 1 to residue 108; the C_L region comprises residues 109–214.

purity and relatively low molecular weight, the amino acid sequences of a number of these urinary proteins have been determined in a number of different laboratories (for a review, see Edelman and Gall, 1969). Comparison of the sequences shows that the light chain consists of a variable or V_L region and a constant or C_L region each ranging in length from 106 to 108 residues. The sites of amino acid variation and some examples of the types of substitution found are also shown in Fig. 3. In a sample of 15 proteins compared, there are 43 positions in the V_L region in which replacements have been observed. On the average, there are somewhat more than two replacements per variable position but certain positions (e.g. 96) have as many as five amino acid substitutions. Other positions in V_L show no variation, and so far no half-cystine residue has been found to be replaced by another amino acid. Most of the substitutions can be accounted for by single base changes in the codons corresponding to the amino acids replaced. Note that in the C_L region there is one site of variation at position 191. This is known to be due to genetic polymorphism at the so-called Inv locus (see Martensson, 1966; Kelus and Gell, 1967). The evidence suggests that this Inv marker is inherited in a simple Mendelian fashion and is specified by alleles of a single gene.

Milstein, Milstein and Feinstein (1969) have pointed out that the variations of human κ chains fall into three subgroups of sequences which cannot be specified by less than three separate genes; furthermore, these genes are non-allelic; that is, everybody has sequences corresponding to all three subgroups. As we shall see later, the presence of V region subgroups in chains which have the same C region sequences has important consequences for theories of diversity and control.

Only a few heavy chains have been studied in detail but the sequence of the Eu heavy chain (Gottlieb et al., 1968; Rutishauser et al., 1969; Edelman et al., 1969) has been compared with two other partial sequences, Daw (Press and Piggot, 1967), and He (Cunningham, Pflumm and Edelman, 1969). Our studies (Edelman et al., 1969) suggest that the variable region of the heavy chain (V_H) is about the same size as that of the light chain and thus the constant region (C_H) is about three times as large. Trypsin cleaves the heavy chain at lysyl residue 222 to produce the Fc fragment. Isolation of a single glycopeptide indicated that the polysaccharide portion of the molecule (see Fig. 2) is attached at aspartyl residue 297. By comparison with the data of Thorpe and Deutsch (1966) we have suggested (Rutishauser et al., 1969) that glutamyl residue 356 and methionyl residue 358 are associated with the so-called Gm specificity and the genetic polymorphism of the Fc region. Thus, like the C_L region of the light chain, the C regions of γ1 heavy chains contain a small number of variants which are inherited in a classical Mendelian fashion (Martensson, 1966).

Together, the light and heavy chain of protein Eu make up 214 and 446 residues and the symmetry of the molecule gives a total of 1320 residues in the entire structure. The half-molecule is the largest protein unit for which a complete amino acid sequence is known. The sequence determination proves the arrangement of chains in γG immunoglobulin, and the exact location of the disulphide bonds (Gall and Edelman, 1969) provides additional details which are important in relating the structure and function of the molecule. The disulphide bonds have been included in Fig. 2, which shows that the molecule contains a number of strikingly linear arrangements. From their NH_2-termini to half-cystines V, the light and heavy chains are aligned in register in the primary structure. The intrachain disulphide bonds are linearly and periodically disposed. In accord with the alignment of light and heavy chains, the corresponding (similarly numbered) intrachain disulphide bonds are in similar positions and the disulphide loops are of approximately the same size. All of the loops in the constant regions (C_H and C_L) vary from 57 to 61 residues. Those in the variable regions range from 66 residues in V_L to 75 residues in V_H.

The covalent chemistry compels conclusions about the symmetry in the three-dimensional structure. Thus there is a two-fold rotation axis in the Fc region which runs right through the interchain disulphide bonds. Moreover, homology relationships among the chains (see below) and work on the antibody combining site (Singer and Doolittle, 1966) suggest that there is a pseudosymmetry axis between the light and heavy chains.

With these facts in mind, we may ask a number of questions about the relation between the structure and the function of the antibody molecule in the selective immune response. What is the origin of diversity in V_H and V_L regions, both of which appear to be necessary for the antigen recognition function? What portions of the molecule serve effector functions? Are there clues in the structural data to the mechanism of differentiation and clonal commitment of antibody-forming cells?

Before considering these questions, it is profitable to ask about the evolutionary origin of the chains by comparing various portions of the sequence of the protein Eu with each other and with portions of other molecules. Two criteria may be used for this purpose: direct visual inspection of any two portions of the sequence for stretches of similar or identical residues, and comparison of polynucleotide sequences corresponding to the amino acid sequences, using a computer according to the method of Fitch (1966).

Visual comparison of V regions of the light and heavy chains of protein Eu shows a number of regions of similarity but no striking homology (Gottlieb et al., 1968). Nonetheless, Fitch's programme shows reasonable homology although comparison with another heavy chain V region indicates that there is no closer relationship between chains from the same molecule than between chains from different molecules (Gottlieb et al., 1968). A comparison of the C_L and C_H regions of Eu shows that there are three regions of equal length in C_H that are strikingly homologous to each other and to C_L. No convincing homology could be found between V and C regions in Eu (Rutishauser *et al.*, 1969; Edelman *et al.*, 1969). These relationships are summarized in Fig. 4.

How may we interpret these relationships? Hill and co-workers (1966) have proposed that both V and C regions of immunoglobulins evolved by successive duplication of a single precursor gene sufficient in size to specify one homology region (110 residues). Our results completely substantiate this hypothesis, although its details may have to be modified, particularly because there is no convincing homology between V and C regions. This may be either because of early evolutionary divergence of genes specifying these regions or because of the particular mechanism by which the V region is diversified. I would favour the idea that a V gene

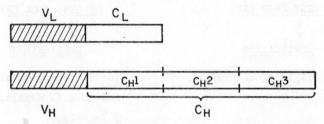

FIG. 4. Regions showing internal homologies in the structure of $\gamma G1$ immunoglobulin. Variable regions V_H and V_L are homologous. The constant region of the heavy chain (C_H) is divided into three regions, C_H1, C_H2, and C_H3, which are homologous to each other and to the C region of the light chain (C_L). The Fc portion of the heavy chain contains homology regions C_H2 and C_H3 (see Fig. 2).

and a C gene diverged early to serve the two major groups of quite different functions: the selective antigen recognition function (ARF), and the effector functions (EF), such as complement fixation. As shown in the hypothetical scheme of Fig. 5, each gene is supposed to have then been duplicated separately. The C genes duplicated in tandem to provide constant homology regions in the heavy chains (such as C_H1, C_H2, C_H3, in Figs. 2 and 4). The immunoglobulin classes also evolved by duplication (Hill *et al.*, 1966) although the size of the duplicating unit is not clear. Studies on the phylogeny of antibodies suggest (Marchalonis and Edelman, 1968) the possibility that γ chains may have originated from genes specifying μ chains rather than by duplication of a small precursor gene. V genes also must have duplicated to give rise to V_L and V_H regions. In addition, however, gene duplication may have been an essential step in producing the necessary diversity of immunoglobulins.

This type of evolution would be compatible with regional differentiation of function in the antibody molecule. We have suggested the hypothesis (Edelman *et al.*, 1969) that each of the pairs of homologous regions $V_H + V_L$, $C_L + C_H I$, $C_H 2 + C_H 2$, and $C_H 3 + C_H 3$ may be folded in a compact domain to serve one of the functions of the molecule. Obviously the details of folding are completely unknown, but the overall relationships of the compact domains may be represented as shown in Fig. 6. The V region domain contains the antigen-binding site which is formed by both V_H and V_L. $C_L + C_H I$ may serve to stabilize the interaction between

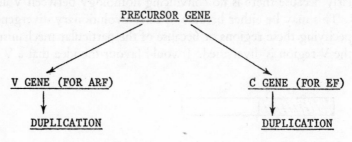

```
                    PRECURSOR GENE



     V GENE (FOR ARF)                 C GENE (FOR EF)
            |                                |
            v                                v
        DUPLICATION                      DUPLICATION

a.  for origin of diversity?     a.  for different domains in
                                     molecule (1 domain, 1 site,
                                     1 function?)

b.  for evolution of classes     b.  for evolution of classes


         EVOLUTION OF MECHANISM TO JOIN ARF AND EF

              (Translocation of V to C)
```

FIG. 5. Hypothetical scheme for evolution of antigen recognition functions (ARF) and effector functions (EF).

the V regions but could also serve a different function. Similarly, the other C_H regions are involved in complement fixation and skin fixation but their exact binding sites are unknown. $C_H 2 + C_H 2$ contains the carbohydrate which plays a role in the excretion of the molecule (Moroz and Uhr, 1967; Melchers and Knopp, 1967). A comparison of human and rabbit Fc fragments, which include homology regions $C_H 2$ and $C_H 3$, shows that there are three regions of sequence which are nearly identical (Rutishauser *et al.*, 1969). The possibility exists that these regions may be concerned with the function of active sites in neighbouring domains. The simplest domain hypothesis is to suppose that one paired homology region

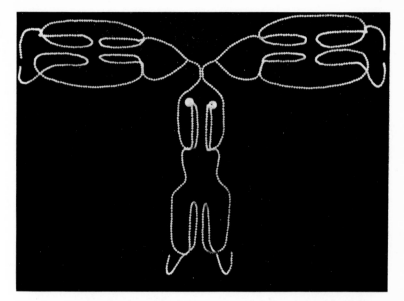

Fig. 6. Poppit bead model of γG1 immunoglobulin. (Compare with Fig. 2). Each bead represents one amino acid residue. Large balls represent the carbohydrate. Disulphide loops are represented as facing each other. In C$_H$2 and C$_H$3 there is a two-fold rotation axis between loops; this axis passes through the disulphide bonds linking the heavy chains. In the Fab portions of the structure, the homology regions and S–S loops are related through a pseudosymmetry axis. Other details of three-dimensional structure are unknown and the model is not intended as a representation of the detailed shape or folding, but rather as a representation of domains of folding.

To face page 312

forms one compact domain with one site and one function (Edelman *et al.*, 1969). Different classes of immunoglobulins such as μ chains may contain extra domains as well as variations of the same domains.

The evolutionary scheme shown in Fig. 5 requires that a mechanism exist to join the V genes and C genes so that ARF and EF are carried out by the same immunoglobulin molecule. This would require that some kind of gene translocation occur in the germ line or soma. There is a compelling reason to suggest translocation in somatic cells as a fundamental mechanism in the antibody system. You will remember that the C region Inv marker in human κ chains is allelic and is specified by a single gene. Since the C regions of the three subgroups are alike (Milstein, Milstein and Feinstein, 1969) and in every individual each non-allelic subgroup has a different V gene, we conclude that each chain must be specified by one of the V genes and one C gene. Because the chains appear to be synthesized from a single starting point (and since we have no evidence of recombination of messenger RNA) we deduce that genetic translocation in somatic cells is likely (Gottlieb *et al.*, 1968; Edelman and Gall, 1969).

The hypothesis I prefer (Edelman, 1967; Edelman and Gally, 1969) is that a V gene episome has a homology region with a C gene and forms a complete VC gene by insertion and crossing over. This provides a reasonable explanation for clonal responses because according to this mechanism at most two of the variants of V regions can be expressed in one cell. In fact, only one is expressed, and in a heterozygous individual with κ chains having Val and Leu at position 191 (see Fig. 3), only Val *or* Leu is seen in immunoglobulin molecules from a single cell. This unusual phenomenon of allelic exclusion can be explained by the translocation hypothesis, for it is highly unlikely that both V genes from each parent would be simultaneously translocated. This would be compatible with the observation that a committed cell has only one specificity (Mäkelä, 1967). Thus, translocation would guarantee that a lymphoid cell would retain its specificity upon stimulation by antigen and it also assures that the cell is most efficient in amplifying the synthesis of its particular antibody molecules.

The actual mechanism whereby antigens trigger mitosis (see Fig. 1) is unknown. Perhaps the simplest hypothesis is to invoke either a change in the conformation of the antibody, or an interaction between two or more antibody molecules or V regions on the cell surface. There may be an actual site on the antibody molecule (perhaps in $C_L + C_H I$) which serves the purpose of triggering mitosis and maturation.

The domain hypothesis and the translocation hypothesis may account in part for the regional differentiation of function in antibodies as well as for the irreversible clonal differentiation required in a selective response. They do not, however, account for the origin of the diversity of V regions. At the gene level, there are three theories to explain this. One theory, proposed in its most complete form by Dreyer and Bennett (1965), is that a separate V gene has evolved for each different V region. This means that the number of V genes in a selective system would have to be very large. Although vertebrates have sufficient amounts of DNA for this purpose, the ticklish point is to explain how so many duplicated genes remained so alike, for obviously in a selective immune response natural selection cannot be directed entirely by the antigen.

A second hypothesis proposed by Brenner and Milstein (1966) is that there is one or at most a few V genes which undergo somatic point hypermutation. The difficulty with this theory is that it fails to account for the necessarily rapid somatic selection which would be required to favour just those mutants which will form folded chains. Cohn (1968) has suggested some ways in which this might be accomplished.

A third theory (Edelman and Gally, 1967) is that there is a small number of genes which evolved to contain a certain set of variants. This pre-evolved set then recombines in the lymphoid cell to form the various sequences. By this mechanism, at least 2^{43} variants can be generated from 43 positions in which mutations have occurred (see Fig. 3). Smithies (1967) modified this theory so that only two genes are assumed to recombine, but this form of the theory appears to be excluded by the presence of three amino acids in one position in mouse light chains (Hood et al., 1967) and by the existence of non-allelic V region subgroups (Milstein, Milstein and Feinstein, 1969).

In any case, so far none of these theories has been tested rigorously. A certain amount of confusion has arisen from the assumption that certain theories are easier to test than others. The obvious critical experiments are to count immunoglobulin genes by DNA–RNA hybridization techniques and to search for a genetic marker in V regions. Unfortunately, these experiments are difficult to carry out. If a homozygous marker is found, Dreyer's theory (Dreyer and Bennett, 1965) and ours (Edelman and Gally, 1967, 1969) would be severely weakened because this finding would suggest that there is only one or at most a few V genes.

The major purpose of this discussion has been to show how facts on the structure of immunoglobulins suggest hypotheses on the basic mechanisms of specificity and control in a selective immune response. At present it is not clear whether the mechanisms of cellular maturation and clonal commitment of antibody-forming cells will have a general bearing on differentiation processes in eukaryotic organisms. The indication that eukaryotic genomes contain multiple copies of genes (Britten and Kohne, 1966) and the long-standing example of transposition of genes in maize (McClintock, 1965) have certain resemblances to some of the unusual features of the antibody system. If these processes and antibody formation are related to differentiation in eukaryotes, then a study of events at the DNA level of somatic cells should be particularly rewarding. Inasmuch as somatic genetics is still difficult, analyses (Lindahl and Edelman, 1969a, b) of enzymic mechanisms of DNA replication, repair and recombination in eukaryotic cells may provide useful information both on the immune response as well as on other specialized manifestations of cellular differentiation.

SUMMARY

Antibodies or immunoglobulins have two major functions in the immune response: (1) they combine with a great variety of chemically distinguishable antigens, and (2) after binding antigens, they mediate a number of essential immune reactions such as complement fixation. The primary structure and arrangement of disulphide bonds of an entire γG immunoglobulin (molecular weight 150 000) has recently been determined and compared with portions of molecules studied by other workers. The molecule consists of two identical light polypeptide chains and two identical heavy chains. Each light chain consists of a variable half (V_L) in which amino acid sequence variations can occur, and a constant half (C_L) in which the sequence is nearly identical from molecule to molecule. The heavy chain variable region (V_H) is homologous to V_L and has approximately the same size. The constant portion of the heavy chain consists of three linearly connected regions, C_H1, C_H2, and C_H3, which are homologous to each other and to C_L. Functional studies indicate that V regions are concerned with antigen recognition functions (ARF) and C regions with chain interaction and effector functions (EF). The data on the amino acid sequences of the molecule can be accounted for by the hypothesis that one or two precursor genes duplicated successively to form antibody molecules with definite regional differentiation of ARF and EF.

A number of theories have been proposed to explain the diversity of V regions and the constancy of C regions. There is evidence that each chain is specified by two genes, V and C, which may be translocated to form a competent VC gene. This can account for the phenomenon of allelic exclusion in antibody-forming cells and is consistent with the notion that one cell produces only one antibody in a selective immune response. At the gene level, there are three theories to account for the diversity of V regions: (1) each region has a single V gene in the germ line, (2) there is a small number of V genes which somatically recombine in lymphoid cells, (3) a single V gene undergoes somatic mutation and somatic selection.

Selective antibody formation can be accounted for by diversification of amino acid sequence, by an antigen-trapping mechanism and by high-gain amplification of antibody synthesis through clonal maturation and division of antigen-stimulated cells. Further structural analysis of antibodies should provide essential clues to the molecular mechanisms of specificity and control which operate at each one of these stages.

Acknowledgement

The work of the author described in this paper was supported by grant GB 6546 from the National Science Foundation and by grant AM 04256 from the National Institutes of Health.

REFERENCES

BRENNER, S., and MILSTEIN, C. (1966). *Nature, Lond.*, **211**, 242–243.

BRITTEN, R. J., and KOHNE, D. E. (1966). *Yb. Carnegie Instn Wash.*, **65**, 78–106.

BURNET, F. M. (1959). *The Clonal Selection Theory of Acquired Immunity.* Nashville, Tennessee: Vanderbilt University Press.

CAIRNS, J. (ed.) (1967). *Cold Spring Harb. Symp. quant. Biol.*, **32**.

COHN, M. (1968). In *Nucleic Acids in Immunology*, pp. 671–715, ed. Plescia, O. J., and Braun, W. New York: Springer.

CUNNINGHAM, B. A., PFLUMM, M., and EDELMAN, G. M. (1969). Unpublished data.

DREYER, W. J., and BENNETT, J. C. (1965). *Proc. natn. Acad. Aci. U.S.A.*, **54**, 864–869.

EDELMAN, G. M. (1967). In *Gamma Globulins, Structure and Control of Biosynthesis*, Nobel Symposium 3, pp. 89–108, ed. Killander, J. Stockholm: Almqvist and Wiksell.

EDELMAN, G. M., BENACERRAF, B., OVARY, Z., and POULIK, M. D. (1961). *Proc. natn. Acad. Sci. U.S.A.*, **47**, 1751–1758.

EDELMAN, G. M., CUNNINGHAM, B. A., GALL, W. E., GOTTLIEB, P. D., RUTISHAUSER, U., and WAXDAL, M. J. (1969). *Proc. natn. Acad. Sci. U.S.A.*, **63**, 78–86.

EDELMAN, G. M., and GALL, W. E. (1969). *A. Rev. Biochem.*, **38**, in press.

EDELMAN, G. M., and GALLY, J. A. (1962). *J. exp. Med.*, **116**, 207–227.

EDELMAN, G. M., and GALLY, J. A. (1967). *Proc. natn. Acad. Sci. U.S.A.*, **57**, 353–358.

EDELMAN, G. M., and GALLY, J. A. (1969). *Brookhaven Symp. Biol.*, **21**, 328–344.

EISEN, H. N., LITTLE, J. R., OSTERLAND, C. K., and SIMMS, E. S. (1967). *Cold Spring Harb. Symp. quant. Biol.*, **32**, 75–81.

FITCH, W. (1966). *J. molec. Biol.*, **16**, 9–16.

GALL, W. E., and EDELMAN, G. M. (1969). Unpublished data.

GOTTLIEB, P. D., CUNNINGHAM, B. A., WAXDAL, M. J., KONIGSBERG, W. H., and EDELMAN, G. M. (1968). *Proc. natn. Acad. Sci. U.S.A.*, **61**, 168–175.

HABER, E. (1964). *Proc. natn. Acad. Sci. U.S.A.*, **52**, 1099–1106.

HABER, E., RICHARDS, F. F., SPRAGG, J., AUSTEN, K. F., VALLOTTON, M., and PAGE, L. B. (1967). *Cold Spring Harb. Symp. quant. Biol.*, **32**, 299–310.

HILL, R. L., DELANEY, R., FELLOWS, R. E., JR., and LEBOVITZ, H. E. (1966). *Proc. natn. Acad. Sci. U.S.A.*, **56**, 1762–1769.

HOOD, L., GRAY, W. R., SANDERS, B. G., and DREYER, W. J. (1967). *Cold Spring Harb. Symp. quant. Biol.*, **32**, 133–145.

JERNE, N. K. (1955). *Proc. natn. Acad. Sci. U.S.A.*, **41**, 849–857.

JERNE, N. K. (1966). In *Phage and the Origins of Molecular Biology*, pp. 301–312, ed. Cairns, J., Stent, G. S., and Watson, J. D. New York: Cold Spring Harbor Lab.

KELUS, A. S., and GELL, P. G. H. (1967). *Prog. Allergy*, **11**, 141–184.

KILLANDER, J. (ed.) (1967). *Gamma Globulins, Structure and Control of Biosynthesis*, Nobel Symposium 3. Stockholm: Almqvist and Wiksell.

KOSHLAND, M. E., and ENGLBERGER, F. M. (1963). *Proc. natn. Acad. Sci. U.S.A.*, **50**, 61–68.

LINDAHL, T., and EDELMAN, G. M. (1969a). *Proc. natn. Acad. Sci. U.S.A.*, **61**, 680–687.

LINDAHL, T., and EDELMAN, G. M. (1969b). *Proc. natn. Acad. Sci. U.S.A.*, **62**, 597–603.

MÄKELÄ, O. (1967). *Cold Spring Harb. Symp. quant. Biol.*, **32**, 423–430.

MARCHALONIS, J. J., and EDELMAN, G. M. (1968). *J. exp. Med.*, **127**, 891–914.

MARTENSSON, L. (1966). *Vox Sang.*, **11**, 521–545.

McCLINTOCK, B. (1965). *Brookhaven Symp. Biol.*, **18**, 162–184.

MELCHERS, F., and KNOPP, P. M. (1967). *Cold Spring Harb. Symp. quant. Biol.*, **32**, 255–262.

MILSTEIN, C., MILSTEIN, C. P., and FEINSTEIN, A. (1969). *Nature, Lond.*, **221**, 151–154.

MOROZ, C., and UHR, J. W. (1967). *Cold Spring Harb. Symp. quant. Biol.*, **32**, 263–269.

PORTER, R. R. (1959). *Biochem. J.*, **73**, 119–126.

PRESS, E. M., and PIGGOT, P. J. (1967). *Cold Spring Harb. Symp. quant. Biol.*, **32**, 45–51.

RUTISHAUSER, U., CUNNINGHAM, B. A., BENNETT, C., KONIGSBERG, W. H., and EDELMAN, G. M. (1969). *Proc. natn. Acad. Sci. U.S.A.*, **61**, 1414–1421.

SINGER, S. J., and DOOLITTLE, R. F. (1966). *Science*, **153**, 13–25.

SMITHIES, O. (1967). *Cold Spring Harb. Symp. quant. Biol.*, **32**, 161–166.

THORPE, N. O., and DEUTSCH, H. F. (1966). *Immunochemistry*, **3**, 329–337.

DISCUSSION

Miller: Were there not indications from M. E. Koshland's data (1967) that some allotypic differences in the N-terminal quarter (V region) of the rabbit H chain exist?

Edelman: Dr Koshland isolated peptides from homozygous rabbits differing in genetic markers and found a correlation between the amino acid composition and the genetic type (1967). If the interpretation of her data is correct we have two alleles with multiple substitutions, which would mean that recombination certainly would have had to be prevented in the germ line. Furthermore, if her hypothesis stands up as she interprets it, there cannot be a very large number of V genes.

Bhargava: How does your model account for the fact that variation in the V region is not quite random?

Edelman: Our model can answer that quite nicely because the mutations have been selected for in evolutionary times. The difficulties arise with Dr Cohn's model.

11*

Bhargava: There are no difficulties with Dr Cohn's model because it allows the undesirable variants to be selected out.

Edelman: Yes, in exactly one month *in utero*. The point is that in our theory selection has already occurred in evolution. These mutations have been selected for in the ensemble, and the recombination of pre-evolved and selected mutations simply rearranges the structure.

Bhargava: Surely variation in the recombination frequency at different points wouldn't be that great or random, as is suggested by the known sequences. For example, you don't find an amino acid like tryptophan being replaced.

Edelman: This model allows you to select for protein structure; any evolutionary model will do that. And the problem of the size of the set needed for a selective theory compared to the size of our sample makes it impossible to conclude anything by inspection. We have about 30 or 40 proteins with sufficient data, and there may be millions, so to conclude anything about that or about the details of recombination at this stage is simply not possible. All one can say is that nothing in the data proves recombination but the data are not inconsistent with it. I shan't speculate about the mechanism of recombination, particularly in the soma, because recombination itself has not been worked out in molecular terms.

Monod: One cannot explain the structural facts and also some of the other facts other than by the episome theory, particularly the fact that there must be a time in the life of the clone when it is generating variability and another time when it is producing antibody and must not generate variability. We know that in bacteria an episome may be in the cytoplasm where it multiplies freely, sometimes to an enormous number, which of course would allow a great deal of mutation to occur even in a theory like yours which also includes recombination. And we also know that when the episome gets implanted on the chromosome the cytoplasmic episomes cannot multiply any more. So it explains at the same time the fact that you have to have these two periods in the life of the clone, and the peculiar nature of the translocation that you have.

Edelman: I find the episome idea extremely attractive and feel that it is probably the explanation for the so-called phenomenon of allelic exclusion, namely that in a heterozygote with Val and Leu at position 191 no immunoglobulin-producing cell produces both types of light chain. Allelic exclusion could be explained by the point that simultaneous translocation of both V genes is highly unlikely.

Harris: At first sight, a period during which you have to generate recombination followed by a period during which you have to shut it off seems to bring a difficulty into the argument. But in *Neurospora* we do have genes which regulate the frequency of allelic recombination in other genes (Catcheside, Jessop and Smith, 1964). These regulatory genes must be subject to phenotypic modulation: you can change the conditions to get either a higher or lower recombination rate. But I'm not sure that you really *need* a period of imposed recombination and a period of suspended recombination. Pictures of clones of cells forming antibody always show many more cells than are actually engaged in the synthesis of the antibody. The clone is formed by extensive cell multiplication, but only a small proportion of the cells are actually synthesizing antibody. The rest are multiplying and perhaps making other things that you are not looking for. You could say that recombination was continuing at a constant rate and that you were still generating diversity in the clone.

Edelman: I believe, although there is no real evidence, that the numbers of different variants involved must be very great indeed, so I am worried about the kinetics of the process. I think that it is much more conservative spatially to distribute all these variants and commit the cells that are there early on, rather than keep them turning over. But since we lack actual data it may be that your temporal model, as in *Neurospora*, could account for a good proportion of the variation.

Harris: It is a great oversimplification to regard mammalian cells in clones as anything but genetically identical, and perhaps they are not even that. They are always phenotypically diverse. When you get a clone going from one cell, the cells on the outside of the clone may be quite different from those on the inside; and if you look at the formation of a specific product, you may find intraclonal variation which has nothing to do with sectoring or mutational events of that kind. Antibody production may be an extreme example of this kind of variation in a mammalian cell clone, where only a few of the cells are doing what you are interested in. The rather rigid concept of a clone comes from bacterial colonies where phenotypic variation is relatively small.

Monod: I wonder, Dr Edelman, why you reject what palaeontologists call orthogenetic selection; why do you want the final product that you observe, a highly avid antibody, to be selected in just one step? Is it not possible that there is evolution in the sense of the evolution of the horse during the Tertiary period?

Edelman: In a selective theory there are many genes specifying antibodies that could interact with a particular antigen, and some of those genes would be silent. So drift could occur. Furthermore, it is clear on a selective theory that arsanilic acid is not exerting an enormous selective pressure, or any other artificial antigen!

If you wish to put the selection as Dr Cohn does, in the soma, so that you have a neo-Darwinian process in the soma, the short time involved implies to me that there must be hypermutation and therefore there must be a most efficient selection demon to exert pressure against all the bad variants and kill those cells. So far, I have not found any acceptable molecular explanation for a selection demon.

Monod: You postulate a single C gene and yet I understand that in a myeloma clone the differential rate of protein production is something like 30 per cent. That would seem to require more than a single gene, on grounds of kinetics. Presumably the gene, whatever it does, has to be extremely active, or translation has to be very active. One wonders whether there might not be local gene reduplication, as occurs in insects. And if that is the case, then the chances of hybridization might be much better.

Edelman: Gene amplification is certainly theoretically permissible. If the number of C genes produced this way equals the number of V genes, then you have a much better chance of hybridization.

Cohn: What does a comparison of the constant regions of the Fd fragments from μ, γ and α reveal? Are they clearly different?

Edelman: That is not yet known.

Tata: Do you think that sugars contribute to the attenuation or exaggeration of the different variabilities?

Edelman: No, for the following reason. Recently it has been shown by Moroz and Uhr (1967) and Melchers and Knopp (1967) that the excretion of the immunoglobulin may depend on the carbohydrate, and therefore it is most likely that the sugars have to do with interaction with membranes and transport from the cell. You also can cleave off the Fc fragment with the sugar without any change in specificity of the Fab fragment.

REFERENCES

CATCHESIDE, D. G., JESSOP, A. P., and SMITH, B. R. (1964). *Nature, Lond.,* **202,** 1242.
KOSHLAND, M. E. (1967). *Cold Spring Harb. Symp. quant. Biol.,* **32,** 119–127.
MELCHERS, F., and KNOPP, P. M. (1967). *Cold Spring Harb. Symp. quant. Biol.,* **32,** 255–262.
MOROZ, C., and UHR, J. W. (1967). *Cold Spring Harb. Symp. quant. Biol.,* **32,** 263–269.

REGULATORY MECHANISMS OF CELLULAR PROLIFERATION IN A PROTEIN-DEFICIENT ORGANISM

M. G. Deo and V. Ramalingaswami

Department of Pathology, All-India Institute of Medical Sciences, New Delhi

We turn now from sub-cellular and molecular events in cells exposed to precisely definable influences to events in an intact multicellular organism, with its metabolic fluidity and its hierarchy of regulatory processes, when it is exposed to a deficiency of protein in the diet.

HUMAN KWASHIORKOR

Protein deficiency in young growing children, widely known by the African tribal name *Kwashiorkor*, is a widespread disorder throughout the developing world. In its established form, it manifests itself as a clearly defined syndrome with several components. Retardation of growth is the earliest sign, followed by failure of growth. Generalized oedema develops. Gastrointestinal disturbances are prominent; diarrhoea may be a persistent feature whether or not intestinal pathogens are isolated. Susceptibility to infections is enhanced. Variable skin changes develop, ranging all the way from simple atrophy to pigmentary disturbances and weeping cracked areas. The hair loses its texture and is sparse and brittle. The bones are radiolucent and poorly trabeculated. The activity of several plasma and tissue enzymes is depressed. Moderate anaemia occurs. There are profound disturbances in the psyche. If untreated with proteins and calories, there is a high mortality (Trowell, Davies and Dean, 1954; Brock, 1961; Ramalingaswami, 1964).

EXPERIMENTALLY INDUCED PRIMATE KWASHIORKOR

The multifactorial background against which kwashiorkor arises makes a sequential analysis of the reaction to protein deficiency in man a rather difficult task. Reproduction of the syndrome in all its essential aspects in experimental animals under controlled conditions enables such a study to be made of the pathobiology of protein deficiency. This has now been accomplished in a number of animals by several workers. In our laboratory, the syndrome has been reproduced in its totality in young growing Rhesus monkeys by force-feeding them a diet deficient in proteins but adequate in calories and all the other essential nutrients (Deo and Ramalingaswami, 1960; Ramalingaswami, Deo and Sood, 1961; Deo, Sood and Ramalingaswami, 1965; Sood, Deo and Ramalingaswami, 1965; Ramalingaswami and Deo, 1968; Jha, Deo and Ramalingaswami, 1968). The primate syndrome so induced is characterized by growth retardation, oedema and a fall in the concentrations of serum albumin, cholesterol and transferrin. The liver shows periportal fatty change to begin with which later becomes generalized; in the pancreas there is a reduction in the zymogen granules and shrinkage and atrophic crowding of acinar elements. Likewise, the salivary glands are atrophied. The germinal centres in the spleen are depleted of cellular elements and a moderate degree of normochromic normocytic anaemia develops. Endochondral and appositional bone growth is slowed down. There is a well-marked mucosal atrophy all along the gastro-intestinal tract. In the later stages, the skeletal muscle loses protein and undergoes extensive atrophy; even the cardiac muscle does not escape the process of muscle breakdown.

We thus have a similar structural frame of kwashiorkor in man and monkey, but the functional picture is still missing. A look at the reactions to protein deficiency shows that although a number of organs are affected, they are not all affected at the same time nor at the same rate. Some organs show a rapid and extensive response to low protein diets. These are organs with a high protein turnover, be they organs that actively synthesize and secrete proteins, such as the liver and pancreas, or organs that have a high rate of cell renewal, such as the small intestine and bone marrow.

CELLULAR PROLIFERATION IN PROTEIN DEFICIENCY

In mammalian systems, large numbers of cells are produced and lost every day. Their birth, differentiation and death are delicately balanced so as to maintain a steady state; the maintenance of this balance is one of the remarkable phenomena of physiological regulation (Patt and Quastler, 1963). In the *continuous cell renewal systems* such as the mucosa of the small intestine, cell production in the progenitor compartment of crypt cells balances cell loss in the differentiated absorptive compartment in the villi. In the *conditional cell renewal systems* such as the liver, the cells which normally have a low rate of proliferation and are long-lived can be stimulated to a burst of proliferative activity by the simple expedient of partial hepatectomy (Bucher, 1967). This stimulated growth, brought out on demand, is also precisely regulated and ceases as soon as the liver size is restored to normal.

Dividing cells pass through a cell cycle made up of recognizable phases —G_1, S, G_2 and M—described by Howard and Pelc (1953). The use of tritium-labelled thymidine together with high-resolution autoradiography has greatly facilitated the study of cell generation cycles and turn-over rates of cell renewal systems (Quastler and Sherman, 1959; Stohlman, 1959; Lamerton and Fry, 1963). By *in vivo* labelling of nuclei of cells in the DNA synthetic phase among the crypt cells of the small intestine and following the labelled cells on autoradiographs through mitosis, through the differentiating compartment and up the villus as a function of time until they are finally cast off, we have gained valuable information on their generation time, maturation time and transit time. With the use of this technique, our earlier work on the primate model indicated that there was a marked slowing down of the cell generation cycle of the crypt cells of the small intestine in protein deficiency (Deo and Ramalingaswami, 1965). There was also evidence in these studies of a significant slowing down associated with functional absorptive defects. Prolongation of transit time is characteristic of the intestinal epithelium of the mouse as age advances (Lesher, Fry and Kohn, 1961). Cell renewal time and cell transit time up the villus are prolonged in four days of total starvation in mice (Brown, Levine and Lipkin, 1963).

In order to characterize further the defect in the cell generative cycle in protein deficiency, two experiments were made in rats, one on the mucosa of the small bowel as an example of a continuous cell renewal system and the other on the liver after partial hepatectomy as an example of a conditional cell renewal system. In the former experiment performed by K. Verma in our laboratory (Verma, Deo and Ramalingaswami, 1969), a single dose of 200 R of X-ray irradiation was given to the whole body of two groups of rats, one on a high and one on a low protein diet. This is a small dose and is known to lead to a temporary post-irradiation "freeze" of the various phases of the cell cycle, of which the flow of cells into mitosis is the most sensitive (Patt and Quastler, 1963). Mitotic counts in the crypt cell population just before irradiation and 40 minutes later revealed a sharp drop in both groups, indicating an effective block from G_2 to M, but the counts at the end of 40 minutes were higher in the deficient group than in the controls (Fig. 1). As this dose of 200 R would result in a block of flow of cells from G_2 to M, the higher mitotic counts in the deficient animals may be indicative of a prolongation of the mitotic phase in protein deficiency.

Further indication of a disturbance in the cell cycle in protein deficiency came in an experiment in which the resumption of mitosis up to 4 hours after irradiation was studied. In control animals, with a dose of 200 R, one would expect the post-irradiation "freeze" of the transition from G_2 to M to be thawing between 2 and 4 hours. Verma demonstrated a striking difference in the number of mitoses among the crypt cells between the animals on the protein-deficient and control diets (Fig. 1). It is

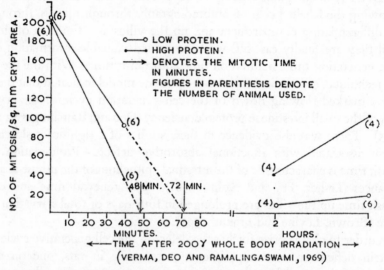

Fig. 1. Mitotic counts in intestinal crypts of rats on high and low protein diets after exposure to 200 R whole-body irradiation.

clear that whereas the control rats fed protein-rich diets were recovering from the G_2-to-M block, the deficient rats had not shown any recovery (Verma, Deo and Ramalingaswami, 1969). Verma's studies have thus established that in protein deficiency: (a) there is a prolongation of the mitotic phase in the continuously dividing population of crypt cells of the small intestine, and (b) there is a delay in the recovery of crypt cells from the G_2-to-M block induced by a small dose of irradiation. To the extent that a higher dose of irradiation to the control animals would have increased the duration of the G_2-to-M block, it may be said that irradiation and protein deficiency act synergistically in delaying the rate of entry of cells into mitosis.

M. Mathur in our department investigated the effect of protein deficiency on a conditional cell renewal system, the regenerating liver following
partial hepatectomy (Deo, Mathur and Ramalingaswami, 1967). The
standard 68 per cent hepatectomy was performed by the Higgins–Ander-
son procedure in protein-deficient and control rats at the same time of day.
A single pulse of [³H]thymidine was administered 22 hours after hepatectomy to both groups. Two hours and 6 hours later, animals were
sacrificed from both groups and labelling indices of hepatocytes estimated
on autoradiographs of liver sections (Fig. 2). It will be seen that whereas
the labelling index rose from 107 to 157 in the 4-hour period in the
control animals, there was hardly any change in the index in the protein

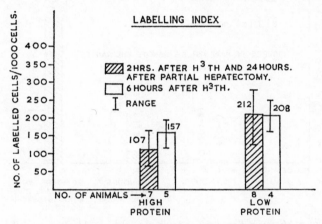

FIG. 2. Labelling indices of hepatocytes from rats on high and low protein
diets after a single pulse of [³H]thymidine.

deficient animals. Thus, 50 new cells were derived from a population of
107 cells in 4 hours in the controls; no significant change in the labelled
cell population occurred in protein deficiency, indicating poor cellular
proliferation.

Colchicine was administered to deficient and control groups 24 hours
after partial hepatectomy and the animals sacrificed 4 hours later. Studies
with colchicine for the past three decades have shown that at certain dosage
levels and within a certain period of time, colchicine acts to arrest cells in
metaphase only, without affecting the rate of entry of cells into mitosis.
By its judicious use in proper concentrations, the mitotic rates, the duration of mitosis and turnover times can be readily quantitated (Hooper,
1961). Striking differences were found in the rates of accumulation of
blocked mitoses between the two groups (Fig. 3). Whereas the mitotic

index rose from 0·5 to 43 in 4 hours of colchicine block in the control animals, there was far less rise in this index in the deficient animals, indicating, in this system also, a marked delay in the flow of cells into mitosis in protein deficiency. The higher mitotic index 28 hours after partial hepatectomy without colchicine in the protein-deficient animals gives a false impression of a higher rate of proliferation in this group (Fig. 3); the mitotic compartment is in fact expanded as a result of the cells staying longer in mitosis.

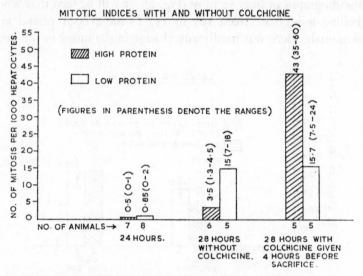

FIG. 3. Mitotic indices, with and without colchicine block, in partially hepatectomized rats on high and low protein diets.

THE HYPOPLASTIC CYTOPENIA OF PROTEIN DEFICIENCY

Whether it is a continuous cell renewal system or a conditional one, the effects of protein deficiency are similar. The rate of flow of cells into mitosis is slowed down and the mitotic phase itself is prolonged (Fig. 4). The DNA synthetic phase is probably also prolonged. The effect on the post-mitotic pre-DNA synthetic phase has not yet been evaluated. It is, however, clear that impairment of cell production is a basic lesion in protein deficiency. In a continuous cell renewal system, this must result in a reduction in the size of the generative compartment, and a hypoplastic cytopenia would be the resulting lesion in all such systems. Hence (1) the atrophy of intestinal crypts, leading to diminished differentiated cell production and reduction in the height of the villi; (2) the hypoplasia of the bone marrow, leading to a normocytic normochromic anaemia;

(3) the delayed sequences in the proliferation and maturation of cells of the epiphyseal cartilage and osteoblasts, leading to a virtual arrest of endochondral bone formation and longitudinal bone growth; (4) the reduced production of antibody-forming cells in the spleen in response to antigenic stimulation, the spleen being in some respects more sensitive to lack of protein than the liver (Kenney et al., 1968). In this study protein deficiency reduced antibody output to the same extent as it reduced the formation of antibody-synthesizing cells; (5) the poverty in fibroblastic proliferation in response to injury, so that wounds do not heal and cirrhosis does not develop after repeated hepatic necroses induced by carbon tetrachloride (Bhuyan et al., 1965); (6) the delayed renewal of

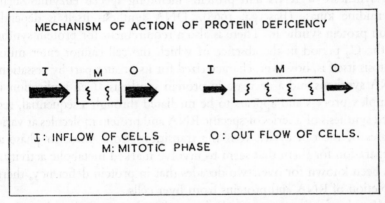

FIG. 4. The effects of protein deficiency on the rate of flow of cells into mitosis.

germinative hair cells and of epidermis resulting in epidermal atrophy, slowing of the rate of hair formation and the formation of thin and mechanically deficient hair (Bradfield, 1968); and (7) the deficiency in the myelinization of the central nervous system on account of diminished proliferation of oligodendroglia in the growing brain (Dickerson, 1968).

IRREVERSIBILITY OF GROWTH FAILURE

The elegant experiments of McCance and Widdowson (1962) had clearly shown that the earlier the animal is a victim of "malnutrition" (in this case calorie restriction), the greater is the likelihood of stunting in growth to become permanent. Although it is by no means proved, severe calorie deprivation during the first year of life of human infants is likely to lead to permanent stigmata in height and head size (Thomson, 1968). Normal growth in all organs of the rat between 0 and 21 days of age is by acquisition of new cells; between 21 and 65 days it occurs by both hyperplasia and hypertrophy of cells and beyond 65 days, by hypertrophy alone (Winick and Noble, 1965). Winick and Noble (1966) made the interesting suggestion that the cellular effects of malnutrition (calorie restriction) depend upon the phase of growth of the animal at the time of onset of malnutrition. Very early in life, malnutrition would impede cell division and organ growth and differentiation and may result in permanent stunting. Later in life, it would lead to changes in cell composition and cell size from which the animal could recover.

REGULATORY MECHANISMS

The flow of continuously dividing cells from G_1 to S seems to require the synthesis of RNA and proteins including special enzymes such as thymidine kinase (Baserga, 1968). DNA synthesis itself is dependent upon protein synthesis. There is also a requirement for protein synthesis in the G_2 period in the absence of which the cell cannot enter mitosis. Mitosis itself is, however, characterized for its greater part by cessation of RNA synthesis and a decrease in protein synthesis. Cell replication is a complex process and appears to be mediated through a sequential, regulated synthesis of a series of specific RNA and protein molecules at various phases of the cell cycle. It is the transitions from phase to phase and preparation for them that seem to involve marked metabolic activity. It has been known for over two decades that in protein deficiency, there is depletion of RNA and proteins from liver cells.

Waterlow's pioneering work shows that the metabolic adaptation to protein deficiency consists of a series of changes in the pattern of protein synthesis and catabolism (Waterlow, 1968a). Immediately upon being placed on a protein-deficient diet, the animal shows a fall in the synthetic rate for albumin (James and Hay, 1968; Kirsch et al., 1968). Later, compensatory adjustments in the catabolic rate of albumin set in, in an effort to maintain the total circulating albumin mass within normal limits. Even in the later stages, the capacity to return to the normal rate of synthesis is well maintained so that an immediate response is possible when the supply of amino acids improves. Other adaptations include a greater reutilization of amino acids along with diminished deamination brought about by a reduction in the activity of urea cycle enzymes (Gaetani et al., 1964; Schimke, 1962; Stephen and Waterlow, 1968). Protein deficiency leads to a rapid loss of RNA from the liver (Munro, 1968) and to a diminished synthesis of DNA in the intestine (Munro and Goldberg, 1964). Munro suggests that the influx of amino acids stimulates protein synthesis, reduces protein breakdown and causes more free ribosomes to aggregate to polysomes. When amino acids are in short supply, the process is reversed and there is a breakdown of both RNA and protein in the cell (Munro, 1968). Circulating plasma amino acids and amino acids in portal blood reaching the liver would be expected to be low in protein deficiency. A reduction in the levels of essential amino acids in circulating blood has been repeatedly demonstrated in kwashiorkor (Whitehead and Dean, 1964). It has been suggested that the action of dietary amino acids on polysome profiles is achieved through a cytoplasmic mechanism not requiring nuclear mediation (Munro, 1968). In view of the fact that protein deficiency leads to loss of cellular components, cytoplasmic RNA and protein whose sequential synthesis is necessary for movement to take place from phase to phase in the cell generation cycle, it is conceivable that the generation cycle will be profoundly disturbed in protein deficiency.

It is known that cellular regeneration can be influenced by the hormones of the pituitary, adrenal and thyroid glands but none of them is essential to the process (Bucher, 1967). The blood levels of both cortisol and pituitary growth hormone are elevated in human malnutrition (Alleyne and Young, 1967; Pimstone et al., 1966) but thyroid function is within normal limits (Montgomery, 1962). The only disturbance in the thyroid we observed in protein-deficient rats was the inability of these animals to replicate cells as effectively as the controls when exposed to a goitrogenic stimulus for growth, as in a conditional renewal system (Ramalingaswami et al., 1965). Cortisol is known to produce an increase in catabolism and growth hormone a decrease in the fractional catabolic rate of albumin (Rothschild et al., 1958; Grossman, Yalow and Weston, 1960; Gabuzda, Jick and Chalmers, 1963). An increase in protein intake produces a fall in the level of the growth hormone (Gabuzda, Jick and Chalmers, 1963) and a marked fall also occurs in 1–2 weeks of administration of a high protein diet to malnourished children. Levels of insulin are reportedly low in malnutrition. As insulin is known to promote amino acid incorporation into muscle protein, it is possible that reduced protein synthesis in muscle is related to insulin lack. These hormonal changes are interesting but complex and it is difficult to know which are primary and which secondary (Waterlow, 1968b).

SUMMARY

A syndrome closely resembling human kwashiorkor has been induced in young growing Rhesus monkeys by feeding them a diet deficient in protein but adequate in calories and other essential nutrients. Every cell and organ in the body is affected but they are not all affected at the same rate nor at the same time. There is a pattern in the response to protein deficiency.

Organs with a high protein turnover show rapid and extensive metabolic and structural alterations. Among these are organs such as the liver and pancreas which actively synthesize and secrete proteins; others such as the small intestine, bone marrow and growing ends of bone have an active cell renewal system.

In a continuous cell renewal system such as the mucosa of the small intestine and in a conditional cell renewal system such as the regenerating liver after partial hepatectomy, a marked slowing down of cell generation was observed in protein deficiency in rats. These studies were made with the use of blocking agents of cell generation such as X-irradiation and colchicine.

A slowing down of cell generation appears to be a basic alteration in response to protein deficiency. Its mechanism is not clear. Diminished influx of amino acids and the associated breakdown of cellular RNA and proteins may be related to the disturbance in cell generation.

REFERENCES

ALLEYNE, G. A., and YOUNG, V. H. (1967). Clin. Sci., 33, 189–200.
BASERGA, R. (1968). Cell Tissue Kinet., 1, 167–191.
BHUYAN, U. N., NAYAK, N. C., DEO, M. G., and RAMALINGASWAMI, V. (1965). Lab. Invest., 14, 184–190.
BRADFIELD, R. B. (1968). In Calorie Deficiencies and Protein Deficiencies, pp. 213–221, ed. McCance, R. A., and Widdowson, E. M. London: Churchill.
BROCK, J. F. (1961). Fedn Proc. Fedn Am. Socs exp. Biol., 20, suppl. 7, 61–65.
BROWN, H. O., LEVINE, M. L., and LIPKIN, M. (1963). Am. J. Physiol., 205, 868–872.
BUCHER, N. L. R. (1967). New Engl. J. Med., 277, 686–696, 738–746.
DEO, M. G., MATHUR, M., and RAMALINGASWAMI, V. (1967). Nature, Lond., 216, 499–500.
DEO, M. G., and RAMALINGASWAMI, V. (1960). Lab. Invest., 9, 319–329.
DEO, M. G., and RAMALINGASWAMI, V. (1965). Gastroenterology, 49, 150–157.
DEO, M. G., SOOD, S. K., and RAMALINGASWAMI, V. (1965). Archs. Path., 80, 14–23.
DICKERSON, J. W. T. (1968). In Calorie Deficiencies and Protein Deficiencies, pp. 1329–1340, ed. McCance, R. A., and Widdowson, E. M. London: Churchill.
GABUZDA, T. G., JICK, H., and CHALMERS, T. G. (1963). Metabolism, 12, 1–10.
GAETANI, S., PAOLUCCI, A. M., SPADONI, M. A., and TOMASSI, G. (1964). J. Nutr., 84, 173–178.
GROSSMAN, J., YALOW, A. A., and WESTON, R. E. (1960). Metabolism, 2, 528–550.
HOOPER, C. E. S. (1961). Am. J. Anat., 108, 231–244.
HOWARD, A., and PELC, S. R. (1953). Heredity, Lond., suppl. 6, 261–273.
JAMES, W. P. T., and HAY, A. M. (1968). J. clin. Invest., 47, 1958–1972.
JHA, G. J., DEO, M. G., and RAMALINGASWAMI, V. (1968). Am. J. Path., 53, 1111–1125.
KENNEY, M. A., RODERUCK, C. E., ARNRICH, L., and PIEDAD, F. (1968). J. Nutr., 95, 173–178.

KIRSCH, R., FRITH, L., BLACK, E., and HOFFENBERG, R. (1968). *Nature, Lond.*, **217**, 578–579.

LAMERTON, L. F., and FRY, R. J. (ed.) (1963). *Cell Proliferation*. Oxford: Blackwell Scientific Publications.

LESHER, S., FRY, R. J. M., and KOHN, H. I. (1961). *Expl Cell Res.*, **24**, 334–343.

McCANCE, R. A., and WIDDOWSON, E. M. (1962). In *Protein Metabolism*, pp. 109–117, ed. Gross, F. Berlin: Springer.

MONTGOMERY, R. D. (1962). *Archs Dis. Childh.*, **37**, 1229–1235.

MUNRO, H. N. (1968). *Fedn Proc. Fedn Am. Socs exp. Biol.*, **27**, 1231–1237.

MUNRO, H. N., and GOLDBERG, D. M. (1964). In *The Role of the Gastro-intestinal Tract in Protein Metabolism*, pp. 189–198, ed. Munro, H. N. Oxford: Blackwell Scientific Publications.

PATT, H. M., and QUASTLER, H. (1963). *Physiol. Rev.*, **43**, 357–396.

PIMSTONE, B. L., WITTMANN, W., HANSEN, J. D. L., and MURRAY, P. (1966). *Lancet*, **2**, 779–780.

QUASTLER, H., and SHERMAN, F. G. (1959). *Expl Cell Res.*, **17**, 420–438.

RAMALINGASWAMI, V. (1964). *Nature, Lond.*, **201**, 546–551.

RAMALINGASWAMI, V., and DEO, M. G. (1968). In *Calorie Deficiencies and Protein Deficiencies*, pp. 865–873, ed. McCance, R. A., and Widdowson, E. M. London: Churchill.

RAMALINGASWAMI, V., DEO, M. G., and SOOD, S. K. (1961). In *Meeting the protein needs of infants and pre-school children*, pp. 365–375. Publication No. 843, National Academy of Sciences, National Research Council, Washington D.C.

RAMALINGASWAMI, V., VICKERY, A. L., JR., STANBURY, J. B., and HEGSTED, D. M. (1965). *Endocrinology*, **77**, 87–95.

ROTHSCHILD, M. A., SCHREIBER, S. S., ORATZ, M., and McGEE, H. L. (1958). *J. clin. Invest.*, **37**, 1229–1235.

SCHIMKE, R. T. (1962). *J. biol. Chem.*, **237**, 459–468, 1921–1924.

SOOD, S. K., DEO, M. G., and RAMALINGASWAMI, V. (1965). *Blood*, **26**, 421–431.

STEPHEN, J. M. L., and WATERLOW, J. C. (1968). *Lancet*, **1**, 118–119.

STOHLMAN, F., JR. (1959). *The Kinetics of Cellular Proliferation*. New York: Grune and Stratton.

THOMSON, A. M. (1968). In *Calorie Deficiencies and Protein Deficiencies*, pp. 289–299, ed. McCance, R. A., and Widdowson, E. M. London: Churchill.

TROWELL, H. C., DAVIES, J. N. P., and DEAN, R. F. A. (1954). *Kwashiorkor*. London: Arnold.

VERMA, K., DEO, M. G., and RAMALINGASWAMI, V. (1969). In preparation.

WATERLOW, J. C. (1968a). *Lancet*, **2**, 1091–1097.

WATERLOW, J. C. (1968b). In *Calorie Deficiencies and Protein Deficiencies*, pp. 61–73, ed. McCance, R. A., and Widdowson, E. M. London: Churchill.

WHITEHEAD, R. G., and DEAN, R. F. A. (1964). *Am. J. clin. Nutr.*, **14**, 1313–1319.

WINICK, M., and NOBLE, A. (1965). *Devl Biol.*, **12**, 451–466.

WINICK, M., and NOBLE, A. (1966). *J. Nutr.*, **89**, 300–306.

DISCUSSION

Harris: Professor Ramalingaswami, is it possible to reverse all these changes that you observe in protein-deficient animals by amino acids only, or do you have to give protein?

Ramalingaswami: In our experimental model we have not given a mixture of amino acids; if we give whole protein in the form of milk protein we can reverse all these changes. In human kwashiorkor, it has been shown by Brock and his colleagues in South Africa that a mixture of essential amino acids is able to initiate the process of recovery (Hansen, Howe and Brock, 1956).

Harris: When I was playing with the analysis of cell cycles about a decade ago (Harris, 1959) I found that if I interfered with the synthesis of proteins by giving certain amino acid analogues during the G_1 phase, then the cell would not enter the S phase. If I interfered in exactly the same way during the S phase, synthesis of DNA went on. This might explain some of the DNA labelling patterns that you see in your protein-deficient animals.

Ramalingaswami: I think this is possible.

Fortier: I wonder if the time at which hormonal activity is assessed in these cases of protein deficiency may not be important. In the early stages of protein deficiency, ACTH, glucocorticoids and growth hormone secretion might increase as a result of non-specific stress, whereas at later stages overall pituitary deficiency might account for some of the effects which you have described. It may be recalled, in this connexion, that ovarian atrophy is part of the "pseudohypophysectomy" syndrome which results from chronic under-nutrition in the rat and that observations in this species have led to the conclusion that the starved rats reacted as if they had been hypophysectomized, in that ovaries, uteri and vaginal epithelium underwent marked atrophy and could be restored to normal by exogenously administered gonadotropins (Mulinos and Pomerantz, 1941; Maddock and Heller, 1947).

Ramalingaswami: The changes in the cell generation cycle in protein deficiency that I described have some resemblance to the effect of hypophysectomy on cell renewal systems; the DNA label stays for a long time in the cell as if the cell is not able to proceed further with the process of division. I cannot interpret an elevated level or a depressed level of a hormone in terms of what is happening in the organs. The studies that have been made so far seem to indicate that the tissues of the kwashiorkor child are being perfused with blood that contains high levels of growth hormone (Pimstone *et al.*, 1966); but whether this is effective in eliciting a response in target organs is a different story. If you feed the kwashiorkor child with protein, the level of growth hormone goes down; this may be due to its rapid utilization, but in this field we have no definite information to connect the cellular events with hormonal events. In partially hepatec-tomized rats, if you give growth hormone you can influence the rate of regeneration to some extent but the presence of the hormone is not critical for regeneration (Bucher, 1967).

White: We demonstrated some years ago (White and Dougherty, 1947) that one of the earliest hormonal responses in mice to either restricted caloric intake or total starvation is the activation, as Professor Fortier has indicated, of the pituitary–adrenocortical mechanism. The data established that with reduced caloric intake lymphoid tissue was more sensitive to degenerative changes (protein loss) than the liver and that the mobilization of amino acid nitrogen was initially not from the liver cell, as had been thought, but from lymphoid tissue. Whereas there is rapid involution of lymphoid tissues in a short period of fasting, this does not occur if the experimental animal is adrenalectomized before fasting (Adams and White, 1950). Subsequently Hoberman (1950) showed that the mechanism of this action of adrenal steroids is apparently at the level of the pool of free amino acids in non-hepatic tissues and the corticoids inhibit amino acid utilization for protein synthesis. In view of what we now know about the primary role of lymphoid cells in a variety of immune phenomena, one wonders whether, very early in protein deprivation, before serious and obvious clinical symptoms are apparent, one might be able to detect an impairment of an immune response, either of the direct or of the secondary, delayed hypersensitivity type, which could alert one to the possibility of impending caloric or protein deficiency.

Sarma: Professor Ramalingaswami, could the reduced antibody response in rats be a secondary effect of some other change? Have you looked at the excretion of vitamins, particularly B vitamins, in these protein-deficient rats? Vitamin B6 and others which are implicated in the antibody response may be excreted because of the lack of apoenzyme and this could then be a secondary effect of protein deficiency.

Ramalingaswami: There are, and there must be, a number of secondary consequences of protein deficiency that affect the metabolism of vitamins, minerals and other nutrients. Dr S. K. Sood in our department has shown for example that the metabolism of iron is altered in protein deficiency (Sood, Deo and Ramalingaswami, 1965). The carrier protein transferrin is markedly diminished in the blood in protein deficiency. Serum iron falls and anaemia develops. In later stages, a vicious cycle develops where analysis of exact events becomes difficult. It would certainly be worth seeing if there is a larger excretion of B vitamins in protein deficiency.

Gopalan: Professor Ramalingaswami has presented interesting observations on the changes in rate of cell renewal in different organs. During the past two decades we have had excellent descriptions of the clinical and morphological changes occurring in protein-calorie malnutrition. All these changes are obviously the end-results of certain regulatory mechanisms which are brought into play in response to the stress of malnutrition. We still do not have much light on these regulatory mechanisms themselves. The amazing thing in malnourished populations is that so few clinical manifestations of nutritional deficiency are seen though the diets are in fact deficient by accepted standards. Advanced types of nutritional deficiency diseases constitute only 1 to 2 per cent of the population even in the worst nourished groups. This raises the whole question of regulatory mechanisms involved in the adaptation of the human organism to nutritional deprivation.

Waterlow (1959) has some observations on this point. His work has produced some evidence of adaptive changes in protein-calorie malnutrition. In the face of dietary protein deficiency, there is an attempt to conserve protein losses and preferentially channel amino acids to the more important organs at the expense of less important ones. There is a pronounced drop in the urinary nitrogen excretion. There is apparently an increased rate of turnover in some tissues and a decreased rate in others. There is evidence of increased incorporation of labelled amino acids in liver and decreased concentration in muscle. On the hormonal side, there is an increase in the circulating growth hormone accompanied by an increase in non-esterified fatty acids (Pimstone et al., 1968). The changes in insulin concentration are apparently not consistent.

The meaning of these changes is still not clear though they are all obviously concerned in regulatory mechanisms which are involved in under-nutrition. It will be interesting if the cell renewal defects in the intestinal villi can be explained on the basis of any of these changes.

Livingston: Professor Ramalingaswami emphasized the effects of protein deficiency on rapidly reproducing tissues like gut lining and liver. It has also been suggested that important organs may be relatively protected. I wonder about the nervous system, an obviously vital organ. Neurons do not replicate, yet they have a very high rate of protein production, several times higher than that of other cells. Neuroscientists are keen to know what this protein may be doing, and specifically, whether protein production may play a role in learning and memory. Professor Ramalingaswami, have you information, and if not, could you and your colleagues be encouraged to investigate the influence of protein deprivation on learning capabilities in your animal studies?

Human beings in deprivation are difficult to study because we seek to restore their intake as soon as possible and because they are so apathetic. One could perhaps establish how rapidly recovery from apathy takes place and whether there are any detectable recoverable or residual learning and memory deficits when deprived people are resupplied with protein. If supplies are limited, for whom does one provide the available rations? Concentrating on survival of children may be heartbreaking because children are likely to be the most drastically affected. Protein deprivation during infancy, when the brain is growing at a high rate, can be devastating to mental development. Dr Joaquin Cravioto, a Mexico City paediatrician, found that protein deprivation in infancy leads to underdeveloped intelligence (Cravioto, 1968). Dr Myron Winick, working with *post-mortem* brains of infants who died of malnutrition in Chile, found subnormal numbers of brain cells for given ages (Winick, 1968). Dr John Bosley and Dr Davis H. Calloway, in tests on normal young adult male volunteers with carefully controlled diets containing practically no protein (less than 500 mg of nitrogen), detected mood changes and a reduction in verbal learning after only about two weeks of such deprivation (personal communication). Naturally, they terminated that experiment immediately, for fear of causing brain damage. These data, from human studies, underline the importance of controlled animal experiments to determine the role of protein in brain functions. Some experiments on rats and pigs point in the same direction but are not so valuable as would be experiments that Professor Ramalingaswami might do on monkeys.

Ramalingaswami: With monkeys recovery is prompt once you give protein, and the protein-synthesizing machinery seems to be intact. The whole aspect of the animal changes and within days one can demonstrate changes in organs in the direction of normality. We have not however studied the central nervous system and behaviour of our monkeys in any detail. We have no parameters, no standards of normality of primate behaviour, available to us. I agree that this is the most crucial aspect of protein malnutrition.

Anand: The question of whether dietary deficiencies, especially of proteins, affect the nervous system and learning and memory came up at the IBRO Symposium held by UNESCO in 1968, and evidence was provided that brain function is affected (Canosa, 1968). In experimental animals malnutrition during the critical phases of development affects the weight and composition of the central nervous system, and causes alterations in behaviour and memory. Studies among malnourished children have also demonstrated poorer mental performance of such children, compared with their siblings without any history of malnutrition.

Livingston: Is there evidence that the brain is protected in terms of maintaining protein production at the expense of other tissues, during acute or chronic deprivation of protein?

Ramalingaswami: No critical work has been done on the brain, but a so-called vital organ which we used to think is protected against protein deprivation—heart muscle—is in fact not so protected (Chauhan, Nayak and Ramalingaswami, 1965).

Gopalan: With current methods of clinical management, nearly 90 per cent of cases of advanced protein-calorie malnutrition can be treated successfully. By successful treatment I mean that immediate fatality can be avoided. The mortality in kwashiorkor, given ideal conditions of clinical management, is around 8 per cent. Within about four weeks, dramatic manifestations like oedema and skin changes disappear; but the return of serum albumin levels to normal levels takes considerable time. Body weights remain at sub-standard levels for several months and, in fact, barely catch up with normal levels.

We have recently examined the possible deleterious effects of protein-calorie malnutrition in childhood on subsequent mental growth and development (Champakam, Srikantia and Gopalan, 1968). Intelligence tests and tests for inter-sensory organization were given to children who had been successfully treated for kwashiorkor 7 to 8 years earlier and their performance was compared with a group of normal matched children.

These studies showed that there was a clear-cut difference between the control subjects and children who had recovered from kwashiorkor. Retardation was particularly pronounced in perceptual and abstract abilities. We, however, like to interpret these results with great caution because factors other than malnutrition could have contributed to this difference. Further, the children who had recovered from kwashiorkor had been hospitalized for long periods and this might have resulted in considerable loss of "learning time". The possible effects of emotional factors like anxiety and fear incidental to hospitalization must also be considered.

Secondly, it is well-known that not all children who subsist on inadequate diets develop kwashiorkor; an important factor may be the level of intelligence, motivation and resourcefulness of the mothers. It is possible that children dependent on mothers who are particularly lacking in these qualities are the ones that develop kwashiorkor. If, indeed, this were so, these very same parental factors which would have contributed to the development of kwashiorkor, could have also contributed to the poor mental performance.

REFERENCES

ADAMS, E., and WHITE, A. (1950). *Proc. Soc. exp. Biol. Med.*, **75**, 590.

BUCHER, N. L. R. (1967). *New. Engl. J. Med.*, **277**, 686.

CANOSA, C. A. (1968). *IBRO Bulletin*, **7**, 69–74.

CHAMPAKAM, S., SRIKANTIA, S. G., and GOPALAN, C. (1968). *Am. J. clin. Nutr.*, **21**, 844.

CHAUHAN, S., NAYAK, N. C., and RAMALINGASWAMI, V. (1965). *J. Path. Bact.*, **90**, 301.

CRAVIOTO, J. (1968). In *Malnutrition, Learning and Behavior*, ed. Scrimshaw, N. S., and Gordon, J. E. Cambridge, Mass.: Massachusetts Institute of Technology Press.

HANSEN, J. D. L., HOWE, E. E., and BROCK, J. F. (1956). *Lancet*, **2**, 911.

HARRIS, H. (1959). *Biochem. J.*, **72**, 54.

HOBERMAN, H. D. (1950). *Yale J. Biol. Med.*, **22**, 341.

MADDOCK, W. O., and HELLER, C. G. (1947). *Proc. Soc. exp. Biol. Med.*, **66**, 595–597.

MULINOS, M. J., and POMERANTZ, L. (1941). *Endocrinology*, **29**, 558–567.

PIMSTONE, B. L., BARBEZAT, G., HANSEN, J. D. L., and MURRAY, P. (1968). *Am. J. clin. Nutr.*, **21**, 482.

PIMSTONE, B. L., WITTMANN, W., HANSEN, J. D. L., and MURRAY, P. (1966). *Lancet*, **2**, 779.

SOOD, S. K., DEO, M. G., and RAMALINGASWAMI, V. (1965). *Blood*, **26**, 421–432.

WATERLOW, J. C. (1959). *Nature, Lond.*, **184**, 1875.

WHITE, A., and DOUGHERTY, T. F. (1947). *Endocrinology*, **41**, 230.

WINICK, M. (1968). *Nutr. Rev.*, **26**, 195–197.

SUBCELLULAR FRACTIONATION TECHNIQUES IN THE STUDY OF CHEMICAL TRANSMISSION IN THE CENTRAL NERVOUS SYSTEM

V. P. Whittaker

Department of Biochemistry, University of Cambridge, and Department of Neurochemistry, Institute for Basic Research in Mental Retardation, Staten Island, New York

Of all biological control mechanisms, the nervous system must be regarded as the most highly evolved and the most rapid in its action. Even in non-vertebrate forms, such as the cephalopods, it provides, in conjunction with the special senses, especially the large, complex eye, a device whereby the animal can acquire extremely complex responses to its environment, including tactile and visual recognition of forms and rapid movements for purposes of predation or escape.

We have little idea of how the central nervous system functions *in toto*. It is often regarded as a "biological computer"; however, man-made digital computers provide an extremely poor model. It is true that certain concepts derived from the automated control systems can be legitimately applied to aspects of nervous system function; thus muscle spindle afferents and their associated γ-efferents form a fine control system for the regulation of muscle movement that has direct analogies in control engineering. At a higher level, the "wiring diagram" of the cerebellum has recently been worked out in considerable detail, and computer simulation of the complex networks involved may well assist in solving, in detail, the specific role that this part of the brain plays in the control of posture and muscular movement (Fig. 1). Other extremely important central functions such as the integration of "drives", affective states, consciousness, dreaming, memory and learning, respond less readily to this approach, and it seems likely that radically new theories will be necessary before much further progress is made. For example, it has recently been suggested (Longuet-Higgins, 1968; Chopping, 1968) that the optical holograph may provide a model for the non-localized character of the memory trace; just as destruction of part of a holograph plate does not destroy a specific part of the image (but merely reduces image definition as a whole)

338

so cortical ablation does not abolish specific memory traces, which clearly have multiple representation in some form throughout the cortex.

If central nervous system function as a whole still largely remains an enigma, much more is known about the functioning of the constituent units, the nerve cells or neurons, and about the way in which they communicate with each other and with their effector cells. Within the

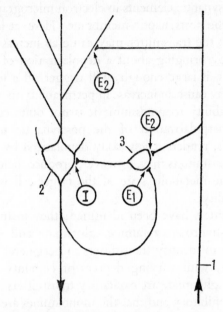

FIG. 1. A simplified version of a type of input–output network embodying a "negative feedback" loop that is found in many areas of the central nervous system. Incoming fibres (1) synapse with the dendrites of a large cell (2) which provides an output through its axons. An interneuron (3) activated both by the input (1) and recurrent collaterals from cell (2) exerts an inhibiting effect on cell (2). Excitatory transmitters E_1 and E_2 and an inhibitory transmitter I are involved at the various synapses as indicated. The arrangement stabilizes the output of cell (2).

neuron, information is transmitted from the dendrites and neuron somata (cell bodies) along the axons to the axon terminations by means of a propagated action potential generated by a transient change in the sodium permeability of the neuronal plasma membrane and sustained by differences in the sodium concentration between the inside and the outside of the cell. Such action potentials are superimposed on and are opposite in sign to the resting membrane potential resulting from the difference in potassium concentration across the cell membrane.

At the functional contacts (synapses) between neurons or between

neuron and effector cell the mode of transmission of information changes
—with a few interesting but unimportant exceptions discussed later—
from being electrochemical to being chemical in nature (reviewed by
Whittaker, 1968). Specific chemical transmitter substances are released,
by the arrival of the action potential, from a presynaptic store or stores;
they diffuse across the synaptic cleft—the empty-looking space between
the pre- and postsynaptic elements in electron micrographs of synapses—
to interact with the postsynaptic membrane. Here, at least two types of
effect may occur: the transmitter may cause an increase in permeability
to cations, thereby bringing about a depolarization of the postsynaptic
membrane which, if large enough, will trigger off a propagated action
potential; or it may cause an increase in permeability to chloride ions (or a
decreased permeability to potassium; it is not quite certain which) resulting in a hyperpolarization of the postsynaptic membrane which
inhibits it—that is, renders it less easily depolarized by excitatory transmitters. Some transmitters may, however, produce their effects indirectly
by influencing the metabolic state of the target cell which secondarily
affects its excitability.

Several transmitters have been identified; they include acetylcholine,
noradrenaline, 5-hydroxytryptamine, glutamate and γ-aminobutyrate.
It has been easier to identify transmitters in peripheral locations than in
central; however, with varying degrees of certainty we can say that
acetylcholine and glutamate are excitatory transmitters, γ-aminobutyrate
and glycine are inhibitory and that the monoamines are probably excitatory to some cells and inhibitory to others. It is doubtful, however, if
this list of putative transmitters accounts for as many as half the synapses
in the mammalian central nervous system, and the identity of the remainder is one of the most urgent problems in neurobiology.

An unexpected feature of transmitter release is its "quantized" character.
This is especially clear in the case of the neuromuscular junction (Katz,
1966). In the resting junction random postsynaptic potentials occur which
are similar to, but much smaller than, the full-scale postsynaptic potentials
(psp's) which signal transmission. The known electrogenic effect of
acetylcholine shows that these miniature psp's cannot be accounted for
by the diffusion of acetylcholine from the presynaptic nerve terminal
molecule by molecule, but require that the transmitter is being released
randomly in packets or quanta of between 1000 and 10 000 molecules at
a time. Such "synaptic noise" has been observed in central synapses also,
and for other transmitters, but it is uncertain if transmitter release is
always quantized in this way.

The morphology (Fig. 2) of the presynaptic nerve terminal is consistent with its role in chemical transmission: the terminal cytoplasm contains numerous synaptic vesicles 500 Å in diameter which are thought to be the ultimate storage sites of transmitter and the basis for its quantized release; the synaptic cleft appears to provide a low resistance path for action currents, thus suppressing electrotonic stimulation of the postsynaptic membrane by presynaptic action potentials.

Chemical transmission is the Achilles heel of the neuron; it is a function which is readily interfered with by chemical analogues of transmitters that are able to gain access to the central nervous system. There is reason

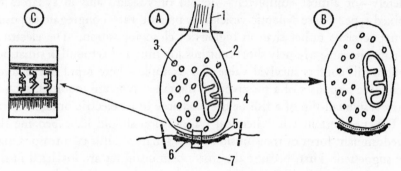

FIG. 2. Diagram illustrating (A) main morphological features of the synapse (1, preterminal axon; 2, sac-like enlargement of preterminal neuronal plasma membrane; 3, synaptic vesicles; 4, intraterminal mitochondrion; 5, synaptic cleft; 6, postsynaptic membrane with postsynaptic thickening; 7, portion of postsynaptic cell); (B) synaptosome formed by cleavage along thick lines in (A) with morphological features 2–6 retained; (C) enlargement of portion of synaptic cleft in (A) enclosed by the rectangle, showing possible interaction of "coded" carbohydrate side-chains of glycoproteins present in the pre- and post-synaptic membranes.

to believe that many of the drugs that have potent central effects, such as the hallucinogens and tranquillizers, owe this property to their ability to interfere at one point or another with the synthesis, storage, release, postsynaptic action and subsequent inactivation of transmitters involved in pathways responsible for affective states and consciousness. Thus the powerful hallucinogen lysergic acid diethylamide (LSD) is a structural analogue both of noradrenaline and of 5-hydroxytryptamine; mescaline is an analogue of noradrenaline, the tranquillizer reserpine is an indole derivative and atropine, a hallucinogen in high doses, blocks the central action of acetylcholine. All these transmitters are involved in ascending pathways in the limbic or reticular activating systems, believed to be concerned with affective states and consciousness.

As exceptions that prove the rule, the discovery, in certain locations, of electrical synapses and of synapses of mixed chemical and electrical type has helped to strengthen the view that synapses conforming to the morphological description already given utilize chemical transmission, even when the identity of the transmitter is unknown (Furukawa and Furshpan, 1963; Robertson, Bodenheimer and Stage, 1963). Electrical synapses in which transmission includes the electrotonic invasion of the postsynaptic membrane by the presynaptic action potential do not have a synaptic cleft; there is apposition of the pre- and postsynaptic membranes as in "tight junctions" between epithelial cells. The cytoplasm is completely—or almost completely—devoid of vesicles, and in synapses of mixed type, where synaptic vesicles are present, they congregate opposite synaptic clefts rather than in regions of close apposition. The electrical synapse has an extremely short transmission time and is found in situations where this confers survival value: for example where rapid offensive or defensive responses of a massive and ungraded type are required, such as the evasive tail-flip of a fish or the discharge of an electric organ.

It is not certain why chemical transmission should have become the predominant form of transmission across synapses. But two reasons may be suggested. First, because the stores of transmitter are localized in the presynaptic terminal, chemical transmission is a guarantee of the unidirectionality of synaptic transmission, an essential prerequisite for the orderly processing of information in the central nervous system. Action potentials may travel in either direction along the axon, but are presumably unable to traverse a synaptic gap antidromically because of the absence of postsynaptic stores of transmitter. Secondly, chemical transmission may provide a kind of chemical coding of neurons, of advantage in suppressing unwanted "cross-talk" between intertwined but functionally distinct pathways.

Although the function of the synapse may be defined as the transmission of information from one excitable unit to another, the "information" transmitted almost certainly consists of more than the rapid depolarizations and hyperpolarizations already mentioned. The coding of information in the central nervous system is unexpectedly complex (Bullock, 1968): trophic influences are exerted by neurons upon their neighbours; memory and learning may well involve changes in synaptic contacts; and in the developing brain extremely large numbers of specific connexions must be made and coded for. The presynaptic nerve terminal is also the receiving station for the protein synthesized by the cell body and passed down the axon by the process of axonal flow (reviewed by Schmitt,

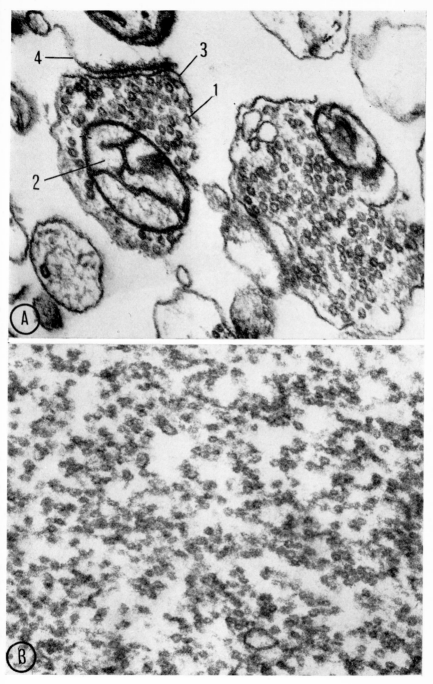

FIG. 3. Electron micrographs of (A) synaptosomes showing 1, synaptic vesicle; 2, intraterminal mitochondrion; 3, external presynaptic membrane; 4, postsynaptic membrane (note cleft material); (B) isolated synaptic vesicles (Fraction D, Figs. 5 and 6). Magnification factor, 70 000; permanganate-fixed.

To face page 342

1968). This flow of materials may stand in a similar relation to the slow time-course events at the synapse as the propagated action potential does to fast synaptic events. Molecular models for these aspects of synaptic function can at present be elaborated only in the sketchiest way. As far as the formation of specific contacts is concerned, it is possible that the lightly staining material seen in the synaptic cleft in permanganate-fixed electron micrograph sections represents informational macromolecules, possibly glycoproteins, anchored to the pre- and postsynaptic membranes whose carbohydrate chains are coded for specific interactions (Fig. 2).

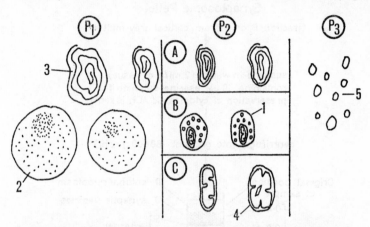

Fig. 4. Diagram illustrating the separation of synaptosomes (1) from the other main constituents of sucrose homogenates of brain tissue, namely nuclei (2), myelin (3), mitochondria (4) and microsomes (5). Differential centrifuging at approximately 10^4, $2-10 \times 10^5$ and 6×10^6 g. min gives three particulate fractions P_1, P_2 and P_3 consisting largely of nuclei and large myelin fragments (P_1), small myelin fragments, synaptosomes and free mitochondria (P_2) and microsomes (P_3). Density gradient separation of fraction P_2 resolves it into three subfractions consisting largely of myelin (A), synaptosomes (B) and free mitochondria (C).

In recent years, the classical electrophysiological and morphological techniques for studying chemical transmission have been supplemented by newer biochemical methods which permit the isolation of presynaptic nerve terminals and their contained organelles by subcellular fractionation techniques (reviewed by Whittaker, 1965 and Marchbanks and Whittaker, 1968).

An unexpected property of the presynaptic terminal is its superior mechanical strength compared with that of surrounding cell processes and the preterminal regions of axons. It is thus relatively easy to find homogenization conditions that cause extensive disruption of glial and neuronal cell bodies, dendrites and axons without breakdown of the presynaptic terminal region. This is simply detached from its axon and postsynaptic attachment and converted into a 'sealed bag with all the morphological features and most of the chemical properties of the original terminal preserved intact (Figs. 2 and 3A). The detached presynaptic nerve terminals have been called *synaptosomes*, a name that emphasizes their cellular origin and particulate character. Synaptosomes may be isolated from other constituents of brain homogenates (cell

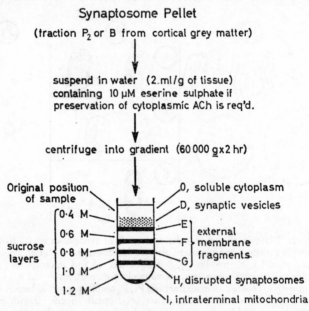

Synaptosome Pellet

(fraction P_2 or B from cortical grey matter)

suspend in water (2.ml/g of tissue) containing 10 µM eserine sulphate if preservation of cytoplasmic ACh is req'd.

centrifuge into gradient (60 000 g x 2 hr)

Original position of sample

0, soluble cytoplasm
D, synaptic vesicles
E }
F } external membrane fragments.
G }
H, disrupted synaptosomes
I, intraterminal mitochondria

sucrose layers
0·4 M
0·6 M
0·8 M
1·0 M
1·2 M

FIG. 5. Separation of hypo-osmotically disrupted synaptosomes into their constituent organelles.

debris, nuclei, mitochondria, microsomes, myelin and membrane fragments) by a combination of differential and density gradient centrifuging (Fig. 4).

Synaptosomes are osmotically sensitive (an indication that they truly are sealed structures) and if suspended in water they burst, discharging a proportion of their contents. If the disrupted synaptosomes are subjected to density gradient separation, a series of fractions is obtained that consists (reading from top to bottom of the gradient) of soluble cytoplasm diluted with water, mono-dispersed synaptic vesicles (Fig. 3B), fragments of external membranes of increasing sizes, incompletely disrupted synaptosomes and intraterminal mitochondria (Fig. 5). Analysis of these fractions permits conclusions to be drawn about the intraterminal location of transmitters and enzymes metabolizing them and also regarding the general enzymic and chemical makeup of the preterminal region (Whittaker, Michaelson and Kirkland, 1964) (Fig. 6).

The transmitter that has been most completely studied so far is acetylcholine. Unfortunately, only a proportion of synaptosomes in any one

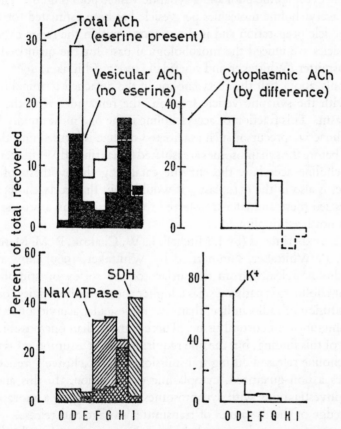

FIG. 6. Composition of fractions prepared as shown in Fig. 5. Note the recovery of the soluble cytoplasmic marker K+ (bottom right diagram) mainly in fraction O, of the external membrane marker NaK-activated ATPase (bottom left) in fractions E, F and G and of the mitochondrial marker succinate dehydrogenase (SDH) in fraction I, vesicular acetylcholine (ACh) (top left) mainly in fraction D and cytoplasmic ACh (top right) mainly in fraction O. Fraction H, consisting of incompletely disrupted synaptosomes, contains all the markers except the very diffusible potassium and cytoplasmic ACh. The distribution of markers is thus consistent with the morphological identifications of the fractions listed in Fig. 5.

preparation are derived from cholinergic neurons (neurons utilizing acetylcholine as a transmitter) since non-cholinergic neurons outnumber cholinergic in all parts of the central nervous system. Thus quantitative estimates of, for example, acetylcholine concentration have to be corrected for the presumed proportion of cholinergic neurons in the total population and there is no way of estimating this very accurately.

Acetylcholine is found to exist within the synaptosome in two pools, a soluble cytoplasmic pool and a synaptic vesicle pool (Fig. 6). The number of acetylcholine molecules per vesicle has been estimated for the isolated vesicle preparation and is about 2000—within the range expected if the vesicles are indeed the morphological basis for the quantized release of transmitter (Whittaker and Sheridan, 1965; Whittaker, 1966). However, only about half the acetylcholine in the detached terminal is associated with the synaptic vesicle fraction: the remainder is in the soluble cytoplasm. This fraction of acetylcholine is the one most rapidly labelled if a radioactive precursor such as tritiated choline is infused into the brain *in vivo* before the synaptosomes are isolated (Chakrin and Whittaker, 1969). Since choline acetylase, the enzyme catalysing the synthesis of acetylcholine, is also in the cytoplasm, it would seem that acetylcholine is first synthesized there and later transferred to the vesicles by a process that so far has not been duplicated *in vitro*.

Recent experiments (by J. Mitchell, L. W. Chakrin, R. M. Marchbanks and V. P. Whittaker, summarized by Whittaker, 1969) show that the acetylcholine released from the surface of the cortex on stimulation of afferent cholinergic pathways has a higher specific radioactivity, one hour after infusion of radiocholine, than that of vesicular acetylcholine isolated from the subjacent cortical tissue. There is more than one possible explanation of this finding, but the one requiring fewest assumptions is that the acetylcholine released during transmission is not exclusively vesicular but includes a non-quantized, cytoplasmic contribution. In this and other ways, investigations with synaptosomes are leading to a more detailed knowledge of the dynamics of transmitter storage and release.

Synaptosomes can be regarded as miniature non-nucleated cells: they possess, as we have seen, an external membrane, a cytoplasm, with its full complement of glycolytic enzymes, and one or more small mitochondria. On incubation in a saline medium fortified with co-enzymes and a suitable energy substrate (e.g. glucose) they respire and synthesize high-energy phosphate compounds (ATP and creatine phosphate) (Marchbanks and Whittaker, 1967). Their external membranes possess carrier-mediated uptake and extrusion properties. Thus, metabolizing synaptosomes take up choline (Marchbanks, 1968), noradrenaline (Colburn *et al.*, 1968) and certain amino acids (R. Graham-Smith, personal communication) by sodium-dependent, ouabain-sensitive mechanisms, and extrude sodium ions (Ling and Abdel-Latif, 1968). The choline uptake is also hemicholinium-sensitive. Thus synaptosomes provide an alternative to whole cell preparations such as tissue slices, erythrocytes or ascites tumour cells in which to study membrane properties, with the advantages of greater simplicity and direct relevance of the results to nerve-cell function.

Synaptosomes are metabolically competent in other ways too: thus they incorporate radioleucine into protein (Morgan and Austin, 1968). This process is dependent on the presence of an energy source such as glucose and oxygen and an intact external membrane carrier system for the uptake of amino acids (L. Austin, personal communication). Synaptosomes also provide samples of terminal axoplasm in studies of axonal flow. *In vivo* most of the radioactive synaptosomal protein resulting from radioleucine administration appears to arise from axonal flow rather than from local synthesis; however, the carbohydrate moiety of glycoproteins is synthesized locally (Barondes, 1966, 1968).

Synaptosomes are often seen to have portions of postsynaptic membrane adhering to them (Fig. 3A). They can thus be used as sources—albeit very impure ones—of postsynaptic membrane and thus of transmitter receptors (De Robertis, Fiszer and Soto, 1967) and also of synaptic cleft material which, as we have seen, may well specify synaptic contacts. In this connexion, it is interesting that several enzymes concerned with the synthesis of the carbohydrate side-chains in glycoproteins and glycolipids are localized in the external synaptosome membrane in the developing chick (S. Roseman, personal communication).

Under proper homogenization conditions, a high proportion of endings are converted into synaptosomes (Clementi, Whittaker and Sheridan, 1966). Advantage may eventually be taken of this fact to avoid the sampling difficulties which occur when attempts are made to assess quantitatively changes in the number of synapses in developmental or learning studies. An important advantage of the synaptosome technique is that it enables us to compare the chemical composition and the metabolic properties of the presynaptic terminals of normal animals with those of animals which have been pretreated in various ways, for example by the administration of drugs or by psychological conditioning. "Split-brain" preparations yield control and pretreated synaptosomes from the same animal (M. Cuénod, personal communication). The synaptosome will clearly have a valuable part to play in extending our knowledge of synaptic function in its broadest sense.

SUMMARY

One of the most complex of all control processes in multicellular organisms is that effected by the central nervous system. We have very little understanding of the way in which the nervous system as a whole works or how information is coded or processed. However, it is clear that one of its key components is the synapse, the specialized structure through which nerve cells make functional contact with their neighbours.

Transmission of information at synapses is believed to be brought about, with a few interesting but relatively unimportant exceptions, by the release of specific chemical transmitter substances from the presynaptic nerve terminal. On the arrival of a nerve impulse the transmitter is released into the synaptic cleft, diffuses to the postsynaptic membrane and there causes either depolarization (excitation) or hyperpolarization (inhibition) of the postsynaptic cell membrane.

In recent years the development of several new techniques has greatly increased our knowledge of synaptic transmission. These include histochemical techniques by which the transmitter or enzymes involved in its metabolism are rendered visible, micro-electrode techniques by which the effect of putative transmitters on single neurons can be compared with synaptic activation, and biochemical methods by which the metabolism, storage and release of the transmitter and the effect of drugs and toxins on these processes can be elucidated. A technique described in detail is one developed in the author's laboratory by which, by homogenizing brain tissue in sucrose, the club-like presynaptic nerve terminals can be detached from their axons and isolated as sealed structures by a combination of differential and density gradient centrifuging. These detached terminals or *synaptosomes* retain the morphology and, as far as is known, the chemical composition of the presynaptic nerve terminal and provide a new type of preparation for the investigation of synaptic function *in vitro*.

REFERENCES

BARONDES, S. H. (1966). *J. Neurochem.*, **13**, 721–727.
BARONDES, S. H. (1968). *J. Neurochem.*, **15**, 699–706.
BULLOCK, T. H. (1968). *Proc. natn. Acad. Sci. U.S.A.*, **60**, 4–14.
CHAKRIN, L. W., and WHITTAKER, V. P. (1969). *Biochem. J.*, **113**, 97–107.
CHOPPING, P. T. (1968). *Nature, Lond.*, **217**, 781–782.

CLEMENTI, F., WHITTAKER, V. P., and SHERIDAN, M. N. (1966). *Z. Zellforsch. mikrosk. Anat.*, **72**, 126–138.
COLBURN, R. W., GOODWIN, F. K., MURPHY, D. L., BUNNEY, W. E., and DAVIS, J. M. (1968). *Biochem. Pharmac.*, **17**, 957–964.
DE ROBERTIS, E., FISZER, S., and SOTO, E. F. (1967). *Science*, **158**, 928–929.
FURUKAWA, T., and FURSHPAN, T. S. (1963). *J. Neurophysiol.*, **26**, 140–176.
KATZ, B. (1966). *Nerve, Muscle, and Synapse.* New York: McGraw-Hill.
LING, C. M., and ABDEL-LATIF, A. A. (1968). *J. Neurochem.*, **15**, 721–729.
LONGUET-HIGGINS, H. C. (1968). *Nature, Lond.*, **217**, 104.
MARCHBANKS, R. M. (1968). *Biochem. J.*, **110**, 533–541.
MARCHBANKS, R. M., and WHITTAKER, V. P. (1967). *Abstr. 1st int. Congr. int. Soc. Neurochem.*, Strasbourg, p. 147.
MARCHBANKS, R. M., and WHITTAKER, V. P. (1968). *Biol. Basis Med.*, **5**, 39–75.
MORGAN, I. G., and AUSTIN, L. (1968). *J. Neurochem.*, **15**, 41–51.
ROBERTSON, J. D., BODENHEIMER, T. S., and STAGE, D. E. (1963). *J. Cell Biol.*, **19**, 159–196.
SCHMITT, F. O. (1968). *Proc. natn. Acad. Sci. U.S.A.*, **60**, 38–47.
WHITTAKER, V. P. (1965). *Prog. Biophys. molec. Biol.*, **15**, 39–96.
WHITTAKER, V. P. (1966). In *Mechanisms of Release of Biogenic Amines*, pp. 147–162, ed. Euler, U. S. von, Rosell, S., and Uvnäs, B. Oxford: Pergamon Press.
WHITTAKER, V. P. (1968). *Proc. natn. Acad. Sci. U.S.A.*, **60**, 27–37.
WHITTAKER, V. P. (1969). *Prog. Brain Res.*, **31**, 211–222.
WHITTAKER, V. P., MICHAELSON, I. A., and KIRKLAND, R. J. A. (1964). *Biochem. J.*, **90**, 293–303.
WHITTAKER, V. P., and SHERIDAN, M. N. (1965). *J. Neurochem.*, **12**, 363–372.

DISCUSSION

Paintal: Your experiment on the cortex in which you collected the released acetylcholine is most fascinating. Have you any idea of the amount of activity in the cortex before you stimulate it? The cortex contains certain cells which are active at one time and when you stimulate certain other cells, these first cells become silent. So one would expect the activity in the cortex to remain about the same.

Whittaker: The release of acetylcholine from the surface of the cortex has been studied in some detail by Dr Mitchell and his co-workers and there is also much information about the distribution of cholinergic fibres in the cortex as detected by a histochemical method based on staining for cholinesterase, which is believed to be a marker for cholinergic fibres (Mitchell, 1963; Collier and Mitchell, 1966; Silver, 1967). It is generally considered that the cell bodies of most of the cholinergic fibres in the cortex are lower down in the various afferent pathways, and cholinergic neurons are represented in the cortex mainly by their nerve terminals. If the cells which become silent as a result of stimulation are not cholinergic it would not make any difference from our point of view. There is evidence that, for example, some of the pyramidal cells have a

12*

large number of cholinergic terminals on them, and it doesn't matter what happens to the other cells so long as the afferent cholinergic pathways have been activated and the acetylcholine has been released from them. Certainly Mitchell has found that the activation of cholinergic afferents to the cortex results in about a fivefold increase in acetylcholine output.

Paintal: I wasn't aware that the evidence for acetylcholine as the main transmitter in the central nervous system was so clear. Dr K. Krnjević (Krnjević, Randić and Straughan, 1966) believes that GABA is a very important transmitter in the central nervous system. One would expect that by stimulating the cortex you would get an equal amount of GABA and acetylcholine, because if they are both transmitters they should be produced together, in a heterogeneous population.

Whittaker: Attempts have been made to detect an increase in the GABA concentration in cups placed on the cortex when the inhibitory pathways of the cortex are activated, which can be done by stimulating the surface of the cortex at rather low frequencies (not the deep structures, as we were doing). It has not so far been possible to demonstrate an unequivocal increase in the amount of GABA released. However, we know that GABA is rapidly taken up by nerve terminals; this can be shown both in slices and in synaptosome preparations, which has been done recently, and it may be that acetylcholine is exceptional among transmitters in being hydrolysed and then the hydrolysis product reabsorbed. We know that both GABA and noradrenaline, for example, are readily taken up by nerve endings. The typical way of terminating transmitter action may be re-uptake by the presynaptic nerve terminal. In that case one would not expect to be able to demonstrate this increase of output unless one blocked the uptake process.

Edelman: Have you tried to fractionate the proteins in your synaptosome preparations and compare them with the soluble brain proteins by, say, acrylamide gel electrophoresis?

Whittaker: We haven't done work of this kind, but H. R. Mahler's group is working on the proteins of synaptosome membranes, determining their turnovers and also comparing them with the soluble cytoplasmic proteins (Cotman, Mahler and Hugli, 1968).

Edelman: I ask because your preparation provides an opportunity to compare some well-characterized fraction of the brain with the overall brain protein pattern. We know that that overall pattern differs from one species to another; soluble brain proteins can be fractionated now on acrylamide gel even though one may not be able to identify many of them by any functional test. This might provide the opportunity to locate some of the brain proteins.

Whittaker: A very interesting thing to do would be to compare the proteins which are labelled with radioactive leucine in the Austin and Morgan type of system where protein synthesis is going on in the isolated terminal, with what is labelled by axonal flow when one injects radio-leucine into the cortex and isolates the synaptosomes 96 hours later. There may be all sorts of differences here waiting to be found out.

Of course there is the problem that it is difficult to get absolutely pure subcellular fractions. This doesn't matter so much when one is studying a substance like acetylcholine, where one knows *a priori* that it is likely to be in nerve terminals anyway. But if one is interested in something as non-specific in its localization as protein, one has to be sure that one is dealing with absolutely homogeneous membrane preparations, and this problem has not been entirely solved. I think we have solved it for the synaptic vesicles but perhaps not completely for the external membranes. Also, the vesicles account for only 0·1 per cent by weight of brain, which is very little material to work with. However, I think that the problems of both resolution and yield may be solved by zonal centrifugation.

Monod: How do you interpret your results on the distribution of acetylcholine between the vesicles and cytoplasm? If the release is quantal, it is hard to see how this could be done effectively unless acetylcholine was released essentially as vesicles.

Whittaker: I still think that our results in general support the quantal hypothesis. The amount of acetylcholine per vesicle, for example, is within the range which electrophysiologists have given for the size of a quantum, which incidentally they can't do more accurately than by a power of ten. It is possible that the quantal phenomenon is superimposed on a tonic background release of acetylcholine, and there is some evidence of this. For example, Mitchell and Silver (1963) followed the resting release of acetylcholine from an innervated nerve–diaphragm preparation using a pharmacological assay and also estimating the number of quanta—that is, the number of miniature end-plate potentials (minips). They also varied the ionic composition of the medium in such a way as to vary the minip frequency. They found with the pharmacological estimates about 100–200 times more acetylcholine than could be accounted for by the minips. Furthermore, this endogenous pharmacological release was not affected by changes in ionic composition which greatly changed the minip frequency. This is rather good independent evidence that there must be two components of acetylcholine release.

Lynen: You found that the specific activity of cytoplasmic acetylcholine was higher than that of the vesicle. I wonder whether your vesicles are a mixed population, some of which rapidly turn over their acetylcholine and others are rather slow in the turnover process?

Whittaker: One could make the assumption that there is a population of rapidly turning over vesicles which is also more labile to hypo-osmotic shock, and that what we define operationally as the cytoplasmic fraction is this labile, metabolically active fraction of vesicles. We cannot exclude this but there is also a third fraction of acetylcholine which is present in the supernatant of the brain fraction, if the brain homogenate is made in the presence of an anticholinesterase. The specific activity of this fraction is much lower than for either of the other two. On your hypothesis that all the acetylcholine is vesicular it must either arise from the breakdown of synaptosomes, or represent vesicles present in the cytoplasm of the nerve cell bodies. And they must be very labile, because that acetylcholine is always recorded as free acetylcholine, even in the original iso-osmotic sucrose homogenate. So there is the paradox of one fraction of acetyl-choline, the synaptosomal cytoplasmic fraction, in which you have to postulate that the structures are metabolically very active and labile, and a second free pool, the supernatant free pool, in which you have to assume that the vesicles are metabolically sluggish and even more labile. One has to make rather a large number of assumptions if one wants to retain the idea of a single, vesicular pool.

Livingston: Is it not possible that some acetylcholine is on its return path, so that it will be captured in the supernatant and not be found either in the vesicles or even inside cell boundaries?

Whittaker: This is equivalent to saying that there are just two pools, a non-synaptosomal and a synaptosomal pool, whereas we find three, on the basis of the specific activity measurements.

Tata: What do you think is the site of protein synthesis in synaptosomes? Does the organelle secrete protein?

Whittaker: Morgan and Austin measured incorporation of radioactive leucine into the protein. They did not demonstrate net synthesis of protein. Part of that incorporation is attributable to the intraterminal mitochondria. The other part has all the characteristics of cytoplasmic incorporation. However, it is true that nobody has ever seen ribosomes or rough-surfaced endoplasmic reticulum in synaptosomes. When I challenged Dr Austin on that point he said, "Nobody saw ribosomes in mitochondria until protein synthesis had been demonstrated in them". It may be a question of looking harder.

Secretion of protein by synaptosomes has not been demonstrated, although in another secretion mechanism which has certain features in common with chemical transmission, namely the release of adrenaline from catecholamine granules in the adrenal medulla, there is a specific binding protein called chromogranin which is released. There is another binding protein in the hypophysis which is also released, neurophycin, and it is an intriguing possibility that there are similar substances in nerve endings.

Talwar: Dr Whittaker's method offers an extremely interesting experimental approach to work on a fraction which is functionally important. Is anything known about the heterogeneity of synaptosomes prepared from various cell types and from various regions of the brain?

Whittaker: We mostly work now with cortex which has been scraped to get rid of myelin. This is a great help in getting homogeneous fractions, because myelin interferes with the separations. Within the cortex there are variations in nerve endings, but not gross variations. However, in the cerebellum where there are giant mossy fibre endings, 2–5 μm in diameter, and very small endings in the molecular layer where there are parallel fibre endings, there is a much greater range of variation and we find that synaptosomes formed from the large terminals come down in the nuclear fraction. Again, the microsomal fraction is heavily contaminated with free synaptic vesicles which presumably derive from the breakdown of a certain proportion of these large endings which don't lend themselves readily to synaptosome formation. So certainly there is heterogeneity. Furthermore, within the cortex itself, only 10–30 per cent perhaps of the terminals are derived from cholinergic cells, so that we have heterogeneity with respect to transmitter type.

Talwar: What are the similarities and differences between the membranes that you can get out of the synaptosome preparations and nerve tissue membranes in general?

Whittaker: I can summarize this by saying that as far as our analyses have gone we can classify nerve membranes into three main types: one is typified by myelin, which has a molar cholesterol-to-phospholipid ratio of 1:2, and has considerable quantities of cerebroside. The second type of membrane is typified by synaptic vesicles and interestingly enough microsomes, which have a very similar composition. These have a relatively higher phospholipid and a lower cholesterol content. External synaptosome membranes form the third type, which is intermediate in composition. Like myelin it has a relatively high cholesterol content, but like microsomes and synaptic vesicles and mitochondria it has a very low, probably negligible, cerebroside content. This composition fits what is known of differences between internal and external membranes in other cells. We know that cholesterol is a characteristic external membrane component. Furthermore, ganglioside is present in the external synaptosome membrane but not in vesicles. There are also the differences in enzyme composition that I mentioned in my paper. And finally the mitochondria have a unique type of phospholipid, namely cardiolipin, which interestingly enough only has analogies in bacteria. It is completely absent from all other membranes in eukaryotic cells, as far as we know. It is present in brain mitochondria in quite a high concentration, about 14 per cent of the total lipid.

Salganik: Is the protein synthesis in synaptosomes sensitive to puromycin and to actinomycin D?

Whittaker: It is partially sensitive to cycloheximide.

Tata: Dr T. M. Andrews at Mill Hill has been examining protein synthesis in brain microsomes, which are of course contaminated quite heavily with synaptosome-derived material. Unlike liver rough microsomes, puromycin will discharge all the nascent protein elaborated by brain ribosomes bound to membranes.

Whittaker: Nerve cells of course have huge amounts of rough-surfaced endoplasmic reticulum and ribosomes. Although incapable of replication they make protein for export down the axon (axoplasmatic flow).

Paintal: You said that you measured the metabolism of the synaptosomes. What was the QO_2?

Whittaker: The uptake in the presence of glucose is about 600 μmole/ g/hr at 37° C, which is high, comparable to brain slices. We believe that much of the metabolism of brain slices, particularly tangential slices through the cortical neuropile which is what biochemists very often use, is attributable to their content of nerve endings. It is a region of intense metabolism.

Edelman: Have you looked at synaptosomes from cortex in which you have induced spreading cortical depression?

Whittaker: No.

Lynen: You mentioned the three types of membrane. Is there potassium-activated ATPase in the external membrane or the internal membrane fraction?

Whittaker: The sodium-potassium activated ATPase is a characteristic of the external membrane. But brain microsomes are also a source of this ATPase; the microsomes are undoubtedly a mixture of fragments of internal and external membranes and we don't know the proportion of the two. But the specific activity of the sodium-potassium activated ATPase in the relatively pure external membrane isolated from synaptosomes is 2–3 times higher than that of the classical brain microsome fraction.

Tata: Perhaps it would be useful to look at 5′-nucleotidase activity, which is possibly the most specific plasma membrane enzymic marker.

Whittaker: Yes; we shall certainly look into that.

REFERENCES

COLLIER, B., and MITCHELL, J. F. (1966). *J. Physiol., Lond.,* **184,** 239.

COTMAN, C. W., MAHLER, H. R., and HUGLI, T. E. (1968). *Archs Biochem. Biophys.,* **126,** 821–837.

KRNJEVIĆ, K., RANDIĆ, M., and STRAUGHAN, D. W. (1966). *J. Physiol., Lond.,* **184,** 78.

MITCHELL, J. F. (1963). *J. Physiol., Lond.,* **165,** 98.

MITCHELL, J. F., and SILVER, A. (1963). *J. Physiol., Lond.,* **165,** 117.

SILVER, A. (1967). *Int. Rev. Neurobiol.,* **10,** 57.

REGULATION OF VISCERAL ACTIVITIES BY THE CENTRAL NERVOUS SYSTEM

B. K. ANAND

Department of Physiology, All-India Institute of Medical Sciences, New Delhi

DURING recent years much experimental data has accumulated on the involvement of various central nervous structures in the regulation of visceral activities, which thereby adjusts the homeostasis of the body. This prompted Professor Fulton (1951) to remark that "among the most significant developments in modern neurology—if not in all scientific medicine—has been the gradual recognition of the existence of vast areas in the forebrain subserving autonomic functions."

HISTORICAL

Langley (1921), in his celebrated monograph, defined the autonomic nervous system as a group of myelinated and unmyelinated fibres which innervate blood vessels and glands. This system was envisaged by him to be a purely peripheral motor system and he did not think of its functional relationship with other parts of the brain and spinal cord. The beginning of our more intimate knowledge of the part played by the forebrain in the integration of autonomic function came with the studies of Karplus and Kreidl (1910) who observed striking autonomic effects after stimulating the hypothalamus. These studies were further extended by Hess (Hess, Brugger and Brucher, 1945; Hess, 1947, 1948, 1949); Bard (1929, 1940); Bard and Mountcastle (1947); Bailey and Bremer (1938); and many colleagues and pupils of Ranson (Ranson, Fisher and Ingram, 1937; Ranson, 1939, 1940). Attention thus came to be focused on the hypothalamus, which for a time came to be designated by Sherrington (1947) as the "head ganglion" of the autonomic nervous system, and was considered to be the highest seat of control and integration for this system.

More recently, knowledge of the functional and anatomical interactions between the hypothalamus and other areas of the brain has increased rapidly. Broca (1878) described the great limbic lobe surrounding the hilus of the brain as a common denominator in the brains of all mammals, and Papez (1937) suggested its involvement in a variety of viscero-somatic and emotional reactions in mammals. Phylogenetic and cytoarchitectural studies, together with recent physiological investigations, have suggested that these higher nervous regions represent an early neural development involved in the higher control of the autonomic nervous system and affectively determined behaviour (MacLean, 1949) (Fig. 1). The structures

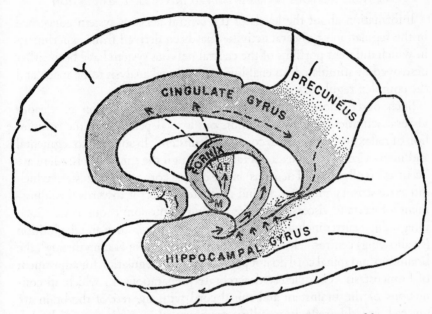

FIG. 1. Diagram of the medial aspect of the brain. The shaded area of the cortex represents what was formerly known as the limbic lobe of Broca. It also shows schematically some of the circuits in the limbic system emphasized by Papez. It corresponds to what is referred to as the visceral brain. M, mammillary body, the hypothalamic recipient of the bulk of of the fornix projecting from the hippocampus; AT, anterior thalamic nuclei, the way-station between the mammillary body and cingulate gyrus. (From MacLean, 1949.)

comprising the anatomical limbic lobe have been shown to have a functional unity and have therefore been collectively designated the "limbic system". MacLean (1949) has given the name "visceral brain" to these structures, as they mostly regulate the visceral functions of the body.

A number of experimental observations (see below) suggest that the limbic system of the brain (including the hypothalamus) is to a large extent responsible for maintaining homeostatic conditions in the body, through its regulation of autonomically determined visceral activities, of endocrine secretions, and of affectively determined behaviour. Enough experimental evidence is also available to suggest that the pattern of nervous regulation of visceral activities, which determine the constancy of the *milieu intérieur*, is similar to the pattern of nervous regulation of somatic activities, which adjust an individual's reactions to the external environment.

CENTRAL NERVOUS LEVELS INVOLVED IN VISCERAL REGULATION

Information about the levels of the central nervous system concerned in the regulation of visceral activities has been derived from experiments in which different portions of the central nervous system have been either destroyed or stimulated, so enabling a functional analysis to be made and the complex regulatory mechanisms to be localized.

Such observations have brought out that the regulation of various visceral activities through autonomic outflows depends upon the completion of reflex arcs in the spinal cord and brainstem. In addition to segmental and inter-segmental reflexes operating through the spinal levels, there are "centres" (collections of neurons) in the central core of the brainstem which integrate sensory information and ultimately direct it towards the adjustment of certain visceral activities through autonomic outflows. Such integrating centres include cardiac, vasomotor, respiratory and deglutition (swallowing) centres. However, reflex mechanisms operating through the brainstem and spinal cord do not provide efficient regulation for adjustment of homeostasis. Completely decerebrate preparations, in which all connexions of the brainstem and spinal cord with the rest of the brain are severed, are able reflexly to adjust cardiovascular, respiratory and other activities but cannot integrate the functioning of these different systems to maintain homeostatic conditions under diverse environmental situations. Such a preparation cannot even maintain normal body temperature (Bard and Woods, 1962).

If, however, the connexions of the hypothalamus with the brainstem and hypophysis are left intact but its connexions with the rest of the brain are severed, homeostatic mechanisms are maintained under basal conditions. This implies that the hypothalamus not only regulates and adjusts the activities of different visceral and endocrine systems but also integrates these with each other so that the sum total of all these integrated responses is reflected in the maintenance of body homeostasis. In addition, hypothalamic integration results in the production of certain drives, resulting in specific motivational behaviours, which are also directed towards maintenance of homeostasis. However, such hypothalamic preparations fail to maintain homeostatic conditions in certain stressful situations or changed environmental conditions. Such a preparation cannot efficiently maintain normal body temperature when exposed to high or low environmental temperatures (Keller and McClaskey, 1964). In such a preparation cardiovascular and respiratory adaptation to hypoxia also does not occur efficiently (Davis, 1951; Turner, 1954). Even the drives and motivated behaviour produced in response to homeostatic changes are abnormal; such a preparation can show states of hunger and satiation but these result in the animal eating everything presented to it without any discrimination between edible and inedible objects (Anand et al., 1959; Chhina, 1960). The motivated behaviour produced in response to the sex drive is also abnormal (Anand, 1964).

On the other hand, if connexions of the limbic system with the hypothalamus and the rest of brainstem and spinal cord are left intact and only neocortical regions are removed, the animal can adjust its homeostatic

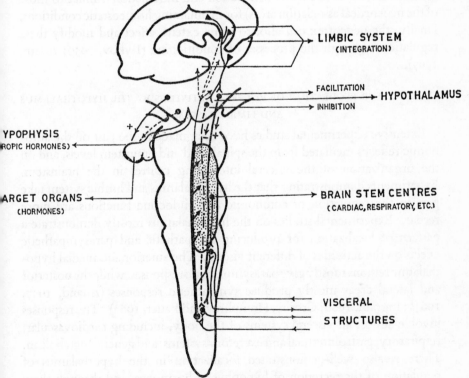

FIG. 2. Diagram of the organization of central nervous regulation of visceral activities.

mechanisms efficiently in diverse situations (Anand, Dua and Chhina, 1957, 1958).

Thus the regulation of visceral activities is functionally organized at different levels of the central nervous system (Anand, 1963a) (Fig. 2). Sensory inputs from various internal organs form the basis of reflexes which are completed through spinal and brainstem levels. The afferents also relay in the brainstem centres for special visceral activities. Finally, most of these afferents project to the hypothalamus, and possibly some into limbic regions (Nauta and Kuypers, 1958). The hypothalamus (like the motor areas) brings about the different visceral and endocrine activities which ultimately regulate the *milieu intérieur* by providing facilitatory and inhibitory controls over autonomic outflows, and adjusting the secretion of hormones through the hypophysis. Certain drives, leading to specific motivated behaviours, also result from integration in the hypothalamus. The hypothalamus is further interconnected with various limbic structures (MacLean and Pribram, 1953; Pribram and MacLean, 1953) which possibly subserve functions of further integration and modulation (similar to those of the neocortical association areas) for maintaining homeostatic conditions. Finally, the neocortex can also, to some extent, affect and modify these regulations through the processes of conditioning (Bykov, 1959; Adam, 1967).

EVIDENCE FOR REGULATION OF VISCERAL ACTIVITIES BY THE HYPOTHALAMUS AND LIMBIC SYSTEM

Extensive experimental studies have been made in the past on the autonomic reflexes mediated from the spinal cord and brainstem levels, and on the organization of the visceral integrating centres in the brainstem. Experiments demonstrating that the hypothalamus and limbic system take part in the regulation of autonomic and endocrine functions are more recent. Experimental studies on the hypothalamus mostly demonstrate a pattern of localization for producing sympathetic and parasympathetic effects on the activities of different viscera. The anterior and medial hypothalamic regions mostly give parasympathetic responses, while the posterior and lateral areas mostly produce sympathetic responses (Anand, 1957, 1963b; Ingram, 1960; Crosby, Humphrey and Lauer, 1962). The responses involve almost all visceral systems of the body, including cardiovascular, respiratory, gastrointestinal and excretory systems, and general metabolism. There is also well-demonstrated localization in the hypothalamus of regulation of the secretion of hypophyseal hormones, and through these of the other endocrine functions. Precise localization for certain types of motivated behaviours, like hunger, thirst and sexual activity, can also be shown.

Experimental studies on the limbic system, by contrast, provide a picture of unspecificity (MacLean, 1954) in terms of these responses. Both stimulation and ablation of various limbic structures evoke responses from the various visceral systems but there is no specific localization for the different functions in the different limbic structures. This shows that the "specific" localization in the hypothalamus is something similar to what is observed in the motor areas of the neocortex, and the "unspecific" integration of autonomic functions in the limbic system is similar to what happens in the neocortical association areas for somatic activities.

I shall highlight here some of the many studies made, in order to stress the involvement of these higher nervous structures in the regulation of various types of visceral activity.

Regulation of cardiovascular activities

Experimental procedures applied to the hypothalamus and limbic structures give rise to changes of heart rate, blood pressure, tone of peripheral blood vessels, and level of blood flow through different structures.

Hypothalamus. Generally speaking, stimulation of posterior and lateral regions of the hypothalamus produces a rise in blood pressure and increase in heart rate, and occasionally fibrillation of the heart (Kabat, Magoun and Ranson, 1935; Magoun *et al.*, 1937; Bronk, Pitts and Larrabee, 1940; Redgate and Gellhorn, 1953; Dua, 1956). Such stimulation has also been shown to release adrenaline and noradrenaline (Magoun, Ranson and Hetherington, 1937; Redgate and Gellhorn, 1953; Folkow and von Euler, 1954; Cross, 1955). Stimulation of anterior and medial hypothalamic regions including the preoptic and septal areas, on the other hand, results in a fall of blood pressure and slowing of heart rate (Wang and Ranson, 1939, 1941; Hess, 1947; Dua, 1956). Bronk, Pitts and Larrabee (1940) demonstrated that hypothalamic stimulation results in changes in the frequency of firing in the sympathetic nerves to the heart.

Lesions and stimulations in the preoptic region and anterior hypothalamus occasionally lead to pulmonary oedema (Gamble and Patton, 1953; Maire and Patton, 1954). Changes in the total blood volume, as well as its distribution in the systemic and pulmonary vascular bed, also occur on stimulation and ablation of hypothalamic regions (Gulati, 1967).

The acceleration of the heart rate and rise of blood pressure produced by stimulating medullary regions can be abolished when the brain is sectioned caudal to the hypothalamus (Piess, 1960), demonstrating that the hypothalamic influences are mediated through the medullary centres. Wilson and co-workers (1961), however, believe that the hypothalamic influences are not mediated through medullary baroceptor reflexes.

Cardiovascular responses during muscular exercise are also possibly mediated through the hypothalamus, since lesions of the posterior hypothalamus very nearly abolish such responses (Smith *et al.*, 1960). They can be elicited by hypothalamic stimulation even when muscles are paralysed with relaxants (Wilson *et al.*, 1961).

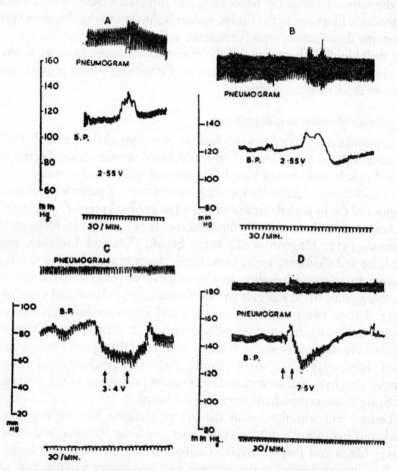

FIG. 3. Blood pressure and respiratory responses on stimulation of limbic structures in unanaesthetized animals with square-wave pulses of 30/sec frequency, 0·2–1 msec duration, and voltages ranging from 1–5 V. A. Stimulation of orbitomesial frontal cortex. B. Stimulation of cingulate gyrus. C. Stimulation of anteromedial group of amygdaloid nuclei. D. Stimulation of hippocampus. (From Anand and Dua, 1956.)

Anand, Dua and Malhotra (1957) observed that after reserpine is given the threshold for eliciting the pressor response from the posterior hypothalamic area is raised, while the threshold for eliciting the depressor response from the anteromedial region is lowered. This suggests that, in addition to neuronal interaction and integration between the different hypothalamic regions, hypothalamic neurons producing different responses may have somewhat different metabolic parameters.

Limbic system. Spencer (1894) produced blood pressure changes by stimulating fronto-temporal regions. These observations have been confirmed by showing a general increase in blood pressure on stimulation of the orbito-mesial cortex (Livingston *et al.*, 1948; Delgado and Livingston, 1948; Kaada, Pribram and Epstein, 1949; Turner, 1954). Cingulate stimulation mostly results in a fall in blood pressure and slow pulse (Smith, 1945; Ward, 1948), but occasionally a rise with some cardiac acceleration. Stimulation of temporal limbic structures like the amygdala and hippocampus mostly results in a fall in blood pressure (Kaada, 1951; MacLean and Delgado, 1953; Wood *et al.*, 1958). Some reports, however, suggest that blood pressure rises on stimulation of the amygdala (Koikegami, Kimoto and Kito, 1953; Koikegami *et al.*, 1957; Chapman *et al.*, 1954). Anand and Dua (1956) stimulated various limbic regions systematically in unanaesthetized animals and observed that stimulation of frontal lobe structures mostly produces a rise in blood pressure accompanied by peripheral vasoconstriction, whereas stimulation of limbic structures contained in the temporal lobe mostly results in a fall in blood pressure accompanied by peripheral vasodilatation (Fig. 3). Changes in the heart rate, however, are quite variable. On the other hand, when the different limbic regions were stimulated in anaesthetized animals (Anand and Chhina, 1959), stimulation of both frontal lobe and temporal lobe limbic structures resulted in a fall of blood pressure, although vasoconstriction was still produced after frontal lobe stimulation and vasodilatation on temporal lobe stimulation. This emphasizes the fact that studies on visceral responses are best made on unanaesthetized preparations, to avoid variable results.

Surgical lesions involving frontal lobe structures (Anand, Dua and Chhina, 1957) produce a small drop in blood pressure, while similar lesions affecting the temporal lobe result in a slight rise in blood pressure.

Regulation of respiratory activities

Experimental procedures involving the hypothalamus and the limbic system have been shown to change the frequency and the depth of respiration.

Hypothalamus. Like cardiovascular effects, changes in respiratory activity are observed on stimulation or ablation of all the hypothalamic regions. Posterior hypothalamic stimulation generally increases the rate of ventilation (Hess and Muller, 1946; Dua, 1956). Stimulation of the medial part of anterior hypothalamus decreases the rate and amplitude of respiration (Dua, 1956). Stimulation of preoptic and supraoptic regions, on the other hand, generally leads to panting (Ranson, 1939; Hess, 1947).

Most of these respiratory responses accompany the cardiovascular changes and may, therefore, be an attempt on the part of the organism to mutually adjust the functions of these two systems for homeostatic regulation. However some of the respiratory effects, like panting, are related to the mechanism of temperature regulation.

Limbic system. Although variable effects on respiration have been reported after stimulating different limbic regions, most workers have observed either arrest of respiration, or a certain level of inhibition, after stimulating various frontal and temporal limbic structures (Ward, 1948; Kaada, Pribram and Epstein, 1949; Lennox and Robinson, 1951; Turner, 1954). Anand and Dua (1956), working with unanaesthetized preparations, generally observed quickening of respiration from stimulation of frontal lobe structures, and inhibition, including apnoea, from stimulation of temporal lobe structures. However, after an anaesthetic both frontal and temporal lobe stimulation result in inhibition of respiration (Anand and Chhina, 1959). Respiratory activity was slightly inhibited by lesions of both frontal and temporal lobe structures (Anand, Dua and Chhina, 1957).

Regulation of gastrointestinal activities

Stimulation and ablation of the hypothalamus and limbic system produce all kinds of changes in the secretory activity and the motility of the alimentary tract. It is also observed that experimental involvement of hypothalamic and limbic structures invariably leads to the production of petechial haemorrhages in the mucous membrane of the alimentary tract and the presence of blood in the stools. Involvement of certain brain structures, on the other hand, leads to acute haemorrhagic ulcers of various sizes.

Hypothalamus. Although the results are somewhat variable in the hands of different workers, generally speaking stimulation of the anterior hypothalamus produces an increase in gastric secretion, tone and motility, while that of the posterior hypothalamus results in relaxation of the stomach, including its pyloric end, and diminution in its secretory activity (Wang *et al.*, 1940; Eliasson, 1953; Sen and Anand, 1957*a*). Intestinal and colonic motility is inhibited by stimulating the anterior hypothalamus (Wang

VISCERAL REGULATION BY CNS

et al., 1940). Salivation and licking are also observed on hypothalamic stimulation (Hess, 1947; Dua, 1956).

Stimulation of the preoptic area in unanaesthetized animals mostly results in the production of acute haemorrhagic ulcers (Sen and Anand, 1957*a*). On the other hand, the formation of ulcers by ligation of the pyloric end is prevented by lesions of the anterior hypothalamus, and their production is facilitated by lesions of the posterior hypothalamus (Sharma, Dua and Anand, 1963).

Limbic system. A number of experiments demonstrate that stimulation of different limbic structures results in changes in the motility and tone of different parts of the alimentary tract, though the results are variable in different hands (Babkin and Kite, 1950; Kaada, 1951; Eliasson, 1952; Koikegami and Fuse, 1952; Sen and Anand, 1957*b*). Such stimulation and ablation also lead to significant changes in the gastric secretory activity of the stomach (Davey, Kaada and Fulton, 1950; Shealy and Peele, 1957; Sen and Anand, 1957*b*). Acute haemorrhagic ulcers are produced only after stimulation of the amygdaloid nuclei (Sen and Anand, 1957*b*).

Regulation of liver and kidney functions and changes in blood chemistry

In addition to changes in bile flow produced by stimulating the hypothalamus (Birnbaum and Feldman, 1962), stimulation and ablation of both hypothalamic and limbic structures produce changes in the functioning of the liver, as demonstrated by changes in serum bilirubin, alkaline phosphatase, plasma proteins and carbohydrate metabolism (Anand, 1963*b*).

Similarly, such experimental procedures applied to the limbic structures disturb renal function, as shown by changes in urea clearance, glomerular filtration rate and renal plasma flow (Gupta, 1960).

Changes in the liver blood flow have also been observed after hypothalamic stimulation (Surange, 1966), and these are not related to variations in the systemic blood pressure. The observation that lesions in the ventromedial hypothalamus block the development of liver necrosis as a result of massive haemolysis produced by malarial parasites in monkeys (Ray *et al.*, 1958) suggests a relationship of this area with the blood circulation through the liver.

It has also been observed that lesions and stimulation of hypothalamic and limbic areas often disturb the chemical composition of the blood, producing variations in its glucose, sodium, potassium and plasma protein content (Anand and Dua, 1956; Anand, Dua and Chhina, 1957; Anand and Chhina, 1959). Not only is the total quantity of plasma protein affected, but there are disturbances of the albumin : globulin ratio also.

366   B. K. ANAND

All such changes would be related, not only to the changed functional activity of liver, kidneys and other visceral organs, but also to changes in the endocrine secretions.

Changes in other visceral activities

In addition to the specific observations described, stimulation and ablation of the hypothalamic and limbic areas very often produce other autonomically mediated visceral effects, such as pupillary changes, pilo-erection, sweating, urination, changes in urinary bladder tone, defecation and changes in splenic activity (Hodes and Magoun, 1942; Kaada, 1951; Gastaut, 1952; Koikegami and Fuse, 1952; Sloan and Kaada, 1953; Koikegami, Kimoto and Kito, 1953; Hess, 1954; French, Hernández-Peón and Livingston, 1955; Gloor, 1955; Grossman and Wang, 1955; Anand and Dua, 1956; Ingersoll, Jones and Hegre, 1956).

Regulation of feeding and drinking

Experimental studies have provided evidence that there are two opposing mechanisms in the hypothalamus that regulate food intake (Anand and Brobeck, 1951; Anand, Dua and Schoenberg, 1955; Anand, 1961), namely a mechanism in the lateral region of the hypothalamus which initiates feeding and is therefore designated the "feeding centre", and one in the medial part of the hypothalamus which brings about satiety after a meal has been taken and is thus termed the "satiety centre". Feeding behaviour is probably based on feeding reflexes operating through the spinal cord and brainstem levels, put into effect by sensory stimuli which make the animal aware of the presence of food (Brobeck, 1960), which are facilitated and inhibited from the hypothalamic regions (Fig. 4). Thus the activities of these hypothalamic centres provide the basis for hunger and satiety states which result in specific drives leading to a particular type of motivated behaviour. These are further influenced and modulated from the limbic system of the brain in a manner which provides "discriminative appetite" (Anand, 1961). The activities of the hypothalamic and limbic mechanisms are further determined by homeostatic changes.

Similarly, the hypothalamus is involved in the regulation of water intake in two ways: (*i*) the anterior hypothalamus by its regulation of secretion of the antidiuretic hormone determines the amount of water lost through the urine (Verney, 1947) and thus indirectly governs the amount of water intake, and (*ii*) the lateral hypothalamic region directly controls the thirst drive (Andersson and McCann, 1955) and thus the amount of water taken in.

Although the mechanisms regulating hunger and thirst are located in the same anatomical area of the lateral hypothalamus, we have demonstrated that these two mechanisms are independent of each other (Anand and Dua, 1958).

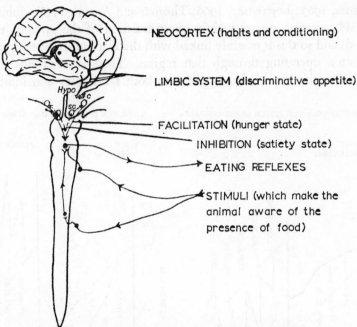

NEOCORTEX (habits and conditioning)

LIMBIC SYSTEM (discriminative appetite)

FACILITATION (hunger state)

INHIBITION (satiety state)

EATING REFLEXES

STIMULI (which make the animal aware of the presence of food)

Fig. 4. An outline of the central nervous mechanisms regulating food intake. Hypo, hypothalamus; FC, feeding centre; SC, satiety centre; →, facilitation; -->, inhibition. (From Anand, 1963a.)

Regulation of endocrine activities

Hypothalamic connexions with the anterior pituitary gland through the portal vessels enable this region to regulate hypophyseal activities, which it does by elaborating releasing factors. In addition it has direct neuronal connexions with the posterior pituitary, so that some of the hormones synthesized in the hypothalamus, like the antidiuretic hormone, are released there. It has been shown by many workers that the hypothalamus regulates the secretion of the tropic hormones of the anterior pituitary—adrenocorticotropin (ACTH), thyrotropin (TSH) and the gonadotropins. Experimental evidence for this is extensive (for recent reviews see McCann, Dhariwal and Porter, 1968; Farrell, Fabre and Rauschkolb, 1968).

The limbic system of the brain would also be expected to influence the secretion of these tropic hormones through the hypothalamus, but such studies on the limbic system are relatively few.

Recent studies by my colleagues, however, show that, just as stimulation of certain hypothalamic regions leads to increased TSH and ACTH output from the hypophysis, stimulation of certain other hypothalamic areas leads to a decrease in the output of both TSH and ACTH below the basal levels (Thomas, 1963; Logawney, 1966; Thomas and Anand, 1968). Inhibition of TSH secretion occurs mainly on stimulation of the preoptic region (Fig. 5), and so this is possibly linked with the mechanism of temperature regulation operating through that region. These observations suggest that just as the hypothalamus provides both facilitatory and inhibitory

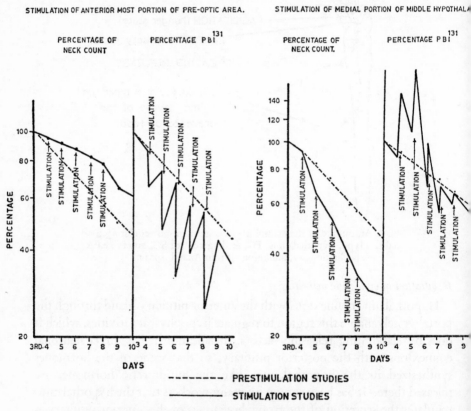

FIG. 5. Turnover rates of ^{131}I from the thyroid to the blood in the form of protein-bound ^{131}I, after subcutaneous injection of ^{131}I, expressed as percentage variation from the counts made on the third day after injection. Stimulation of the medial part of the middle hypothalamus caused a rapid turnover of ^{131}I, shown by a fall in the neck (thyroid) count and a rise in the protein-bound ^{131}I levels on each day of stimulation (*right*), indicating increased secretion of TSH. Stimulation of the anteriormost part of the preoptic area produced the opposite effects (*left*), with a decreased secretion of TSH below the basal level.

mechanisms for the adjustment of autonomic activities, similar reciprocally acting dual mechanisms are available for adjusting the secretory activities of endocrine hormones also.

Regulation of visceral activities by the cerebellum

Although the cerebellum is not part of the limbic system, certain parts of it, especially the palaeocerebellum, are intimately connected with limbic structures (Anand et al., 1959). Since neocerebellar structures which are connected with the neocerebrum (neocortex) influence somatic activities mediated through the neocortex, it is reasonable to expect that the palaeocerebellum influences visceral activities through the limbic system. Cerebellar stimulation produces various autonomic effects like changes in pupil diameter, galvanic skin response, movements of alimentary tract, urinary bladder function, and cardiovascular and respiratory responses (Stella, 1939; Moruzzi, 1947; Bard and Mountcastle, 1947; Wang, Stein and Brown, 1956). Rasheed (1965) observed that stimulation of the palaeocerebellar region in unanaesthetized animals mostly produces a rise in blood pressure, an increase in heart rate and a rise in urinary bladder pressure; if the same stimulation is repeated after giving an anaesthetic, the blood pressure falls and respiratory activity decreases.

Regulation of visceral activities by the caudate nucleus

Certain autonomic responses are also obtained by stimulating the caudate nucleus (Wang, 1960; Rubinstein and Delgado, 1963). Anand and Chhina (1959) and Subberwal, Anand and Singh (1965) mostly obtained a fall in blood pressure and inhibition of respiration by stimulating the rostral part of the head of the caudate.

FEEDBACK ACTION OF HOMEOSTATIC CHANGES INTO THE CENTRAL NERVOUS SYSTEM

The central nervous regions which regulate and integrate the various visceral and endocrine activities described have to be supplied with information as a basis for such regulation. It is proposed that this information is supplied in two ways. Firstly, it is routed via afferent nerves coming from the visceral structures involved in a particular activity. This information, as has been pointed out, is not only essential for completing reflex arcs operating through the spinal cord and brainstem levels but also is ultimately fed into the hypothalamic and limbic integrating mechanisms. For example, sensory afferents bringing information about distension of the stomach

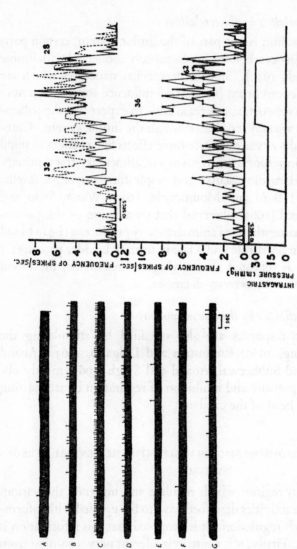

FIG. 6. Unit activity recorded from a neuron in the satiety centre of a starving cat (*left*); and spike frequencies of four more such units (*right*) exposed to increased intragastric pressure produced through indwelling balloons. *Left*: A, control; B, immediately after raising the intragastric pressure to 15 mm Hg; C, immediately after raising the intragastric pressure to 30 mm Hg; D, 5 min after raising the intragastric pressure to 30 mm Hg; E, after deflation of the intragastric balloon; F, after raising the pressure in an intraperitoneal balloon to 30 mm Hg; and G, on raising the pressure in the intragastric balloon after severing the gastric vagal branches. The unit activity increased with increase of intragastric pressure. *Right*: spike frequencies of four satiety units correlated with changes of pressure in intragastric balloon. The roughly linear relationship between the spike frequency and the intragastric pressure is demonstrated. (From Anand and Pillai, 1967.)

(with food), ascending through the vagus nerve, ultimately project to the "satiety" centre in the hypothalamus (Anand and Pillai, 1967) (Fig. 6) and by activating it, cut out further eating.

In addition, the change or changes produced in the internal environment as a result of visceral activity also affect and influence the nervous regulating mechanisms specific for this activity (Anand, 1963a). If the higher nervous integrating and regulating mechanisms in the hypothalamus and limbic system manage to keep the *milieu intérieur* constant, they can perhaps do so if there is a feedback in terms of all changes affecting the internal environment. This means that either there must be specific neurons in the central nervous regulating mechanisms which are sensitive to the various kinds of changes (chemical or physical) taking place in the internal environment; or there are receptors sensitive to the components of the *milieu intérieur* present at the periphery which ultimately project sensory information into the central nervous system; or both. Enough experimental evidence is available to suggest the presence of both these types of sensing mechanisms.

Specifically sensitive mechanisms responding to changes in homeostasis have so far mostly been shown to be present in the hypothalamic region. Thermosensitive (Anand, Banerjee and Chhina, 1966; Cross and Silver, 1966) and osmosensitive (von Euler, 1953; Sawyer and Gernandt, 1956; Jewell and Verney, 1957; Cross and Green, 1959; Nakayama, Eisenman and Hardy, 1961; Nakayama *et al.*, 1963; Brooks, Ushiyama and Lange, 1962) neurons have been demonstrated in the hypothalamus. Feedback by circulating hormones into the hypothalamic regions is an accepted fact (Flerko and Szentagothai, 1957; Hodges and Vernikos, 1958, 1959; Harris, 1960; Lisk, 1960; Halasz, Pupp and Uhlarik, 1962; Fortier, 1963; Anand, 1964; Reichlin, 1964; Michael, 1965; Motta, Mangili and Martini, 1965; Davidson, 1966; Mangili, Motta and Martini, 1966; Lincoln and Cross, 1967; Ramirez *et al.*, 1967; Chhina *et al.*, 1968; McCann, Dhariwal and Porter, 1968; Chhina and Anand, 1969). Neurons sensitive to hypoxic and hypercapnic changes have also been demonstrated (Cross and Silver, 1962a, 1962b, 1963). We (Anand, Dua and Singh, 1961; Anand *et al.*, 1961, 1964; Mayer, 1966; Anand, 1967) have demonstrated neurons in the "satiety" area which are sensitive to changes in the level of glucose utilization in the body (Fig. 7a and b). Similarly, there are neurons sensitive to injection of hypertonic solutions, which result in drinking (Andersson and Larsson, 1957, 1961). In addition, neurons which determine hunger and thirst drives respond to applications of adrenergic and cholinergic substances (Grossman, 1962). Neurons sensitive to other components of the internal environment may be expected to be demonstrated: they will help to explain such experimental observations as the preferential selection by animals with sodium or calcium deficiencies of diets containing the deficient elements (Scott, 1946; Harriman, 1955a, b; Lewis, 1964; Lat, 1967).

As a result of feedback of information through afferent nerves, and directly through the circulating blood, the hypothalamic (and possibly the limbic) integrating mechanisms are activated. These then bring about appropriate responses which correct any homeostatic imbalances. This is achieved by bringing about changes in visceral activities through autonomic outflows, as well as by changes in the secretory pattern of hormones. In addition,

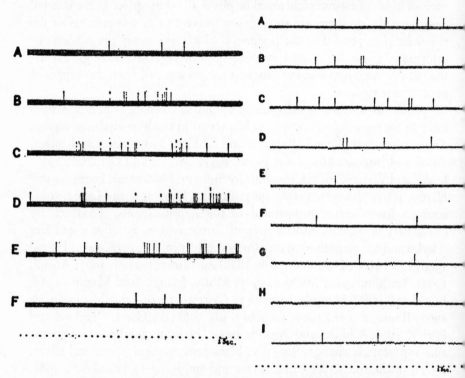

FIG. 7a. Unit activity recorded from neurons of satiety centre after giving intravenous glucose (*left*) and intravenous insulin (*right*). *Left*: A, before giving glucose; and at frequent intervals after intravenous glucose infusion: B, after 10 min, C, after 20 min, D, after 25 min, E, after 35 min and F, after 60 min. The unit activity was increased with injection of blood glucose but tended to return within an hour. *Right*: A, before giving insulin; and at certain intervals after insulin injection: B, after 5 min, C, after 10 min, D, after 15 min, E, after 25 min, F, after 35 min, G, after 45 min, H, after 55 min. Frequency of the spikes was increased initially for 5–10 min but later decreased. (From Anand *et al.*, 1964.)

feedback of information into the hypothalamic region results in the production of certain drives leading to motivated behaviours, which are also directed towards correcting homeostatic deficiencies. Most basic motivational forces arise from some physiological need state, but some compelling drives (such as sexual motivation), which are also important in homeostatic maintenance, cannot be traced to specific needs or deficiencies in the body, and are brought about by particular environmental stimuli. Homeostatic drives such as hunger, thirst and elimination are

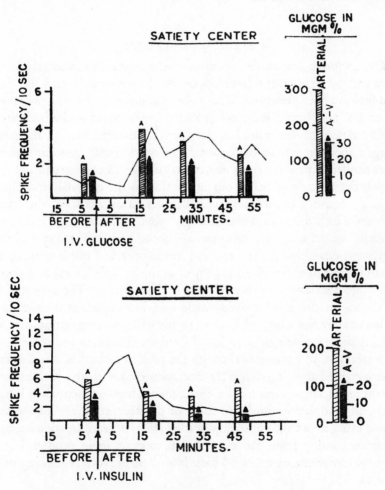

FIG. 7b. Spike frequency of two units from the satiety centre correlated with changes in arterial blood glucose and arterio-venous glucose difference produced by intravenous injection of glucose (*upper*), and intravenous injection of insulin (*lower*). (From Anand *et al.*, 1964.)

elicited and reduced directly by changes in the internal environment, which feed information back into the nervous regions. The non-homeostatic drives, such as sexual or emotional arousal and activity, are typically elicited and reduced by changes in the external environment, though internal environment changes which "prime" the organism are essential for these also. For example, sensory information from the external environment can activate central nervous mechanisms related to sex only when these have been "primed" by the feedback of gonadal hormones (Chhina *et al.*, 1968; Chhina and Anand, 1969) (Fig. 8).

CONCLUSIONS AND SUMMARY

Observations on a number of autonomic, endocrine and visceral responses obtained by experimental studies on the hypothalamus and the limbic system have been presented. These observations do not give us a complete picture of how homeostatic conditions are maintained in the body by the limbic system of the brain, but they provide us with enough material to suggest that the pattern of regulation of our internal environment is on lines similar to the pattern of nervous regulation of our somatic activities, which determine our behaviour in relation to the external environment (Fig. 9). Sensory inputs from various internal organs form the basis of reflexes which are completed through spinal and brainstem levels. The afferents ascend and have relay centres in the brainstem for special reflex activities; examples are the cardiac, vasomotor and respiratory centres. Finally, most of these afferents project to the hypothalamus, and possibly some into limbic regions after relaying in the thalamus. The hypothalamus is like the motor area on the somatic side, in as much as it executes the different activities which ultimately regulate the autonomic outflows, thus modifying reflex activities mediated from the brainstem and spinal levels. The secretion of hormones through the pituitary gland is also regulated. The hypothalamus is extensively interconnected with the various limbic structures and these regions, therefore, may subserve functions of integration and modulation similar to those of the neocortical association areas. These bring about the constancy of the *milieu intérieur*. In addition to the afferents coming from the visceral structures, changes produced in the internal environment also feed back into these nervous regulating mechanisms.

Most of the concepts presented here are not new, but in keeping with the present emphasis on integration in explaining central nervous mechanisms, I have attempted an integration of these concepts. Experimental

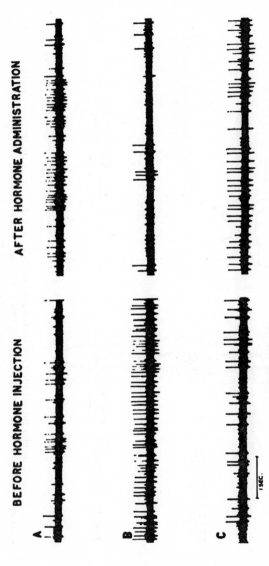

FIG. 8. Unit activity of a single neuron from the preoptic area in a sexually immature female monkey, before (*left*) and after (*right*) an intravenous injection of oestrogen. A, spontaneous discharge; B, during vaginal stimulation; and C, a few minutes after vaginal stimulation. The administration of hormone converted the excitatory response into an inhibitory response. (From Chhina and Anand, 1969.)

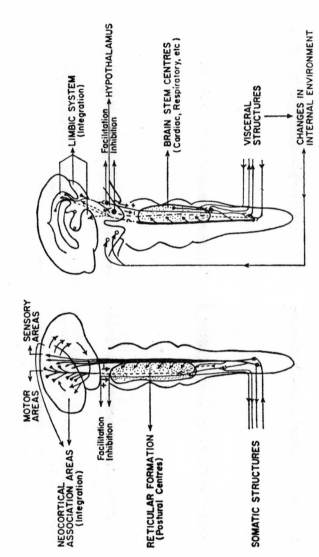

FIG. 9. Diagrams representing patterns of central nervous regulation of the somatic structures which determine activity in relation to the external environment (*left*), and visceral activities which maintain the internal environment (*right*). (Anand, 1963*a*.)

evidence to support these concepts is by no means complete and there are wide gaps in our knowledge, but this presentation may stimulate further experimentation for bridging these gaps.

REFERENCES

ADAM, G. (1967). *Interoception and Behaviour*, pp. 1-152. Budapest: Publishing House of the Hungarian Academy of Sciences.
ANAND, B. K. (1957). *Indian J. Physiol. Pharmac.*, **1**, 149-184.
ANAND, B. K. (1961). *Physiol. Rev.*, **41**, 677-708.
ANAND, B. K. (1963a). In *Brain and Behaviour*, pp. 43-116, ed. Brazier, M. A. B. Washington: American Institute of Biological Sciences.
ANAND, B. K. (1963b). *Indian J. med. Res.*, **51**, 175-222.
ANAND, B. K. (1964). *Bull. natn. Inst. Sci. India*, **27**, 20-27.
ANAND, B. K. (1967). In *Handbook of Physiology*, sect. 6, vol. I, pp. 249-263, ed. Code, C. F., and Heidel, W. American Physiological Society. Baltimore: Williams and Wilkins.
ANAND, B. K., BANERJEE, M. G., and CHHINA, G. S. (1966). *Brain Res.*, **1**, 269-278.
ANAND, B. K., and BROBECK, J. R. (1951). *Yale J. Biol. Med.*, **24**, 123-140.
ANAND, B. K., and CHHINA, G. S. (1959). *Indian J. Physiol. Pharmac.*, **3**, 27-38.
ANAND, B. K., CHHINA, G. S., SHARMA, K. N., DUA, S., and SINGH, B. (1964). *Am. J. Physiol.*, **207**, 1146-1154.
ANAND, B. K., and DUA, S. (1956). *J. Neurophysiol.*, **19**, 393-400.
ANAND, B. K., and DUA, S. (1958). *Indian J. med. Res.*, **46**, 426-430.
ANAND, B. K., DUA, S., and CHHINA, G. S. (1957). *Indian J. med. Res.*, **45**, 345-352.
ANAND, B. K., DUA, S., and CHHINA, G. S. (1958). *Indian J. med. Res.*, **46**, 277-287.
ANAND, B. K., DUA, S., and MALHOTRA, C. L. (1957). *Br. J. Pharmac. Chemother.*, **12**, 8-11.
ANAND, B. K., DUA, S., and SCHOENBERG, K. (1955). *J. Physiol., Lond.*, **127**, 143-152.
ANAND, B. K., DUA, S., and SINGH, B. (1961). *Electroenceph. clin. Neurophysiol.*, **13**, 54-59.
ANAND, B. K., MALHOTRA, C. L., SINGH, B., and DUA, S. (1959). *J. Neurophysiol.*, **22**, 451-457.
ANAND, B. K., and PILLAI, R. V. (1967). *J. Physiol., Lond.*, **192**, 63-77.
ANAND, B. K., SUBBERWAL, U., MANCHANDA, S. K., and SINGH, B. (1961). *Indian J. med. Res.*, **49**, 717-724.
ANDERSSON, B., and LARSSON, B. (1957). *Acta physiol. scand.*, **38**, 22-30.
ANDERSSON, B., and LARSSON, B. (1961). *Acta physiol. scand.*, **51**, 75-89.
ANDERSSON, B., and McCANN, S. N. (1955). *Acta physiol. scand.*, **33**, 333-346.
BABKIN, B. P., and KITE, W. C., JR. (1950). *J. Neurophysiol.*, **13**, 321-334.
BAILEY, P., and BREMER, F. (1938). *J. Neurophysiol.*, **1**, 405-412.
BARD, P. (1929). *Archs Neurol. Psychiat., Chicago*, **22**, 230-246.
BARD, P. (1940). *Res. Publs Ass. nerv. ment. Dis.*, **20**, 551.
BARD, P., and MOUNTCASTLE, V. B. (1947). *Res. Publs Ass. nerv. ment. Dis.*, **27**, 362-404.
BARD, P., and WOODS, J. W. (1962). *Trans. Am. neurol. Ass.*, **87**, 37-39.
BIRNBAUM, D., and FELDMAN, S. (1962). *J. Lab. clin. Med.*, **60**, 914-922.
BROBECK, J. R. (1960). In *Handbook of Physiology*, sect. 1, vol. II, pp. 1197-1250, ed. Field, J., Magoun, H. W., and Hall, V. E. American Physiological Society. Baltimore: Williams and Wilkins.
BROCA, P. (1878). *Revue anthrop.*, ser. 2, **1**, 285-498.
BRONK, D.W., PITTS, R. F., and LARRABEE, A. G. (1940). *Res. Publs Ass. nerv. ment. Dis.*, **20**, 323-341.
BROOKS, C. McC., USHIYAMA, J., and LANGE, G. (1962). *Am. J. Physiol.*, **202**, 487-900.

378 B. K. ANAND

BYKOV, K. M. (1959). In *The Cerebral Cortex and the Internal Organs*, pp. 41–213, ed. Hodes, R. Moscow: Foreign Languages Publishing House.
CHAPMAN,W. P., SCHROEDER, H. R., GEYER, G., BRAZIER, M. A. B., FAGER, C., POPPEN, J. L., SOLOMON, H. C., and YAKOVLEV, P. I. (1954). *Science*, 120, 949–950.
CHHINA, G. S. (1960). *Role of limbic system of brain in the regulation of affective behaviour*, pp. 67–83. Ph.D. Thesis, Punjab University, Chandigarh.
CHHINA, G. S., and ANAND, B. K. (1969). *Brain Res.*, 13, 511–521.
CHHINA, G. S., CHAKRABARTY, A. S., KAUR, K., and ANAND, B. K. (1968). *Physiol. Behav.*, 3, 579–584.
CROSBY, E. C., HUMPHREY, T., and LAUER, E.W. (1962). *Correlative Neuroanatomy*, pp. 331–342. New York: Macmillan.
CROSS, B. A. (1955). *J. Endocr.*, 12, 15–28.
CROSS, B. A., and GREEN, J. D. (1959). *J. Physiol., Lond.*, 148, 554–569.
CROSS, B. A., and SILVER, I. A. (1962a). *J. Endocr.*, 24, 91–103.
CROSS, B. A., and SILVER, I. A. (1962b). *Proc. R. Soc. B*, 156, 483–499.
CROSS, B. A., and SILVER, I. A. (1963). *Expl Neurol.*, 7, 375–393.
CROSS, B. A., and SILVER, I. A. (1966). *Br. med. Bull.*, 22, 254–260.
DAVEY, L. M., KAADA, B. R., and FULTON, J. F. (1950). *Res. Publs Ass. nerv. ment. Dis.*, 29, 617–627.
DAVIDSON, J. M. (1966). In *Neuroendocrinology*, vol. I, pp. 565–611, ed. Martini, L., and Ganong, W. F. New York: Academic Press.
DAVIS, G. D. (1951). Ph.D. Thesis, Yale University.
DELGADO, J. M. R., and LIVINGSTON, R. B. (1948). *J. Neurophysiol.*, 11, 39–55.
DUA, S. (1956). *Electrical stimulation of hypothalamus in unanaesthetised animals*, pp. 34–92. M.Sc. Thesis, All-India Institute of Medical Sciences, New Delhi.
ELIASSON, S. (1952). *Acta physiol. scand.*, 26, suppl. 95, 7.
ELIASSON, S. (1953). *Acta physiol. scand.*, 30, 199–214.
EULER, C. VON (1953). *Acta physiol. scand.*, 29, 133–136.
FARRELL, G., FABRE, L. F., and RAUSCHKOLB, E. W. (1968). *A. Rev. Physiol.*, 30, 577–588.
FLERKO, B., and SZENTAGOTHAI, J. (1957). *Acta endocr., Copenh.*, 26, 121–126.
FOLKOW, B., and EULER, U. S. VON (1954). *Circulation Res.*, 2, 191–195.
FORTIER, C. (1963). *Comp. Endocr.*, 1, 2–24.
FRENCH, J. D., HERNÁNDEZ-PEÓN, R., and LIVINGSTON, R. B. (1955). *J. Neurophysiol.*, 18, 44–55.
FULTON, J. F. (1951). *Frontal Lobotomy and Affective Behavior*, p. 32. New York: Norton.
GAMBLE, J. E., and PATTON, H. D. (1953). *Am. J. Physiol.*, 172, 623–631.
GASTAUT, H. (1952). *J. Physiol. Path. gén.*, 44, 431–470.
GLOOR, P. (1955). *Electroenceph. clin. Neurophysiol.*, 7, 243–264.
GROSSMAN, R. G., and WANG, S. C. (1955). *Yale J. Biol. Med.*, 28, 285–297.
GROSSMAN, S. P. (1962). *Am. J. Physiol.*, 202, 872–882.
GULATI, P. (1967). *Study of hypothalamic control of total blood volume with special reference to redistribution in pulmonary and systemic circuits*. M.D. Thesis, All-India Institute of Medical Sciences, New Delhi.
GUPTA, P. D. (1960). *Influence of limbic system on renal haemodynamics and urea clearance*. M.Sc. Thesis, All-India Institute of Medical Sciences, New Delhi.
HALASZ, B., PUPP, L., and UHLARIK, S. (1962). *J. Endocr.*, 25, 147–154.
HARRIMAN, A. E. (1955a). *J. Nutr.*, 57, 271–276.
HARRIMAN, A. E. (1955b). *J. genet. Psychol.*, 86, 45–50.
HARRIS, G.W. (1960). In *Handbook of Physiology*, sect. 1, vol. II, pp. 1007–1038, ed. Field, J., Magoun, H. W., and Hall, V. E. American Physiological Society. Baltimore: Williams and Wilkins.
HESS, W. R. (1947). *Helv. physiol. pharmac. Acta*, suppl. 4.

HESS, W. R. (1948). *Vegetative Funktionen und Zwischenhirn.* Basel: Schwabe.

HESS, W. R. (1949). *J. Physiol. Path. gén.*, **41**, 61A–67A.

HESS, W. R. (1954). *Das Zwischenhirn; Syndrome, Lokalisationen, Funktionen.* (2nd edn). Basel: Schwabe.

HESS, W. R., BRUGGER, U., and BRUCHER, V. (1945). *Mschr. Psychiat. Neurol.*, **III**, 17.

HESS, W. R., and MULLER, H. R. (1946). *Helv. physiol. pharmac. Acta*, **4**, 339–345.

HODES, R. N., and MAGOUN, H. W. (1942). *J. comp. Neurol.*, **76**, 109–190.

HODGES, J. R., and VERNIKOS, J. (1958). *Nature, Lond.*, **182**, 725–726.

HODGES, J. R., and VERNIKOS, J. (1959). *Acta endocr., Copenh.*, **30**, 188–196.

INGERSOLL, E. H., JONES, L. L., and HEGRE, E. S. (1956). *Anat. Rec.*, **124**, 310.

INGRAM, W. R. (1960). In *Handbook of Physiology*, sect. 1, vol. II, pp. 951–978, ed. Field, J., Magoun, H. W., and Hall, V. E. American Physiological Society. Baltimore: Williams and Wilkins.

JEWELL, P. A., and VERNEY, E. B. (1957). *Phil. Trans. R. Soc. B*, **240**, 197–324.

KAADA, B. R. (1951). *Acta physiol. scand.*, **24**, suppl. 83.

KAADA, B. R., PRIBRAM, K. H., and EPSTEIN, J. A. (1949). *J. Neurophysiol.*, **12**, 347–356.

KABAT, H., MAGOUN, H. W., and RANSON, S.W. (1935). *Archs Neurol. Psychiat., Chicago*, **34**, 931–935.

KARPLUS, J. P., and KREIDL, A. (1910). *Pflügers Arch. ges. Physiol.*, **135**, 401–416.

KELLER, A. D., and McCLASKEY, E. B. (1964). *Am. J. phys. Med.*, **43**, 181–213.

KOIKEGAMI, H., DODO, T., MOCHIDA, Y., and TAKAHASHI, H. (1957). *Folia psychiat. neurol. jap.*, **11**, 157.

KOIKEGAMI, H., and FUSE, S. (1952). *Folia psychiat. neurol. jap.*, **6**, 94–103.

KOIKEGAMI, H., KIMOTO, A., and KITO, C. (1953). *Folia psychiat. neurol. jap.*, **7**, 87–108.

LANGLEY, J. N. (1921). *The Autonomic Nervous System.* Cambridge: Heffer.

LAT, J. (1967). In *Handbook of Physiology*, sect. 6, vol. I, pp. 367–386, ed. Code, C. F., and Heidel, W. American Physiological Society. Baltimore: Williams and Wilkins.

LENNOX, M. A., and ROBINSON, F. (1951). *Electroenceph. clin. Neurophysiol.*, **3**, 197–205.

LEWIS, M. (1964). *J. comp. physiol. Psychol.*, **57**, 348–352.

LINCOLN, D. W., and CROSS, B. A. (1967). *J. Endocr.*, **27**, 191–203.

LISK, R. D. (1960). *J. exp. Zool.*, **145**, 197–208.

LIVINGSTON, R. B., FULTON, J. F., DELGADO, J. M. R., SACHS, E., BRENDLES, S. J., and DAVIS, G. D. (1948). *Res. Publs Ass. nerv. ment. Dis.*, **27**, 405–420.

LOGAWNEY, S. (1966). *Role of hypothalamus in ACTH secretion in monkeys.* M.D. Thesis, All-India Institute of Medical Sciences, New Delhi.

McCANN, S. M., DHARIWAL, A. P. S., and PORTER, J. C. (1968). *A. Rev. Physiol.*, **30**, 589–640.

MACLEAN, P. D. (1949). *Psychosom. Med.*, **11**, 338–353.

MACLEAN, P. D. (1954). *Electroenceph. clin. Neurophysiol.*, **5**, 91–100.

MACLEAN, P. D., and DELGADO, J. M. R. (1953). *Electroenceph. clin. Neurophysiol.*, **5**, 91–100.

MACLEAN, P. D., and PRIBRAM, K. H. (1953). *J. Neurophysiol.*, **16**, 312–323.

MAGOUN, H. W., ATLAS, D., INGERSOLL, E. H., and RANSON, S. W. (1937). *J. Neurol. Psychopath.*, **17**, 241–255.

MAGOUN, H. W., RANSON, S. W., and HETHERINGTON, A. (1937). *Am. J. Physiol.*, **119**, 615–622.

MAIRE, F. W., and PATTON, H. D. (1954). *Am. J. Physiol.*, **178**, 315–320.

MANGILI, G., MOTTA, M., and MARTINI, L. (1966). In *Neuroendocrinology*, vol. I, pp. 297–370, ed. Martini, L., and Ganong, W. F. New York and London: Academic Press.

MAYER, J. (1966). *New Engl. J. Med.*, **274**, 610–616.

MICHAEL, R. P. (1965). *Br. med. Bull.*, **21**, 87–90.

MORUZZI, G. (1947). XVII Int. Physiol. Congr., Oxford, pp. 114-115.

MOTTA, M., MANGILI, G., and MARTINI, L. (1965). Endocrinology, 77, 392-395.

NAKAYAMA, T., EISENMAN, J. S., and HARDY, J. D. (1961). Science, 136, 506-561.

NAKAYAMA, T., HAMMEL, H. T., HARDY, J. D., and EISENMAN, J. S. (1963). Am. J. Physiol., 204, 1122-1126.

NAUTA, W. J. H., and KUYPERS, H. G. J. M. (1958). In Reticular Formation of the Brain, pp. 3-30, ed. Jasper, H. Boston: Little, Brown.

PAPEZ, J.W. (1937). Archs Neurol. Psychiat., Chicago, 38, 725-743.

PIESS, C. N. (1960). J. Physiol., Lond., 151, 225-237.

PRIBRAM, K. H., and MACLEAN, P. D. (1953). J. Neurophysiol., 16, 324-340.

RAMIREZ, V. D., KOMISARUK, B. R., WHITMOYER, D. T., and SAWYER, C. H. (1967). Am. J. Physiol., 212, 1376-1384.

RANSON, S. W. (1939). Ergebn. Physiol., 41, 56-163.

RANSON, S. W. (1940). Res. Publs Ass. nerv. ment. Dis., 20, 342-399.

RANSON, S. W., FISHER, C., and INGRAM, W. R. (1937). Archs Neurol. Psychiat., Chicago, 38, 445-466.

RASHEED, B. M. (1965). Cerebellar influence over certain vegetative functions. Ph.D. Thesis, All-India Institute of Medical Sciences, New Delhi.

RAY, A. P., ANAND, B. K., DUA, S., and SHARMA, G. K. (1958). Bull. natn. Soc. India Malar., 6, 173-175.

REDGATE, E. S., and GELLHORN, D. (1953). Am. J. Physiol., 174, 475-480.

REICHLIN, S. (1964). Ciba Fdn Study Grp Brain-Thyroid Relationships: with Special Reference to Thyroid Disorders, pp. 17-32. London: Churchill.

RUBINSTEIN, E. H., and DELGADO, J. M. R. (1963). Am. J. Physiol., 205, 941-948.

SAWYER, C. H., and GERNANDT, B. E. (1956). Am. J. Physiol., 185, 209-216.

SCOTT, E. M. (1946). J. Nutr., 31, 397-406.

SEN, R. N., and ANAND, B. K. (1957a). Indian J. med. Res., 45, 507-513.

SEN, R. N., and ANAND, B. K. (1957b). Indian J. med. Res., 45, 515-521.

SHARMA, K. N., DUA, S., and ANAND, B. K. (1963). Indian J. med. Res., 51, 708-715.

SHEALY, C. N., and PEELE, T. L. (1957). J. Neurophysiol., 20, 125-139.

SHERRINGTON, C. S. (1947). The Integrative Action of the Nervous System. (2nd edn.) New Haven: Yale University Press.

SLOAN, N., and KAADA, B. R. (1953). J. Neurophysiol., 16, 203-220.

SMITH, O. A., JABBUR, S. J., RUSHMER, R. F., and LASHER, E. P. (1960). Physiol. Rev., 40, suppl. 4, 136-141.

SMITH, W. K. (1945). J. Neurophysiol., 8, 241-255.

SPENCER, W. G. (1894). Phil. Trans. R. Soc. B, 185, 609-657.

STELLA, G. (1939). J. Physiol., Lond., 96, 26.

SUBBERWAL, U., ANAND, B. K., and SINGH, B. (1965). Indian J. med. Res., 53, 1034-1039.

SURANGE, S. G. (1966). A study of the effects of stimulation of different areas of hypothalamus on total liver blood flow. M.D. Thesis, All-India Institute of Medical Sciences, New Delhi.

THOMAS, S. (1963). The hypothalamic regulation of thyroid activity in monkeys. M.D. Thesis, All-India Institute of Medical Sciences, New Delhi.

THOMAS, S., and ANAND, B. K. (1968). XXIV Int. Physiol. Congr., Washington.

TURNER, E. A. (1954). Brain, 77, 448-486.

VERNEY, E. B. (1947). Proc. R. Soc. B, 135, 35-106.

WANG, G. H., STEIN, P., and BROWN, V. W. (1956). J. Neurophysiol., 19, 340-349.

WANG, P. Y. (1960). Scientia sin., 9, 434-444.

WANG, S. C., CLARK, G., DEY, F. L., and RANSON, S. W. (1940). Am. J. Physiol., 130, 81-88.

WANG, S. C., and RANSON, S. W. (1939). J. comp. Neurol., 71, 457-472.

WANG, S. C., and RANSON, S. W. (1941). *Am. J. Physiol.*, **132**, 5-8.
WARD, A. A., JR. (1948). *J. Neurophysiol.*, **11**, 13-23.
WILSON, M. F., CLARK, N. P., SMITH, D. A., and RUSHMER, R. F. (1961). *Circulation Res.*, **9**, 491-496.
WOOD, C. D., SCHOTTELIUS, B., FROST, L. L., and BALDWIN, M. (1958). *Neurology*, **8**, 477-480.

DISCUSSION

Fortier: I am interested by your identification of a TSH-inhibitory centre in the hypothalamus. Where is this centre located and what is the meaning of your different parameters of stimulation with respect to inhibition? Is it to be inferred that stimulation of a given site with certain parameters results in excitation, whereas inhibition is obtained with different parameters?

Anand: I was referring to the well-known fact, which applies to other areas besides the hypothalamus and limbic system, that in a situation where there are different types of neurons you can selectively stimulate some neurons with a certain parameter of stimulation, and when you change the parameter of stimulation you may activate other neurons that lie nearby. That is why if you are investigating the responses from any area, and especially the hypothalamus which has such a convergence of neurons, one has to change the parameters of stimulation to observe the different responses produced.

We have found that a decreased turnover (release) of radioactive iodine from the thyroid into the blood occurs when we stimulate the anterior-most portion of the preoptic area of the hypothalamus. We also know that neurons in this area mostly respond to an increase in the body temperature. So here is a mechanism which integrates the temperature-adjusting responses to increased body temperature with simultaneous inhibition of TSH secretion. We could only get inhibition of the TSH response from stimulation of this area. On the other hand, inhibition of the ACTH response was obtained by stimulating posterior regions of the hypothalamus, mostly the mammillary region.

Fortier: A technical difficulty in establishing the existence of a TSH-inhibitory centre results from the highly unspecific inhibitory response of TSH to any kind of stimulation, with the obvious exception of cold. It is well known, indeed, that, cold excepted, any stimulus resulting in ACTH release concurrently inhibits TSH secretion (see review by Brown-Grant, 1966). Couldn't you possibly have induced non-specific stimulation which would account for the observed TSH inhibition?

Anand: This is an important question, and to test for this non-specificity we tried to study both the ACTH response and the TSH response in the same monkey. I am not saying that this is a specific TSH-inhibiting *centre*, because by this we mean a very concentrated group of neurons producing a specific response. This response can be obtained from a wider area in the anterior-most portion of the hypothalamus. When we stimulate this area there is a diminution in the turnover of radioactive iodine (TSH) without any simultaneous increase in ACTH. When we stimulate the medial portion of the hypothalamus from the supraoptic area right up to the median eminence we specifically increase the TSH turnover; when we stimulate points lateral to these there is no change in TSH. If the stimulation of the median eminence in some animals simultaneously increases ACTH this results in decreased TSH turnover, which is secondary to the ACTH response.

White: You have referred to glucoreceptors; there are examples in mammals in which a particular structure responds with secretory activity to alterations in the concentration of a metabolite arriving at the secretory cell. The classical example is of insulin secretion by the pancreas which is regulated by the level of blood glucose. Another example is the secretion of parathyroid hormone or calcitonin depending on the level of blood calcium. In the case of an ion, for example, Ca^{++}, one thinks of explanations in the area of enzyme activation and inactivation. What general models involving permeability factors and mechanisms of secretion can be developed for other substances affecting cellular secretory activity, such as glucose? What do we mean when we say that there is an "analyser" in the islet cells of the pancreas that tells the pancreatic cell how much blood glucose is present?

Monod: Or equally when we say that there is an analyser in the hypothalamus which promotes male or female behaviour depending on whether the hormone is testosterone or oestrogen?

Anand: It has been shown that the activity changes in the hypothalamic neurons produced by increased glucose utilization are not a result of sensory projections coming from receptors at the periphery, because even after making midcollicular sections these changes can still be produced. Also, the neuronal activity is related to the utilization of glucose rather than to the level of blood glucose. But then, how does glucose utilization specifically increase the activity of these hypothalamic neurons? Is it by action on the cell membrane; or by increased concentration of glucose inside the cell increasing cellular activity; or are there special enzymic reactions which are triggered? We have no specific answer for this.

We have investigated the sensitivity of hypothalamic neurons to the level of circulating gonadal hormones along with sensory stimuli originating from the genital organs, as they interact together to give a specific response. In the preoptic area of the hypothalamus, neuronal activity is inhibited in adult female monkeys by vaginal stimulation. We took sexually immature female monkeys and on vaginal stimulation it was observed that the preoptic neurons discharged at a higher rate. When the same monkey was given one injection of oestrogen, and records were taken from the same neuron, vaginal stimulation now reversed the excitatory response to an inhibitory one. So when we talk of the specific sensitivity of a neuron to some chemical mechanism, possibly other mechanisms are also involved.

Gopalan: What is the evidence that acute anxiety and emotional crisis can increase secretion of ACTH? For example, it has been documented that even the type of anxiety associated with examinations in college students can produce striking negative nitrogen balance. Is there any evidence that this response is mediated through increased adrenocortical secretion?

Anand: In such an anxiety state the bodily responses are mediated through a number of mechanisms operating through the central nervous system. Not only is there an increased ACTH output but many other visceral systems are influenced simultaneously through the autonomic nervous system.

REFERENCES

BROWN-GRANT, K. (1966). In *The Pituitary Gland*, vol. 2, pp. 235–269, ed. Harris, G. W., and Donovan, B. T. London: Butterworths.

SOME GENERAL INTEGRATIVE ASPECTS OF
BRAIN FUNCTION

ROBERT B. LIVINGSTON

Neurosciences Department, University of California at San Diego, La Jolla, California

WHEN I left New York, newspapers carried a Lichty cartoon which I thought prophetic for my part in this symposium. The cartoon showed the chairman of a Speaker's Committee introducing Senator Snort, saying, "In place of the astronauts who were unable to come, we have a speaker today who has often engaged in spectacular flights of fancy!" Professor Christopher Longuet-Higgins was the intended speaker for this final paper; I am therefore substituting for a "real astronaut". We have missed Chris for his many potential contributions throughout this meeting, but never more keenly than now!

Recall first the characteristics of life referred to by Professor Monod: self-reproduction and goal-directedness. These criteria are met by all living systems, from slime mould to man; they are not met by any non-living systems. Quite patently, the goals, goal-seeking behaviour, reproductive mechanisms and fashions of reproduction vary enormously throughout the plant and animal kingdom, yet most of the underlying molecular tricks to sustain life are the same. It is now our purpose to attempt to characterize the neurological mechanisms by which such complicated processes may operate in higher forms of life.

The nervous system has assumed so dominant a role in the goal-directedness of mammalian behaviour that we can concentrate our attention on its integrative processes. Integrative aspects of nervous systems are ultimately dependent on underlying molecular processes similar to those discussed earlier in this meeting. But the processes which organize functions in nerve cells and glia, which in turn organize functions in nervous systems, have introduced into evolution qualitatively new and more powerful overall controls. Consciousness and cultural communication which have emerged through nervous system evolution now exercise influences that are already deleteriously affecting and might even effectively destroy the entire biosphere. This is goal-directedness, to be sure, but with a mean and narrow frame of reference for the goal assumed. Self-reproductiveness has also got out of hand in the largest sense: there is some general lack of appreciation of the distinction between recreation and procreation!

I will attempt first to sketch in a few firm and a few speculative notions as to how the brains of men may work. Then I will attempt to indicate how I believe we can achieve a profounder understanding of the plights and the opportunities of mankind if we will but devote additional efforts toward understanding the integrative processes of the human nervous system as these processes account for perception, value judgements and behaviour. This Ciba Foundation symposium is just such an endeavour. Naturally, there will need to be some "spectacular flights of fancy" at this primitive stage in the development of neurosciences understanding.

EVOLUTION OF NERVOUS SYSTEMS AFFECTING EVOLUTION

Nervous systems have been elaborated from cellular and intercellular functions already present in rudimentary form in non-nervous cells. At the level of cellular differentiation there appear to be no discontinuities in evolution and none are obvious in the initiation of neurons from ancestor cells. Who can venture the limits to such anti-entropic processes?

There has been one notable break in the otherwise smooth curve of evolutionary processes. Some few thousands of years ago (or perhaps a few hundred thousands), during the most recent moments on an evolutionary scale of time, mankind began experimenting and creating a remarkable discontinuity in evolution. Since some period of the murky beginnings of speech and the formation of symbolic language, conscious cultural transmission has evolved rapidly. It has evolved increasingly rapidly during the last few thousand years and with astonishing rapidity during the last few decades. It is possible that the limitations of the speed of light and some other boundary conditions may introduce an asymptote to this evolutionary jump, even if we escape from massive catastrophe.

Conscious cultural transmission is enormously more powerful than biological heredity, because it can acquire advantages from any sources, not only from direct ancestors; and it can distribute advantages unlimitedly, not only to immediate descendants. Conscious cultural transmission coupled to increasingly precise and fundamental control of biological heredity has extraordinary potentialities. Marshall Nirenberg has cautioned that "When man becomes capable of instructing his own cells, he must refrain from doing so until he has sufficient wisdom to use this knowledge for the benefit of mankind. I state this problem well in advance of the need to resolve it, because decisions concerning the application of this knowledge must ultimately be made by society, and only an informed society can make such decisions wisely" (Nirenberg, 1967).

Recent gains in man's control over the environment, recent expansion of the sources of power available, achievements in communication and transportation, and extensions of brain potentialities by resources now becoming available, all provide colossal opportunities for either the enormous improvement or the utter destruction of the biosphere. If we can combine conscious cultural transmission with an increasing measure of control over biological heredity toward the improvement of things that mean the most in human life, we shall have taken an important step from being *Homo faciens*, the doer, the manufacturer, to becoming what we have called ourselves, by aspiration, *Homo sapiens*, the wise one.

RELATIONS BETWEEN MOLECULAR BIOLOGY AND BRAIN CIRCUITRY

There exist bridges between the molecular and chemical control systems and the rather global organ system controls that we are about to consider. These are like the flimsy suspension bridges across mountainous chasms encountered along the footpaths linking Katmandu with the summits of the Himalayas and Karakorams. Clinging to the weak principles suspended between different levels of complexity of organization, using slippery facts for foot treadles, crossing over the roaring torrents of literature cascading beneath, few individuals venture from the cities to the glaciers, while the Sherpas feel at home only in their own rarified and tilted environment.

Built-in codes

The steps leading from molecular controls to information processing in the nervous system can be looked on as "codes". It is clear that there is not one code for the nervous system, as for the genetic code, but many nervous system codes. There is, moreover, no single locus for the action of any of the several different codes (Perkel and Bullock, 1968).

The first point about coding in the nervous system is that there are certain subjective and behavioural operations that are built into limited regions of the brain. There are genetically endowed centres the activation of which generate "like" and "dislike" subjective experiences and organize such contrasting behaviours as approach and avoidance. Within these systems are subsystems characterized as appetitive and satiety regions related to thirst, hunger, sex, freedom from constraint, novelty, and so on. These built-in systems relate to generalized mood and feeling states and are thought to be instrumental for learning and memory, for example, as reinforcement systems (Livingston, 1967).

These mechanisms are difficult to describe except in operational terms but they have been demonstrated in the brains of a wide variety of mammals including man. These built-in controls belong to phylogenetically ancient parts of the nervous system and are likely common denominators for all vertebrate nervous systems. They are closely connected with brain-stem systems for general control of posture and locomotion—for motor readiness, for example. They are also closely connected with hypothalamic-pituitary systems for the general control of the internal milieu, such as visceral readiness. The counterparts of these approach and avoidance systems in invertebrates have not been revealed, but we may reasonably suppose from behavioural observations that something comparable must be governing in the nervous systems of these creatures.

Applied on top of these built-in approach and avoidance controls are systems which express in increasing degree in higher mammalian forms more rapid conduction and more distinctive functions. These relate to special sense receptors, fine discriminative cutaneous receptors, and to the discriminative controls effecting skilled motor acts. These systems are more popular among anatomists and physiologists because their fibre connexions are larger in diameter, more thickly sheathed with distinctive myelin; their nuclear waystations are composed of more uniform populations of cell bodies, more clearly distinguished from the surround; and their transmission of signals is more localized, discrete and specific as to sensory signalling within a given sensory modality and motor signalling within given motor nuclei.

The most impressive coding within such built-in and discrete systems is by means of the "labelled line". The "labelled line" refers to the common-place observation that signals generated along the visual pathway yield visions, auditory pathway activation elicits acoustic experiences, olfactory stimulation smells, and so on. Excitations within the systems relating to appetite, approach, satiety and avoidance induce general moods which in turn yield complex, goal-directed activities. These may be considered as "hedonistic labelled lines". If goal-directed activity is frustrated one way, the individual will seek its realization by another, for as long as the mood persists.

Other codes apply at unitary levels: information may be keyed to the moment of firing, the interval between firings, spatial and temporal arrays of firings of units, and to neural events other than impulses. These latter include the induction of generator potentials in receptors, synaptic potentials at synapses, and spatial, temporal and amplitude distributions of graded potentials. Intercellular events include the release of transmitters and ions, ably described here by Dr Whittaker (pp. 338–349), the elaboration and release of neurosecretions, and electrotonic couplings by means of electrical synapses and through electrotonic interactions among active and passive cells and fibres.

An exemplary evaluation of established and candidate codes in these and other categories of neuronal information transfer is provided by Perkel and Bullock (1968). The point I want to make here is that some tenuous relations are already at hand to identify how to proceed from discussions of molecular events to the control of membrane and intercellular events, and, we hope without lapse of logical continuity, to discussions of information transmission by nervous systems. It is still heady and dangerous work getting across that chasm, but there is a will to find a way.

ORGANIZATION OF MAMMALIAN NERVOUS SYSTEMS

Input–output relations

The nervous system, of course, has receptors, including sensory end-organs in skin, muscle and viscera, as well as special sense receptors in the nose, mouth, eye and ear. Processes, direct and relayed, extending from these sensory receiving stations to enter the central nervous system are clearly *afferent*, sensory in function. Nerves which govern the secretions of glands, contractions of muscle and migrations of pigments are equally clearly *efferent*, motor in function. All other neurons are, strictly speaking, providing transitional transmission between input and output units. Can we be certain whether these are primarily sensory or motor? Among cells arising in the central nervous system and going toward skin, muscle, viscera and special senses (and therefore presumably efferent), are classes of nerves which serve to control the performance of the sensory cells. They can alter the likelihood of firing of the incoming cell. Are such neurons motor? Or sensory?

What about interneurons lying directly between sensory input and motor output? They are receiving afferent signals and would appear to be sensory in that context, but they are delivering signals to efferent units and appear to be motor in that context. This is true for all central neurons whether they are ascending along so-called sensory pathways (from which they may deliver messages surprisingly directly to motor centres) or they are descending along so-called motor pathways (from which they may deliver messages surprisingly directly to sensory relay stations). Even the pyramidal tract, the great cortico-spinal motor pathway, thought to be essential for skilled voluntary performances, sends off collateral branches which have multiple sensory influences along the way. I exaggerate our dilemma slightly, but do so because the older categorization of sensory and motor pathways up and down through the central nervous system has long distracted us from observing the biological processes in strictly operational terms. Professor Walle J. H. Nauta at the Massachusetts Institute of Technology has a good term for the remainder of the nervous system after you subtract the obviously afferent and obviously efferent. He calls this remainder the Great Intermediary Net.

Let us go a step further. Let us consider the example of Professor Longuet-Higgins who is a superb pianist (as well as astronaut). In learning to play the piano he, like any musician, became accustomed to hearing the effect of his own motor performance. After much practice, the nervous system is presumed to build up certain expectancies, programmed anticipations respecting the environmental loop lying between motor performance and sensory feedback resulting from that performance. What we think of as the execution of a motor skill is in reality a motor-sensory skill involving not only the feedback of information from sensory units in limbs but also the acoustic feedback. It is well known that if one provides an erroneous time lag for the acoustic sensory feedback, there will be an involuntary decomposition of the central motor control processes. Analogous decompositions of perceptions can be achieved by interfering with motor performances. So it is helpful to consider motor and sensory mechanisms as *interdependent systems*, interdependent peripherally as well as centrally. We can now look at input–output relations of the nervous system from still another vantage point: sensory perception and motor action can best be appreciated from the point of view of the goal orientation of the individual. We will come back to this consideration later.

Another aspect of input–output relations concerns the numbers of potential channels for information transmission. Difficulties are encountered in counting the numbers of fibres in a nerve bundle, and the numbers of afferent, dorsal root ganglion cell bodies, and the numbers of motor and autonomic nerve ganglion cell bodies. Many counts done using light microscopy will have to be radically revised upwards because electron microscopy has disclosed considerable numbers of nerve fibres that are too small to visualize with light microscopic magnifications. So far there has been little interest in making such counts with electron microscopy, but you can appreciate why. Next, there is the difficulty we encountered earlier of establishing among the nerve fibres which ones are coming in and which are going out, and which among each of these directional categories are really sensory and which motor.

Assuming the correctness of present estimations, there are only a few tens of millions of afferent fibres and probably no more than a few million efferents, give or take fifty per cent. Nevertheless, it is widely acknowledged that there is a very much larger number of neurons that are confined entirely to the central nervous system. There are perhaps ten billion (10^{10}) (the most popular and convenient number, although only a crude guess). What do such numbers imply? That there are at least several thousands of central neurons for every possible effector neuron. There would be already a considerable convergence (perhaps 10:1) going from input units directly to output. Adding the central nervous system into the circuit introduces an enormous obligatory convergence on motor expression (perhaps 5–10 000:1). The only behavioural expressions possible in mammals are through secretions of glands, contractions of smooth, cardiac and skeletal muscles and migrations of pigment granules. Professor Walle J. H. Nauta has likened this nervous system convergence toward effectors to the hypothetical problem of a hotel with five to ten thousand guests but only one bathroom!

The principles of evolution being what they are, it is indubitable that nervous systems have gained survival advantages through the expansion of numbers of sensory intake pathways and pathways for the central control of sensory transmission, in comparison with the dimensions of the action systems available. Expansion to a very much larger extent of the wholly central pathways (presumably engaged in sorting out perceptions and in creating and evaluating various options for action), has also enjoyed conspicuous evolutionary advantage.

Horizontal and vertical organizations

The mammalian nervous system is long, like a sausage or a loaf of French bread. It is therefore simpler to cut across; you might say there is "practically no end" to cutting it lengthwise. Because of this simple longitudinal characteristic, most physiological interventions, and most anatomical studies, have involved cutting across and hence have directed our attention toward the segmental or cross-sectional controls acting at different levels of the neuraxis. As a consequence, we know more about segmental reflexes than we do about longitudinal control systems affecting integration lengthwise along the neuraxis.

What are some of the integrative capabilities of the lower parts of the spinal cord? Patients with complete transection of the spinal cord can recover residual reflex capabilities which are remarkably complex in their composition of biologically useful functions. These operate over considerable bodily regions, for example, both legs and pelvic regions, integrating skeletal muscles, smooth muscles and glandular responses. And they may generate sequences of spontaneous and reflexly evoked activity enduring over considerable intervals of time, measured in minutes or longer. An example is the recovery of periodic emptying of the urinary bladder, provided that the immediately relevant segments of the spinal cord are not destroyed. If sacral and lower lumbar segments are intact in the male, erection can be reflexly elicited. With still higher lesions so that upper lumbar and lower to middle thoracic segments are included, erection and reflex ejaculation, with normal seminal fluid and viable, active normal sperm, are produced. These reflex patterns occupy a space–time co-ordination framework that is quite elaborate: all parts of the complicated chain of reflexly guided subsystems are co-ordinated toward the end of reproduction.

Paraplegics can be taught to recognize when their completely isolated lower neuronal segments are in trouble, for instance, when the urinary bladder needs emptying. They learn that some clue is regularly provided by the somatic or autonomic nervous system controls governing the isolated lower segments. These clues can be perceived and interpreted as useful signals. For example, some paraplegics get a mild headache accompanying the reflex rise in arterial pressure in the legs which in turn is an accompaniment of distension of bowel or bladder. Others detect skeletal muscular jerking or changes of tone in muscles of the legs. These signals are sufficient to alert the patients to respond appropriately. How much of this kind of peripheral loop signalling is present in normal subjects is an interesting but entirely unanswered question.

Propriospinal and spino-bulbar reflexes

In experimental animals, appropriate electrical stimulation of any one of the dorsal roots of the spinal cord gives rise to well-known monosynaptic and polysynaptic flexor reflex responses ipsilaterally, and crossed-extensor reflex responses contralaterally. What is less well known is that polysynaptic responses following such stimulation are detectable from ventral roots on both sides of the spinal cord all the way along the spinal cord and brainstem including all of cranial motor outflow. The output responses show an increasing latency as the impulses travel along the neuraxis from the point of entry of the stimulus. These polysynaptic intersegmental reflex responses represent a diffusely projecting system triggered by the appropriate sensory input (Shimamura and Livingston, 1963). The effective sensory input is a barrage of cutaneous afferent impulses such as might be initiated by an abrupt mild pain stimulus. This diffusely projecting longitudinal sensorimotor reflex signal goes down as well as up the spinal cord and brainstem from whatever point of stimulus input, radiating upwards and downwards from stimuli applied to the face, arm, trunk or leg.

Another reflex effect initiated by the same stimulus input is quite different. This is dependent on relay by cells in the general region of the vital cardiovascular and respiratory centres in the medulla oblongata, the bulb. Impulses enter via the dorsal roots of the spinal cord and sensory roots of the brainstem. They travel, without motor expression, to the bulb whence they are relayed and re-transmitted to all levels of the spinal cord and brainstem. Along each motor outlet impulses are emitted as a reflex motor response. Destruction of the bulbar relay leaves the diffusely projecting direct propriospinal reflex response intact. The propriospinal reflex is controlled by projections widely distributed along both sides of the spinal cord that are relayed back and forth from side to side of the cord so that hemisections are circumvented, even when two hemisections are made from opposite sides of the spinal cord one or two segments apart. This is a secure pathway extending throughout the length of the neuraxis. The spino-bulbar relayed response is reflected both ways from the bulb, up into the brainstem and down the length of the spinal cord as a result of stimulating any appropriate input source (Shimamura and Livingston, 1963).

An interesting feature is that the bulbar relayed reflex travels exactly enough faster than the propriospinal reflex that the long-circuited one meets the direct reflex for combined motor effects at the levels of the upper and lower extremities. A stimulus introduced at the lower limb level elicits a succession of increasingly delayed responses going upwards: these are met at the upper extremity by the spino-bulbar circuit which has gone up to the bulb, relayed there, and returned to the level of the upper limb motor nuclei by the time the propriospinal reflex reaches that level. Similarly, stimuli introduced at the upper limb level yield both propriospinal (slow) and spino-bulbar relayed (fast) reflexes which meet at the level of motor segments for the lower extremity. The goal-seeking advantages of these two different pathways whose reflex responses are synchronized at the point of expression appear to be to mobilize the extremities at the same instant with both direct and relayed messages. The bulb serves as an important relay station because it is the focal point for a wide variety of cortical, basal gangliar, cerebellar, spinal and visceral signallings, the point of convergence of messages from widely distributed important parts of the nervous system. The bulb thereby serves as an important locus for longitudinal integration. The combined reflex response is a generalized flexion. This reflex combination has been demonstrated by Shimamura in the cat, dog, monkey and man (Shimamura et al., 1964).

A physiological basis for the "magnet" response

A somewhat more elaborately goal-directed reflex response is elicited when a cat's foot is pushed upwards from the plantar surface, as in weight-bearing. This represents a convergent sensorimotor control system that involves the bulbar and spinal reticular formation. This reflex is related to vestibular input: whenever the vestibular system is appropriately stimulated there is a projection of impulses, relayed in the bulb, which results in a reflex response via all spinal motor outflows. Because of the bulbar relay there is a dependence of this downward projection on the integrity of the bulbar reticular formation. Now, if at the same time that the vestibular signal is introduced there is an appropriate dorsiflexion of the foot, then the normal vestibular ventral root reflex response is greatly exaggerated (Gernandt, Katsuki and Livingston, 1957). The motor outflow relates exclusively to the foot being dorsiflexed. The net effect is to induce an increased motor discharge affecting both flexors and extensors in that limb only. This results in that limb becoming more pillar-like, more capable of bearing weight.

This vestibulo-spinal effect, magnified locally according to the posture of the foot, assures that the cat does not have to use its forebrain to "decide" exactly when to provide the extra motor strength needed for that extremity to support the body. Both timing and force "decisions" originate locally according to when the limb is dorsiflexed and put into an appropriate position to support the animal's weight. You can readily appreciate the importance of such a local limb-supporting control. This combines with ongoing vestibular contributions at all spinal levels. Each time the cat touches its paws to the ground while walking, jumping, springing or landing, the vestibular outflow will automatically be greatly reinforced locally according to the appropriateness of the posture of each foot. Whatever may be the higher level considerations by the cat in directing its bodily performance, there are local "decisions" which contribute to the strength and stability of the limb as a weight-bearing member. There is no wasted anticipatory discharge, nor any protracted discharge after it is no longer needed for useful action, in respect to any individual limb. Uneven and moving surfaces can be used for support without having to negotiate all controls from the forebrain.

Convergence and divergence for integration

The Scheibels have revealed the extent and pattern of convergence of collateral fibre contributions which invade the brainstem reticular formation from classical sensory and motor pathways (Scheibel and Scheibel, 1967). They have shown overlapping and convergence of inputs and outputs of reticular units throughout the neuraxis. Reticular cells located at different levels possess dendrites which reach out widely to invade the fields within which the major sensory and motor pathways send arborizing collateral terminations. Because of this reaching out, the dendrites of the reticular formation are called iso-dendritic, reaching out uniformly and ubiquitously throughout these territories. When a brainstem microscopic section is stained with silver and held up to the light, one is impressed by the appearance of a sunburst effect radiating outward from the reticular core as a result of light reflected from these radially distributed fibres. These fibres form a signal-sampling device that enables reticular cells to gain access to many modalities of both sensory and motor traffic. The largest of the reticular units have remote axonal connexions which may extend all the way from the forebrain to the spinal cord, with many intervening way-stations (Scheibel and Scheibel, 1967). This input convergence and output divergence allows reticular cells to be in communication with a wide variety of incoming, intrinsic, and outgoing messages relating to a wide spectrum of potential perceptions, value judgements and actions. This is the very essence of integration. The connexions suggest that the role of these neurons is to stitch the nervous system together so that the organism can function in an integrated way, as a whole.

Similar reticular units are widely distributed throughout the bulbar zone which yields generalized inhibitory influences on somatic and visceral motor systems and the pontine and midbrain region which controls generalized facilitation of motor systems and generalized arousal of the forebrain. These elements play an important role in the creation and maintenance of the central tides of wakefulness, sleep, readiness for action and attention, and relaxation. Certain cortical fields project into these systems and can arouse a naturally sleeping animal and in other ways influence general states of the animal. Thus we observe that certain goal-directed coordinations influencing the system as a whole in the government of posture and locomotion, musculoskeletal and visceral preparedness, and corticofugal contributions to these general control processes, provide coordinations indispensable for the operation of the individual as a whole. It is partly from this region and partly through the influence of projections passing through this region that the brain exercises its dynamic control over sensory input (Livingston, 1959).

Limbic mechanisms

Perhaps the most interesting control system is one that is farthest removed from the input–output connexions of the nervous system. We may anticipate that these most remote systems have access to the most highly abstracted information and that they generate the most abstracted commands. The limbic system sits like a double doughnut surrounding either side of the upper end of the brainstem. The brainstem is inserted between these two doughnut rings and there is a heavy penetration of the medial brainstem by limbic projections just at that junction. It is in this location that upward-bound and downward-bound traffic to and from the cerebral hemispheres is vulnerable to limbic influences.

The only direct sensory input to the limbic system is olfactory. But the limbic system does receive indirect and presumably heavily processed information from other sensory systems (Nauta, 1964). The most anterior extensions of neocortex, the frontal lobes, are the most remote (from input and output) regions of the neocortex. They are fed by a combination of long cortico-cortical association pathways from cortical association areas and projections from the limbic system. The frontal lobes project downward through the hypothalamus via a pathway commingling with and sharing limbic system outflow, identified as the medial forebrain bundle. The destiny of this combined projection is to the medial part of the midbrain reticular formation, into the midst of mechanisms for the control of states of consciousness, readiness for action and release of action for approach or avoidance. A further outfall is by way of the hypothalamic-pituitary mechanisms governing neuro-endocrine control of the internal milieu.

Positive and negative reinforcement systems

Jim Olds and Peter Milner, fifteen years ago, initiated a sequence of investigations devoted to the exploration of brain mechanisms involved in motivation. They found that they could obtain positive reinforcement by stimulating selected regions of the brain (Olds and Milner, 1954). These regions were later identified as coinciding largely with the limbic system. In the same year, Delgado, Roberts and Miller discovered a complementary set of loci yielding negative reinforcement, located more medially and posteriorly (Delgado, Roberts and Miller, 1954). By positive reinforcement is meant that an animal will seek out places where it was when such brain stimulation was administered; it will do work, learn to run a maze, cross an electrified grid or eat bitter food to obtain such stimulation. Negative reinforcement has an opposite effect: the animal will avoid places where such stimulation was delivered, will do work, learn a maze, and so on, to turn off such stimulation.

Regions for both positive and negative reinforcement vary in the intensity of the drive states elicited and show differences in preference of reinforcement effects relating to appetite and satiety. The entire limbic, hypothalamic and midline midbrain region seems to be the site of control of moods and motivations and to participate in the subjective experiencing and outward expression of emotional states. An important fact is that other, vast regions, including the bulk of the neocortical mantle, most of the basal ganglia, most of the dorsal thalamus and most of the cerebellum are without reinforcement effects; they are neutral or indifferent in these respects.

Positive and negative reinforcement effects are sometimes referred to as something akin to "reward" and "punishment" systems. This is helpful in visualizing their influence on behaviour. From what has been said already about their discovery, it is obvious that behaviour can be shaped by electrical stimulation of these central reinforcement systems. Similarly, local chemical stimulation is effective. Much effort has been directed to determining the approximate equivalence of external and internal positive and negative reinforcements and to securing hierarchies of preferences. It is likely that these systems play a substantial role in the "like" and "dislike" behaviours of humans and that they provide the drive states essential for learning.

Possible memory storage mechanisms

It has been established that the bilateral loss of about two-thirds of the anterior part of the hippocampus and closely associated temporal lobe structures which make up part of the limbic system is followed by permanent loss of the ability to store new memories (Ojemann, 1966). An individual with such a loss will be able to function apparently normally in a given context but he cannot remember events after a few minutes: he cannot learn the names of new associates, cannot remember what he had for breakfast, cannot recall his own drawings and has an essentially blank slate of memories from the time of his temporal lobe loss. Yet he is able to recall in normal fashion memories of his antecedent experiences. This establishes that the hippocampus is not the site of memory storage but is somehow essential for the process of laying down of new memories.

Functions and interrelations of the limbic and brainstem systems suggest that these may play a sequential role in the process of memory storage (Livingston, 1966). The reticular formation is interpreted as playing a role in the recognition of familiar objects. The experiencing of novelty is associated with strong orientation reactions with reticular and forebrain electrical activation and closely associated synchronized waves (theta rhythms) in the hippocampus. It is supposed then that the limbic system is involved in evaluating whatever is novel or has not become habituated, the evaluation being a prognostication of the need for approach or avoidance in respect to that novelty, "like" or "dislike". The nervous system is apparently designed so that events followed by positive or negative reinforcement, "reward" or "punishment", tend to be stored in memory. The stronger the polarity of the reinforcement, in general, the more enduring and vivid the memory. According to this view, whenever an event is followed by strong reinforcement there will be a strong outflow discharge from the limbic-hypothalamic systems. This discharge goes into the brainstem reticular system and effects a generalized arousal of the forebrain. It is postulated that this arousal is accompanied by a message, probably neurochemical or neurohumoral in nature, which signifies "now store!", an order that has an effect of inducing growth or improvement of synaptic relations among all the circuits just previously activated. In this way the phenomenon of generalization could be explained. That is, non-relevant aspects of the event or environmental context will be remembered as well as relevant ones. It would account for all combinations of connexions: all just previously active circuits would be caused to be improved by growth or synaptic improvement. It would account for the fact that spurious aspects of the event tend to lose behavioural significance because they will not be regularly associated with reinforcement. This has been likened to the fading out of transient and moving objects in a long-exposure film and the continually more deeply engrained configuration of the constancies.

Guidance for individual and species survival

MacLean has concluded from his extensive study of the limbic system that the lateral parts of this system, particularly the amygdaloid complex, are involved in the organization of behaviours important to the survival of the individual: search for, securing and consuming food, establishing and defending territory and fighting for self-maintenance. The medial portion of the system, particularly the septum, he believes has to do primarily with survival of the species: grooming, searching for a mate, securing a mate, mating, fighting for a mate and protecting offspring, nesting, den and home-making behaviours (MacLean, 1958).

SPECULATIONS RELATING BRAIN FUNCTIONS TO THE HUMAN PREDICAMENT

Traditionally the nervous system has been looked upon as a passive system responding to stimuli from the environment, an essentially stimulus-bound system. Increasingly it has been necessary to recognize that the nervous system is spontaneously active, composing itself in relation to sleep and wakefulness cycles, moods, appetites, what it pays attention to, what it recognizes and what it remembers, and emitting actions which lead systematically to the selection and approach or avoidance of stimuli. Recently it has been recognized that the central nervous system is controlling its own afferent input in much the same way that it controls its own motor output. These controls are exerted on the sensory receptors and sensory relay stations and seem designed to shape perceptions according to past experiences, expectations and purposes.

Psychology has long known that these factors affect testimony relating to perception, but it has been supposed from the time of Descartes that such signals are reliably represented centrally and presumably available to consciousness. It is now evident that modulation of sense data can begin at the earliest stages of incoming signals. Sensory controls can expand or contract or exclude or divert sensory signals to channel them into different pathways and combinations according to past experiences and physiological states momentarily obtaining. The central control of sensory transmission is exercised in organized, coherent, goal-directed ways according to the central state of the organism. The nature of the stimulating world is coming to be recognized as having a peculiar relationship to the nature of perceptual experiences, a relationship that is idiosyncratic according to the past experiences, motives and anticipations of the individual. This has been more widely recognized as a phenomenal part of life and accepted from a philosophical point of view in Asia than it has been in Europe and America. Bolstered by scientific assumptions and by experiments designed expressly to establish the regularities of sensory processing, Europeans and Americans have been relatively more resistant to the fact that what each man perceives is something to which he is continuously, involuntarily and actively contributing.

We all recognize that a normal newborn baby can be reared in any society and become fully accepted and behaviourally consonant with that society. But if the child is transplanted after the age of about three years, he exhibits many trace influences of his "likes" and "dislikes", his inner logic and world view, and his attachment to environmental characteristics of his land of origin. If he migrates still later, he carries with him a greater encumbrance of such stigmata. If he makes such a transition after about the age of twenty, he will be an obvious transplant even to a momentary and casual observer, and he will probably never make the full adjustment of internal commitment and identification with his new society. Of course, much depends upon how different the two cultures are. This in general is the commonplace experience. It is evident that the nervous system is involved in the idiosyncratic acculturation commitments exhibited.

It appears that our perceptual and value-judgement systems as well as other mechanisms controlling our behaviour become shaped early in life to a set of relatively stable constancies. We can now visualize some of the brain mechanisms involved in this process. In the past it has been assumed that when two people were witness to the same event and gave conflicting testimony, one or the other or both was advertently or inadvertently distorting the truth. Courts of law are predicated on defining where the truth lies from discrepant testimony.

It has long been assumed that each witness to the same event must have had practically equivalent sensory data from which to establish his convictions and to derive his testimony. From what is now being demonstrated in animals we may have to be more cautious about such conclusions. It appears likely that every man has a distinctive set of perceptually distorting lenses, partly culture bound, partly individualistic. He will have more or less difficulty in getting along with others who have acquired different lenses according to how discrepant the two lens developments have been and whether or not either of the two of them has encountered and already developed enough tolerance for varieties of people with substantially different lens systems to be able to accommodate new interface discrepancies constructively. The main problem is that once we have become committed to a given set of cultural or individual perceptually distorting lenses, we cannot take them off. They have become too strongly and too often reinforced by the particular circumstances in which we were reared for us to be aware of them, much less to be able to disengage ourselves from them. And those who have shared equivalent experiences with us have developed similar lenses—giving us the false impression of veridical perceptual experiences. Patriotism is a cultural force to oblige conformity of perceptual and judgemental experiences and it is dangerous, and counter-constructive, for a society that seeks to orient itself to reality.

It seems to me that these portents emerging from the study of nervous systems have potential bearing on a broad range of problems of nationalism, patriotism, the educational experiences useful for diplomacy and more general requirements for education that are very far reaching. I believe that if we were more fully aware of what is now known in relation to brain and behaviour there would be a greater tendency for us to be tentative about explosive interpersonal and international situations, more tolerant of ambiguities, and better able to create circumstances in which man's conscious cultural advancement can be assured of continuity. Nervous systems have only begun to reveal their potentialities for neighbourly occupation of the earth.

It is obvious that the nervous system does not prepare us for a world of infinite variety. It becomes committed during childhood to what particular experiences we have had. The brain is at once the integrator of physiological, psychological and social transactions. We need urgently to comprehend the fundamental brain mechanisms governing social behaviour. There is some promise that gain of knowledge of the underlying processes may lead to a greater chance for success in the experiment of man eventually becoming humane.

We still aspire to the declaration of Sophocles that: "Of all the wonders, none is more wonderful than man, who has learned the art of speech, of wind-swift thought, and of living in neighbourliness."

REFERENCES

DELGADO, J. M. R., ROBERTS, W. W., and MILLER, N. E. (1954). *Am. J. Physiol.*, **179,** 587–593.

GERNANDT, B. E., KATSUKI, Y., and LIVINGSTON, R.B. (1957). *J. Neurophysiol.*, **20,** 453–469.

LIVINGSTON, R. B. (1959). In *Handbook of Physiology*, sect. 1, vol. I, pp. 741–760, ed. Field, J., Magoun, H. W., and Hall, V. E. American Physiological Society. Baltimore: Williams and Wilkins.

LIVINGSTON, R. B. (1966). *Neurosci. Res. Prog. Bull.*, **4,** 235–347.

LIVINGSTON, R. B. (1967). In *The Neurosciences: A Study Program*, pp. 568–577, ed. Quarton, G., Melnechuk, T., and Schmitt, F. O. New York: Rockefeller University Press.

MACLEAN, P. D. (1958). *J. nerv. ment. Dis.*, **127,** 1–11.

NAUTA, W. J. H. (1964). *Neurosci. Res. Prog. Bull.*, **2,** 1–35.

NIRENBERG, M. (1967). *Science*, **157,** 633.

OJEMANN, R. G. (1966). *Neurosci. Res. Prog. Bull.*, **4,** suppl., 1–70.

OLDS, J., and MILNER, P. (1954). *J. comp. Physiol. Psychol.*, **47,** 419–427.

PERKEL, D. H., and BULLOCK, T. H. (1968). *Neurosci. Res. Prog. Bull.*, **6,** 221–348.

SCHEIBEL, M. E., and SCHEIBEL, A. B. (1967). In *The Neurosciences: A Study Program*, pp. 577–602, ed. Quarton, G., Melnechuk, T., and Schmitt, F. O. New York: Rockefeller University Press.

SHIMAMURA, M., and LIVINGSTON, R. B. (1963). *J. Neurophysiol.*, **26**, 416–431.
SHIMAMURA, M., MORI, S., MATSUSHIMA, S., and FUJIMORI, B. (1964). *Jap. J. Physiol.*, **14**, 411–421.

DISCUSSION

Harris: When you talk about reward or punishment by certain brain areas, what are these areas actually doing? Presumably the area itself doesn't either reward or punish.

Livingston: Quite specifically, it does. Activity in certain parts of the brain *is* intrinsically reinforcing to that individual. Positive and negative reinforcement systems are built into certain identifiable brain circuits. If you locally activate a positively reinforcing area electrically (by implanted electrodes) or chemically (by implanted cannulae), the animal shows satisfaction. It will seek out the locale in its environment where that activation occurred; it will learn to run a maze, pull against a spring, turn a wheel, or push a lever to receive such activation. It will disclose whether, and by how much, it prefers such stimulation to having access to more familiar desiderata, such as water, food or sex, when these are withheld. Similar demonstrations of satisfaction have been evaluated in a wide variety of animals including human subjects who, in addition to pressing levers, say that they feel "content" or "happy" or "groovy" or provide some other revelation of subjective satisfaction.

Control of approach and avoidance behaviour is built into the nervous system. But satisfaction, as Santayana reminds us, is the touchstone of behaviour. Punishment can be used in learning experiments, but it is not equivalent to reward. Different brain regions are affected and different effects on learning are obtained. Punishment, or threat of punishment, is often used as a deterrent, but the outcome is unpredictable. Control of the situation reverts to the victim; the "name of the game" may be abruptly changed. Let a teacher or parent imagine using punishment predominantly and then try to predict any specific outcomes! Try deterrence with an automobile at the next intersection! To be convincing, the threat must be sustained to the point that the outcome, at least within desperate limits, is beyond the control of the initiator. The person who wants to shape behaviour has dramatically lost control. Deterrence is an unpredictable and unsafe international as well as interpersonal policy.

Anand: You have rightly emphasized the important role of the reticular system, and especially the regions in the brainstem which bring about

14*

both arousal upwards and activation of receptors downwards, and you have pointed out that the system has to be activated in terms of sensory inputs. We have made some observations on Yogis, looking for their arousal patterns in terms of electro-encephalographic activity, to see whether the activation or arousal of the reticular system is dependent on peripheral sensory inputs or not (Anand, Chhina and Singh, 1961). We find that when Yogis are not meditating their EEG pattern is similar to a normal awake person or animal, and their alpha activity is desynchronized when any sensory stimulus is applied. But when the Yogis are "meditating" their alpha activity becomes more prominent, and it cannot be de-synchronized by any kind of sensory input. This kind of arousal pattern of the cortex does not seem to be dependent upon activation of the reticular system by the sensory inputs. Could the reticular system be internally conditioned to activate itself?

Livingston: I believe so. There are specific cortical areas which project into the reticular formation (French, Hernández-Peón and Livingston, 1955). These may be presumed to be susceptible to conditioning. "... These cortico-subcortical mechanisms might participate in such aspects of consciousness as voluntary alerting, maintenance of the aroused state, focusing of attention, vigilance or perhaps 'set', and meditation or intro-spection. . . . If it is appropriate to postulate that normal cerebral activity is influenced by these cortical zones, their participation in disordered states of cerebral function might logically be considered . . . common symptoms of mental illness include agitation, sleeplessness, hyperactivity and in-stability of thought content, while other syndromes are characterized by an abnormal reduction in activity, somnolence, inattention, etc. . . ." (*loc. cit.*, pp. 90–91). The same cortical fields have visceral as well as somatic representations, are known to exert control over sensory as well as motor activities, contribute to central (brainstem) excitability and thereby alter background readiness for action as expressed caudally through extra-pyramidal outflow and cephalically through the reticular activating system.

The net results of our learning to behave in our environment are achieved by modulating and entraining central nervous system activities, not by overriding them altogether. The old notion of nervous mechanisms being in a state of rest until activated by stimuli from the outside is super-seded. Central neurons are spontaneously active. Equally clearly, the ner-vous system uses special circuits to seek satisfaction; for example, it seeks activation of those regions through the use of both internal and external pathways and by influencing sensory as well as motor pathways.

Paintal: The importance of intrinsic activity in the nervous system, compared to the effects of external stimuli, is an extremely fundamental point. One wonders how generalized this property is in animals.

Secondly, are there any observations on invertebrates—perhaps insects or cephalopods—with regard to reward and punishment systems?

Livingston: Spontaneous activity is so widespread that it may be assumed to be universal from individual cells in primitive nerve nets and ganglia to the most complex nervous systems. It is not known what systems of reward and punishment exist in invertebrates, but insects and even the lowly medusae display motivations, satisfactions and dissatisfactions and must have specially organized nervous mechanisms to provide for the outward expression of these internal states (Bullock and Horridge, 1965; Van der Kloot, 1968).

Anand: You said that circuits through the limbic system are very important in the storage of memory patterns. What is the evidence for this?

Livingston: If you destroy the hippocampal region in the limbic system bilaterally, you preclude the recording of new memories (Ojemann, 1966). If you cut the fornix, which is one of the main hippocampal outflows, you drastically interfere with the storage of new memories (*loc. cit.*). Yet old memories persist and appear to be unaffected by these procedures. We are obliged to conclude that memories are not stored in the missing parts, but that the remainder of the nervous system, in the absence of those parts, is critically deficient with respect to the processing of subsequent memories. The individual so deprived can hang on to immediate contexts but can't recall them later.

During a learning experience with intact animals, if you stimulate the zone of the reticular formation into which limbic and frontal regions project, you can either enhance or interfere with learning (Fuster, 1958; Olds and Olds, 1961; Ojemann, 1966). Learning can also be enhanced by giving drugs which activate the brainstem reticular formation, even when the drugs are given shortly after the experiences that are presumed to be learned (Quarton, 1967).

Activation of the reticular formation gives rise to a d.c. potential shift in the cortex like that seen when an animal experiences novelty (Rowland, 1963). Associated with this are peculiar changes in rhythmic activity in the hippocampus. If the novel stimulus is repeated a few times, the d.c. response and hippocampal signs disappear *pari passu* with the animal's behavioural disregard of or habituation with respect to that stimulus. Such d.c. potential shift and hippocampal rhythm changes also accompany learning, both classical and instrumental or operant conditioning (Livingston, 1966).

Adey has demonstrated a reversal of wave-front precedence between the hippocampal gyrus (cortex immediately next to the hippocampus) and the adjacent entorhinal cortex (Adey, 1967). The reversal takes place between the time of early learning and consolidation to a well-learned response. The d.c. shift is probably an accompaniment of reinforcement during learning since it occurs with reinforcement alone, whether internally or externally induced. A hungry animal, with food placed in its mouth, will show such a d.c. potential, but when it is satiated, the d.c. shift disappears and at the same time learning ceases to progress. When hunger recurs the d.c. shift is restored. Averaged electrical activity in the "feeding centre" in the hypothalamus increases when a hungry animal sees food or eats but this does not occur when he is satiated (Wyrwicka, 1964).

Talwar: What is known of the way in which this regulatory system works? Most of the interrelations that you indicated seemed to be in terms of facilitation and inhibition. Does the main function of the reticular system consist of the filtering of certain information to the higher centres?

I also wonder whether there is any clear evidence to show that the neuronal cells found in the various central nervous system areas are identical or different in their metabolic characteristics. In other words, are the cells which ultimately receive the sensory information identical in their modality to those in the reticular system and is it merely their junctional relationship that distinguishes their function?

Livingston: The reticular formation, and indeed any sector of the nervous system, can do two things which in turn have at least three kinds of effects. There are electrical phenomena due to controlled fluxes of ions across membranes and to electrotonic forces set up by metabolic systems acting on membranes, and there are chemical phenomena due to secretion or release of chemical compounds such as neurohormones and neurotransmitters. These processes are considered to be electrical and chemical signalling systems. Three effects of such signalling include increased excitability of neighbouring and connected cells, diminished excitability by virtue of stabilized membranes, and improvement in connexions so that previously ineffective circuits become effective. Neuroscientists are only beginning to grasp the fundamentals of these processes and effects.

The relationship between neurons and glia which is structurally so intimate and is probably functionally significant is still pretty mysterious. Most is known about the interdependence of glia and neurons in myelin formation, but this is probably only one of many neuron–glia interdependences. From the way glia swarm around cell bodies and synapses in tissue culture, as seen by time-lapse photography by Dr Charles Pomerat (1964), one suspects that the glia must be creating changing environmental influences. Dr Sanford Palay has speculated, on the basis of electron micrographic evidence, that glia may be making different partitions and groupings of synaptic junctions (personal communication).

The reticular formation is apparently involved in recognition of stimuli (Starr and Livingston, 1963). It is involved, along with the limbic system, in orientation to novelty (Livingston, 1966). It plays a role in control of transmission along both sensory and motor pathways (Livingston, 1959). It very likely is involved in switching incoming signals so that they occupy narrower or wider regions depending on experience (habituation), interest (states of drive, emotional states) and expectations (built up from remote and recent past experiences). Presumably such dynamic functions are governed mainly by facilitatory and inhibitory influences. Neurons have an extraordinary range of chemical specification. This is evident in the precision of embryonal ordering of brain circuitry (Edds, 1967), regenerative specificities (Sperry, 1963), reactions to toxic agents and the considerable variety of neurohormones and neurotransmitters (Kravitz, 1967). On top of this chemical specificity is the elaborateness of connexions. Precisely where cells arise and whither they connect is critical to the outcome of the messages transmitted (Scheibel and Scheibel, 1967; Brodal, 1969).

Talwar: Do you envisage the physical representation of learning as localized or diffused in the brain?

Livingston: The reason I suggest a generalized "now store" order is that previous models of learning and memory processes have left a gap, at least for me, in explanations of how for example, bell circuits become fastened to salivary circuits. Reinforcement somehow stabilizes *that* particular pathway but the explanations must go out of their way to explain the phenomena of generalization, irradiation, and so on. It seems to me simpler to suppose that reinforcement does not have to butter up each particular pathway that is to become conditional (how did it "know" which pathways?) but that it releases a generalized order for all recently active pathways to be strengthened, no matter what concatenation of activities may have just occurred (Livingston, 1964, 1966, 1967). This subsumes all the phenomena of association into one diffusely projecting signal. Every biologically meaningful event (yielding reinforcement for that animal) will tend to be stored. Whenever the event recurs it will be re-stored. Biologically irrelevant events will fade away because they are only randomly stored by reinforcing signals. Evolution has created nervous systems that predict. Whenever there is a regular occurrence of A followed by B, the nervous system predicts (anticipates) B when A occurs. If the same event (even if it is an investigator's causally unrelated bell) repeatedly leads to biological satisfaction (food in the mouth of a hungry animal), the combination, the Gestalt, of those occasions will become stored and re-stored so as to create an habitual predictive pathway.

Anand: When integration takes place at the higher central nervous regions, in addition to facilitation or inhibition of some of the lower circuits it also results in certain "drives" which are based on the experience acquired by way of the reward and punishment systems. Integration at the higher levels not only results in facilitation of certain reflexes; it also creates a feeling or a drive in the individual which determines its further behaviour pattern. If glucose utilization changes in the body, it not only facilitates feeding, but also gives the "feeling" of hunger resulting from activation of certain higher centres. This distinct feeling motivates further behaviour.

Monod: Dr Livingston mentioned that sensory inputs can be processed all along the line including being almost completely blotted out; and we have all had these experiences, particularly with respect to sound. If you are thinking very hard about something you don't hear sounds. This is not true of vision; you cannot blot out vision if your eyes are open. But towards auditory impulses you can be virtually deaf. Is it known at what stage in transmission this occurs?

Livingston: The first stage of auditory control lies with the two middle ear muscles which are governed by separate cranial nerves arising within the brainstem. Contraction of these muscles can attenuate sound input to the receptors by about 20 db. This is a considerable outside curtain against sound. At the hair cell receptors a further control exists, control by the olivo-cochlear bundle which also arises in the brainstem (Rasmussen, 1967). This bundle can drastically shut down the response to acoustic input (Galambos, 1956; Desmedt, 1962). Similar controls are exerted at the successive nuclear relays from cochlear nuclei to auditory cortex (Livingston, 1959). The control is by anticipation, or by feedback from undesired signals, in accordance with the animal's past experiences, expectations and purposes (Starr and Livingston, 1963; Livingston, 1959). Something similar occurs along the visual pathway with the pupil (Stark, 1959), central control of retinal ganglion cell transmission (Granit, 1955) and subsequent relays. You may have had the experience of staring blankly without seeing, or reading a page without recalling the information. It is evident that a shutdown or enhancement of intake can take place at any of a number of levels in any given sensory pathway (Livingston, 1959). We have no very convincing ways to establish how much sense data is lost and exactly where, but the present interpretation suggests that sensory selection by reduction and enhancement, by narrowing and widening the area of distribution of signals, by abstraction and modulation, may take place earlier along the sensory trajectory than had been previously supposed. The data of Maturana and his colleagues (Lettvin *et al.*, 1959) and of Hubel and Wiesel (1961) are especially pertinent here.

Paintal: On a more general point, I wonder whether approaches that include the building of models, although very useful for trying to find out how to answer specific questions about nervous system functioning, are not in the long run misleading because they tend to limit one's imagination?

The other point is the use of new terms. You use the term "store" for memory, and terms like "central activation" for the older terms of central excitatory state.

Livingston: I appreciate your concern. We cannot help making models. If not intentionally we do so inadvertently. Models are assumed whenever we design a particular experiment or discuss a given topic. That seems inescapable. How self-conscious we are about modelling, and how explicit, is a matter of custom, style and discipline.

The use of terms is similarly binding. We have in mind certain metaphors that lie behind the terms we use. As science progresses we have to stretch the terms until a point is reached when the old metaphors no longer suffice and new terminology is warranted. This is a subjective turning point that will vary from individual to individual. I believe that terms like "central excitatory state" and "memory" can stand additional metaphors as to what may actually underlie these complicated processes. I depend on people like Moruzzi and Magoun (1949) and their followers for the metaphor of "activation" and on computer people for the metaphor of "now store". The actual biological processes involved are still unknown—the terms are used to describe phenomena. I think of long-term memory storage as involving predominantly growth, but other processes may be involved as well, leading to conditional alterations in circuitry (Bullock, 1966; Livingston, 1966).

Whittaker: Many of your slides showed what looked like wiring diagrams, and you are evidently thinking in terms of gates, of potentials arriving at certain time-intervals and so on, which could lead to a computer model which could be simulated. I wonder if this is an adequate approach in order to understand the functioning of the nervous system as a whole. In other words, is the nervous system this kind of machine? In this sort of model one would not enquire into the chemical events taking place in synapses because synapses would be regarded simply as switches which can be specified by certain parameters. The molecular activities of the synapse would only be relevant for specifically understanding drug action, or something of this kind.

Livingston: Let me make clear my impression that the whole of the nervous system is controlled by molecular mechanisms: its development, the processes of perception, judgement, emotion, learning, memory and behaviour. No single model of the nervous system is adequate. Descartes related brain functions to the elaborate hydraulic fountain and clock systems of his day; we relate them to computers; both analogies are grossly inadequate. Model-making at different levels of membranes, synapses, circuits, and of processes such as perception and behaviour, helps to focus the dialogue and to make our questions more specific, hopefully to rule out some naively attractive possibilities, to provide testable and refutable explanations. I would point out that there are multiple codes for transmitting information by way of nerve impulses and that there are many sites for this coding that are located elsewhere than at synapses (Perkel and Bullock, 1968). This does not denigrate synapses as critical and as obvious sites for control. They are very likely the main sites, but certainly not the only sites. It would be nice if we could experimentally manipulate and come to understand the functions of one type of neuron or one type of synapse and assign all complex brain functions to such "elementary processes" multiplexed. But it is evident that there are not only many codes operating at different sites, but many kinds of neurons and many kinds of synapses. And when they are put to functioning together as an ensemble, it matters very much exactly how they are connected with one another. It shouldn't surprise us that brains beggar description because they are, after all, the most complicated mechanisms known to man.

There is no question that certain parts of the central nervous system connect with certain other parts, and that the fine structure of these connexions may be exquisitely detailed (Ramón y Cajal, 1909–1911; Scheibel and Scheibel, 1968). There is no doubt that there are numerous specific kinds of localized influence. Yet the nervous system also seems to obey some general principles of regional organization and specialization. To this extent any gross model of the central nervous system must resemble a crude wiring diagram. But such configurations are not intended to indicate, or to limit, what the detailed connexions may signify locally. If you wish to be more precise about a limited part of a circuit, you create a more limited model, such as one synaptic bouton, or all of the synapses on a single cell, or all of the input–output relations of a single cell; you can also model all the input–output relations of a ganglion, or a complex structure such as the cerebellum. I have introduced models of an even more general sort. We are certainly making a horrendous jump, perhaps an inadmissible one, to go from subcellular molecular control systems to the kinds of global generalizations I have made. You need to "go loose" to make the jump from molecular models to models of how the brain as a whole may work, but some of us find it extraordinarily exciting to work at this level of complexity and uncertainty.

Monod: A representation of interactions which we know do *not* take place by way of material circuits, say of intercellular controls by means of inducers, repressors and so on, will nevertheless *look* like a wiring diagram, but that is within the logic of the system; it is the only way of representing a complex control system in which certain things or events are related to other things. Even a mathematical function depicted on paper will look like this. It doesn't say what the connectivities are like, however.

Whittaker: I am trying only to draw a distinction between an approach in terms of a wiring diagram model and a model in more chemical terms. By the term "wiring diagram" I mean not a model which simply expresses connectivity in a purely schematic way, but a way of thinking of the neuron and axon as a kind of wire and the synapse as a kind of switch, and building the system up on that. An alternative possibility is that there is some chemical integration of the environment in the central nervous system. These cells are living in a kind of soup in which various chemical events take place, and this may be important to consider. Certainly there are parts of the nervous system, for example the cerebellum, where you can draw very precise wiring diagrams, based on the synaptic connexions observed, and where there is reason to believe that although the precise functioning of the system is not yet understood, it *could* be understood in terms of fairly simple electrical circuitry. But there are other parts, like the centres that you have described, where the processes are so intertwined and where the system doesn't seem to make sense in terms of a simple logic-unit type of approach.

Livingston: I strongly endorse what you say about the integrative processes involved in the various "soupy" cellular environments. The muco-polysaccharides and hyaluronidases and other compounds on the surfaces of cells, and the interstitial and cerebrospinal fluids alike, probably trap ions and perform spatial–chemical control functions of great importance. We tend to stress the axon and action potentials, but in evolutionary terms, decrementless conduction was invented to meet the requirement of safe long-distance transmission, when cells that needed to be connected to each other had to become too far separated to signal by means of graded, decremental chemical transmission (Bullock and Horridge, 1965). Although some information coding occurs at branching points of neurons, most control is exerted at the business ends of neurons, at the cell body (pacemakers and excitability controls intrinsic to the cell), at the dendrites and at the terminal arborizations and synapses. It is at these subcellular, cellular and intercellular regions that the continuous processes and graded responses and integrated levels of facilitation and inhibition occur. The axonal long-conducting systems simply safeguard the conveyance of coded impulses from one axon hillock to the terminal arborizations of that neuron. Most of the dynamic and plastic controls undoubtedly take place in relation to the connexions between cells.

Talwar: It appears that in the central nervous system every action is more or less a multi-unit event; nothing is performed in isolation, and integration of a large number of cellular elements is necessary for the expression of any act. And while one cannot overlook the fact that the language of communication between these units is that of electrical impulses, specific molecules may still have a role in terms of establishing *new* contacts. The potentialities may be inherent in cells, but experiences induce the expression of those potentialities and create new circuits. Would that fit in with the way you are thinking?

Livingston: Yes, indeed.

Monod: Dr Livingston, I was not quite clear what philosophy you were recommending from your physiological findings! You mentioned hedonism; would it not rather be epicurean, which is a much more intelligent way of being a hedonist?

Livingston: I adhere to the French epicurean interpretations of hedonism! I am not trying to design nor even to advocate a philosophy. I am not a philosopher but a technician working on biological problems which seem to be pertinent to the dialogues among philosophers. What seems to be of philosophically greatest importance in our day of risky international anarchy is that the biological mechanisms we inherit do not prepare us for a world of infinite variety. Something remarkable happens to us early in life. We become committed to certain limited and limiting perceptual judgemental and motor skills. Whatever is built into our nervous system originally is remarkably modified by functional exercise in the finite environment within which we are reared. These culturally bound commitments begin to take form shortly after birth.

The brain doubles in size during the first six months and doubles again by about the fourth year. From then until the age of about twenty it shows only a very modest increase in weight. And from that age on, the brain is shrinking. Much of the increase in weight and bulk is due to the binding of axons with sheaths of myelin, but there is a vast increase in size and complexity of cell bodies and a multiplication and extension of the fine processes of axons, dendrites and synaptic surfaces. Conel has shown week-by-week changes evident by light microscopy alone during the early postnatal period of human life (Conel, 1939–1957). Others have shown this using light microscopy in experimental animals (Altman, 1967). Moreover, there are cells which haven't taken up their final station at the time of birth (Caley and Maxwell, 1968). Such cells migrate to establish connexions during this post-natal period of brain development. Perhaps these cells can be affected in accordance with the earliest imprinting of experiences.

The active completion of construction of the brain during the period of earliest individual experiences may contribute biologically to the remarkable early adaptability of human beings and may initiate the compelling commitments of our earliest experiences to the limitations of this early life. This may contribute to the morphological differences seen in the brains of rats, depending on whether early life took place in an enriched or an impoverished environment (Bennett et al., 1964). This has been found to have an effect on brain weight, on neuronal: glial ratios, and on the concentration of acetylcholinesterases. Regardless of these biological considerations, it is obvious that early experiences affect us in enduring ways: if a person is transplanted culturally after the age of about three years, detectable signs of the original culture remain in evidence; if migration takes place after the age of about seven, these signs are obvious in everyday life; if a person migrates after the age of about twenty, his adoption will be manifestly incomplete, to himself as well as to others. Early experiences affect us in relation to the kinds of perceptual experiences we have, the styles of thinking and logical disposition we exercise, the judgements we apply and, in general, the things we like and dislike. This emphasizes the importance of culturally binding or culturally liberating experiences which might be provided during early childhood. Some enduring commitments are established well before the child begins formal schooling. The lessons from these facts for philosophy and for political science are spectacular.

Because environments are so different in different parts of the world, people growing up in different cultures have different perceptual, judgemental and logical experiences and opportunities. We all grow up culturally idiosyncratic even within a given nationality. One of the problems everyone faces in getting along in the world is that we naturally, naively assume that what we observe around us is what any other person observes. If his report of what he witnesses differs from what we are experiencing, we are inclined to conclude either that he is deliberately lying or that he has slipped into some psychologically dubious ratiocination about his own testimony.

The biological evidence now before us is that animals presented with simple stimuli, but with experimentally different scheduling and type of reinforcements, develop objective differences in the central nervous system responses and corresponding behavioural responses to identical stimuli. Presumably they have different perceptual experiences. Their attachments of value, judgements and actions are conspicuously modified according to their past experiences, expectations and purposes. I infer that another person may be just as honest and open-minded as I think I am being when we share experiences and yet our testimonies are in conflict. How much these experientially committed discrepancies can be exaggerated by remoteness from being able to share the experiences, and by conscious and inadvertent cultural editing of transmitted information, I leave to your imaginations!

When we encounter difficulties in our interpersonal relations and when our nations encounter cross-cultural difficulties, we are inclined to attribute the differences to faults of character on the part of someone other than ourselves and our nation. In our own community we are surrounded by like-thinking individuals so that it is practically impossible and perhaps culturally dangerous to become disengaged from similar surrounding cultural aberrations. Since we become individually committed to those limited perceptual and judgemental experiences with which we have gained familiarity, none of us can see any extended part of the world as it really is; we see only what our limited environmental exposure has permitted us to see. Because these commitments become ingrained so early and are so exceedingly difficult to dislodge or modify, the problem of getting along in the world is not simply one of translation of one language into another; it is much more fundamental. If we knew more about the qualitative and quantitative differences of these processes, and if we were able to recognize existing interpersonal and intercultural gulfs in their true proportions, we would become more tolerant of contrary views and values, we would become more tentative in our own interpretations, we would become more able to sort out ambiguous situations objectively, and we would be more inclined to admit that the other person might be right, that he certainly deserves our recognition that he and we can adhere to contrary views, pending some more objective measure than our or his judgement of the issue. He very well may be telling us perfectly honestly about his experiences; these may be different because he grew up in an appreciably different environment from our own. If we understood this fully, we might become less imperious, less reflexive in our responses, and more able to sort out conflicts constructively. Animals don't have the capacity for cultural commitment that we do, they haven't comparable communications, and they lack access to the systems of power that are at our disposal. As a consequence, they don't run the risks we run in confrontation across different realms of experience. They also lack the opportunity and hence responsibility for self-examination that science is now making available to us for the first time in history. We have always had an external (psychological) appreciation for these things and now we can discover exactly how and where such commitments take place in the nervous system. The "how" and "where" is crucial, philosophically and practically.

Bhargava: The thought that patterns are fixed so early in life is rather dismaying. Is there experimental evidence to support this view?

Livingston: There is much evidence indicating how committing early experiences are. This has to do not only with subtleties of language and social behaviour but also with very simple kinds of perceptual experience. Adelbert Ames, Jr. (1955) made fascinating demonstrations to illustrate these effects. John Dewey described these findings as the most important contributions to psychology during this century. I have attempted here to suggest that the physiological apparatus of sensory control and of limbic and reticular selection may provide the biological substrata for psychologically and socially obvious, urgent and compelling problems. Psychological and social control processes do not exist apart from the nervous system. They are built in, by genetic and experiential endowments.

Anand: Is it possible to change those fixed patterns pharmacologically, surgically, or in any other way?

Livingston: Since memory is a distributed function, it becomes difficult to change these patterns. I agree with Dr Bhargava that this is a discouraging revelation. Yet if these problems do exist (as I perceive them), and if they are taken seriously (as I take it they must be), then in relation to both individual and international conflicts we can justifiably say, "cool it, man!" You recognize that an apparent irreconcilability of views may occur not for want of good character and willingness to be constructive; it may become built into brain circuits according to differences in the circumstances in which individuals spent their earliest years. Who can see the world other than through these biological lenses?

To help us all out, we need to create additional institutional defences and protections for resolving individual and international conflicts and for establishing improved levels of justice. These need to be invented and institutionalized, as ideals have been invented and institutionalized in the past. And they need to be extended throughout and honoured all over the world.

REFERENCES

ADEY, W. R. (1967). *Prog. physiol. Psychol.*, **1**, 1–42.

ALTMAN, J. (1967). In *The Neurosciences: A Study Program*, pp. 723–743, ed. Quarton, G. C., Melnechuk, T., and Schmitt, F. O. New York: Rockefeller University Press.

AMES, A., JR. (1955). *The Nature of our Perceptions, Prehensions and Behavior*. Princeton: Princeton University Press.

ANAND, B. K., CHHINA, G. S., and SINGH, B. (1961). *Electroenceph. clin. Neurophysiol.*, **13**, 452–456.

BENNETT, E. L., DIAMOND, M. C., KRECH, D., and ROSENZWEIG, M. R. (1964). *Science*, **146**, 610–619.

BRODAL, A. (1969). *Neurological Anatomy in Relation to Clinical Medicine*. London: Oxford University Press.

BULLOCK, T. H. (1966). *Neurosci. Res. Prog. Bull.*, **4**, 105–233.

BULLOCK, T. H., and HORRIDGE, G. A. (1965). *Structure and Function in the Nervous Systems of Invertebrates*. San Francisco: Freeman.

CALEY, D. W., and MAXWELL, D. S. (1968). *J. comp. Neurol.*, **133**, 17–43.

CONEL, J. L. (1939–1957). *The Postnatal Development of the Human Cerebral Cortex*. Cambridge, Mass.: Harvard University Press.

DESMEDT, J. E. (1962). *J. acoust. Soc. Am.*, **34**, 1478–1496.

EDDS, M. V., JR. (1967). In *The Neurosciences: A Study Program*, pp. 230–240, ed. Quarton, G. C., Melnechuk, T., and Schmitt, F. O. New York: Rockefeller University Press.

FRENCH, J. D., HERNÁNDEZ-PEÓN, R., and LIVINGSTON, R. B. (1955). *J. Neurophysiol.*, **18**, 74–95.

FUSTER, J. M. (1958). *Science*, **127**, 150.

GALAMBOS, R. (1956). *J. Neurophysiol.*, **19**, 424–437.

GRANIT, R. (1955). *J. Neurophysiol.*, **18**, 388–411.

HUBEL, D. H., and WIESEL, T. N. (1961). *J. Physiol., Lond.*, **155**, 385–398.

KRAVITZ, E. A. (1967). In *The Neurosciences: A Study Program*, pp. 433–444, ed. Quarton, G. C., Melnechuk, T., and Schmitt, F. O. New York: Rockefeller University Press.

LETTVIN, J. Y., MATURANA, H. R., McCULLOCH, W. S., and PITTS, W. H. (1959). *Proc. Inst. Radio Engrs Am.*, **47**, 1940–1951.

LIVINGSTON, R. B. (1959). In *Handbook of Physiology*, sect. 1, vol. I, pp. 741–760, ed. Field, J., Magoun, H. W., and Hall, V. E. American Physiological Society. Baltimore: Williams and Wilkins.

LIVINGSTON, R. B. (1964). *Cited in*: Nauta, W. J. H. (1964). *Neurosci. Res. Prog. Bull.*, **2**, 25–27.

LIVINGSTON, R. B. (1966). *Neurosci. Res. Prog. Bull.*, **4**, 235–347.

LIVINGSTON, R. B. (1967). In *The Neurosciences: A Study Program*, pp. 568–577, ed. Quarton, G. C., Melnechuk, T., and Schmitt, F. O. New York: Rockefeller University Press.

MORUZZI, G., and MAGOUN, H. W. (1949). *Electroenceph. clin. Neurophysiol.*, **1**, 455–473.

OJEMANN, R. G. (1966). *Neurosci. Res. Prog. Bull.*, **4**, suppl., 1–70.

OLDS, J., and OLDS, M. E. (1961). In *CIOMS Symp. Brain Mechanisms and Learning*, pp. 153–187, ed. Gerard, R. W., Fessard, A., Konorski, J., and Delafresnaye, J. R. Oxford: Blackwell.

PERKEL, D. H., and BULLOCK, T. H. (1968). *Neurosci. Res. Prog. Bull.*, **6**, 221–348.

POMERAT, C. M. (1964). *The Dynamic Aspects of the Neuron in Tissue Culture* (Film). Pasadena Foundation for Medical Research. Los Angeles: Wexler Film Productions.

QUARTON, G. C. (1967). In *The Neurosciences: A Study Program*, pp. 744–755, ed. Quarton, G. C., Melnechuk, T., and Schmitt, F. O. New York: Rockefeller University Press.

RAMÓN Y CAJAL, S. (1909–1911). *Histologie du système nerveux de l'homme et des vertébrés.* Paris: Maloine.

RASMUSSEN, G. L. (1967). In *Sensorineural Hearing Processes and Disorders*, pp. 61–75, ed. Graham, A. B. Henry Ford Hospital International Symposium. Boston: Little, Brown.

ROWLAND, V. (1963). In *Brain Function: cortical excitability and steady potentials; relations of basic research to space biology*, pp. 136–176, ed. Brazier, M. A. B. Berkeley: University of California Press.

SCHEIBEL, M. E., and SCHEIBEL, A. B. (1967). In *The Neurosciences: A Study Program*, pp. 577–602, ed. Quarton, G. C., Melnechuk, T., and Schmitt, F. O. New York: Rockefeller University Press.

SCHEIBEL, M. E., and SCHEIBEL, A. B. (1968). In *The Thalamus*, pp. 13–46, ed. Purpura, D. P., and Yahr, M. D. New York: Columbia University Press.

SPERRY, R. W. (1963). *Proc. natn. Acad. Sci. U.S.A.*, **50**, 703–710.

STARK, L. (1959). *Proc. Inst. Radio Engrs Am.*, **47**, 1925–1939.

STARR, A., and LIVINGSTON, R. B. (1963). *J. Neurophysiol.*, **26**, 416–431.

VAN DER KLOOT, W. G. (1968). *Behavior.* New York: Holt, Rinehart and Winston.

WYRWICKA, W. (1964). *Electroenceph. clin. Neurophysiol.*, **17**, 164–176.

CHAIRMAN'S CLOSING REMARKS

PROFESSOR F. G. YOUNG

AT the end of what I think we all agree has been a most stimulating conference I must express gratitude and thanks, on behalf of all the conference visitors to Delhi, for the kindness and help that we have received in this great capital city. We especially appreciated the fact that His Excellency the President of India found time to come to inaugurate our proceedings, and that the Honourable Minister of Health, President of the All-India Institute of Medical Sciences, also addressed us on that memorable occasion. We are greatly indebted to Dr K. L. Wig, Director and Dean of the All-India Institute of Medical Sciences, and to Professor B. K. Anand, the Vice-Dean, for willing and effective cooperation in many aspects of the arrangements for this conference. To the members of the staff of that institute, especially Professor G. P. Talwar, we members of the conference express thanks for most enjoyable hospitality and kindness. We have indeed been treated most generously by many Indian friends and colleagues, too many to name, but to whom our warm thanks are due.

And in expressing the gratitude of the visitors to Delhi I must include thanks to the many members of the staff of the India International Centre, where the visiting members of the conference have lived together, properly and comfortably, during nearly a week of scientific debate. Especially do I include in my thanks the staff whom we do not see, in the kitchens and in the offices. We are most grateful for excellent service in every way.

I must also express, as Chairman of this conference, appreciation to those who read such interesting papers and fostered such lively discussion. Have the contributors never overrun their allotted time? Never? Well hardly ever. But the discussers have spoken at length, and the observers not at all—just as it should be.

And now, having expressed thanks at the end of this most enjoyable conference I assume the privilege of the Chairman of the symposium, and declare that my contribution is not open to discussion.

INDEX OF AUTHORS*

Entries in bold type indicate a contribution in the form of a paper; other entries indicate contributions to the discussions.

*Author and subject indexes prepared by William Hill.

INDEX OF SUBJECTS

Protein—*continued*
  synthesis,
    action of cortisone, 101
    amphibian metamorphosis and, 133, 134, 138, 139, 141, 143–144, 151, 154
    availability of ribosomes, 92–93
    cyclic AMP in, 82
    growth hormone activity, 86–107
    hormonal control, 87–88, 106, 133, 134–135, 151–152
    in bacteria, 103
    in enzyme systems, 18, 26
    inhibition by actinomycin, 116
    inhibition by cortisol, 100, 106
    in nucleoli, 55, 56, 57, 58
    in nucleus, 53, 55, 58, 59, 60, 62
    insulin in, 86–107
    in synaptosomes, 352, 354
    regulation, 59, 133, 134–135
    role of oestrogens, 116
Puromycin, reaction with ribosomes, 95, 96

Recognition,
  in immune system, 292
  of antigens, evolution, 310–313, 317
Recognition elements, 293
  genes as, 255
  in immune response, 293, 301
  in learning, 290, 292
  nature of, 255
Recombination, generating immunoglobulin variants, 314, 318–319
Reflexes, propriospinal and spino-bulbar, 391–393
Regulation, of cellular proliferation in protein-deficient organism, 321–337
Reinforcement systems in nervous system, 395–396, 397, 401
Repressor protein, 8–10, 18
  binding to operator, 12–15
  effect of ageing, 25
  purification of, 6, 22, 23
  structure, 16, 24, 25
Repressor–inducer complex, 6, 16, 24
  membrane filter assay, 4–6, 22
Repressor–operator complex, 8, 10–12, 16, 19, 24
  inducer and anti-inducer interaction, 16
  membrane filter assay, 4–6, 22, 24
  specificity, 16
  structure, 17
Respiration, regulation by hypothalamus and limbic system, 363–364
Responses to unexpected stimuli, mechanism of, 255–257
  in immune system, 258–281, 304
  in learning, 281–293, 386–387, 398–400
Reticular units, 394, 401, 403
Reward and punishment systems, 396, 397, 401
Ribonuclease, uptake by cells, 160, 174, 175
Ribonucleic acid (RNA),
  hybridization with DNA, 134, 138, 147
  intranuclear turnover, 59, 135
  polydisperse, 56, 127–128
  synthesis,
    action of cortisol, 129
    action of growth hormone, 113–114

Ribonucleic acid (RNA)—*continued*
  synthesis—*continued*
    action of steroid hormones, 129
    in cells, 134, 135
    induction by oestrogens, 152, 155, 156
    inhibition by actinomycin, 89
    in nuclei, 53, 55, 58, 59, 60, 62
    in nucleoli, 55, 56, 57, 58
    role of oestrogens, 117
    transcription rate, 125
    transfer to cytoplasm, 56, 57, 134
m-Ribonucleic acid (messenger),
  action of growth hormone, 88–91
  action of insulin, 88–91
  degradation, 90
  recognition, criteria for, 125, 126–127
  synthesis, 88, 124
    action of growth hormone, 113
    role of oestrogens, 115–120, 125
  transport from nucleus to cytoplasm, 90
t-Ribonucleic acid (transfer), 57, 61, 62
  action of growth hormone, 88
  synthesis, role of oestrogens, 120
Ribosomes, activity, effect of insulin and growth hormone, 93–98
  availability of, 92–93
  concentration, 101
  effect of growth hormone, 102
  nuclear, 57, 58
  on m-RNA, 99
  reaction with puromycin, 95, 96
  stability and turnover, 104, 153

"Second messenger" hypothesis, 73, 82, 110
Serine, incorporation, 124–125
Spleen cells, and immunity, 254
Starvation, fatty acid synthesis in, 31, 33, 41, 42, 47, 48, 49
Steroid hormones,
  control of protein synthesis, 152–153
  effect on RNA synthesis, 129
  feedback mechanism, 212
  mode of action of cyclic AMP, 72, 81
Synapses,
  chemical transmission, 342
  electrical transmission, 342
  function, 342–343
  mode of information transmission, 340
  stabilization in learning, 288–289, 397, 405–406
Synaptic proteins, 286–287, 289
Synaptosomes, 344–345
  acetylcholine in, 346
  heterogeneity of, 353
  metabolism, 347, 354
  protein synthesis in, 352, 354

Testosterone, 382
Thymectomy,
  depressing immunological competence, 213
  effects of, 272
    on lymphocytes, 240
  neonatal, consequences of, 213, 220
Thymic extracts, *see Thymosin*
Thymocyte, permeability, 176